Third Edition

Rice Almanac

Source Book for the Most Important Economic Activity on Earth

Edited by J.L. Maclean, D.C. Dawe, B. Hardy, and G.P. Hettel

2002

IRRI

CABI Publishing

WARDA ADRAO

CIAT Centro Internacional de Agricultura Tropical International Center for Tropical Agriculture

FAO FIAT PANIS

CABI Publishing
A Division of CAB International
Nosworthy Way
Wallingford
Oxon OX10 8DE
UK

Tel: Wallingford +44 (0)1491 832111
Fax: +44 (0)1491 833508
E-mail: cabi@cabi.org
Web site: www.cabi-publishing.org

Published in association with:

International Rice Research Institute
DAPO Box 7777, Metro Manila, Philippines

West Africa Rice Development Association
01 B.P. 2551, Bouaké 01, Côte d'Ivoire

International Center for Tropical Agriculture
Apartado Aéreo 6713, Cali, Colombia

Food and Agriculture Organization
of the United Nations
Via delle Terme di Caracalla, Rome, Italy

The International Rice Research Institute (IRRI), the West Africa Rice Development Association (WARDA), and the International Center for Tropical Agriculture (CIAT, its Spanish acronym) are three of 16 Future Harvest research centers funded by the Consultative Group on International Agricultural Research (CGIAR). The CGIAR is cosponsored by the Food and Agriculture Organization of the United Nations (FAO), the International Bank for Reconstruction and Development (World Bank), the United Nations Development Programme, and the United Nations Environment Programme. Its membership comprises donor countries, international and regional organizations, and private foundations. Its mission is to contribute to food security and poverty eradication in developing countries through research, partnerships, capacity building, and policy support, promoting sustainable agricultural development based on the environmentally sound management of natural resources. FAO is the United Nations agency that leads international efforts to defeat hunger and works to help developing countries modernize and expand agriculture, forestry, and fisheries and ensure good nutrition for all.

Catalog entries for this book are available from the British Library, London, UK, and from the Library of Congress, Washington, DC, USA.

ISBN 0 85199 636 1

Cover photo: Rice terraces in eastern Bhutan (Gene Hettel).

Typeset, printed, and bound in China.

Contents

Foreword

The first edition of the Rice Almanac was published in 1993 in response to the long-felt need to bring together general information about rice—its origin, its growth and production, the ecosystems under which it is grown, and opportunities for increased yields.

A second edition was published in 1997, incorporating much updated material on production and geographical data plus new information on a number of countries as well as regional information on West African countries from the West Africa Rice Development Association (WARDA) in Côte d'Ivoire, on the Latin American and Caribbean region and countries from the Centro Internacional de Agricultura Tropical (CIAT) in Colombia, and on European rice production. WARDA and CIAT became copublishers of the second edition with IRRI.

The second edition also gave rise to an Internet site, Riceweb (www.riceweb.org), which contains all the Almanac contents as well as a host of additional information about rice, such as rice testing protocols, recipes, a glossary of rice terminology, and access to rice literature and many other rice-related Web sites from around the world. Riceweb has become a highly visited site, earning acclaim also from several Web-site rating and other organizations, such as the Dow Jones Business Directory, *New Scientist, Natural Selection*, and *USA Today*.

For the third edition, we have doubled the number of countries for which production-related information is provided. This is due to help from the Food and Agriculture Organization of the United Nations (FAO), which agreed to provide material from its rice production country database (CORIFA—country rice facts) and to become a copublisher of the Almanac. The Almanac now covers 64 countries, including all the major producers.

The production and other statistics used herein are derived primarily from FAO, which include official country data (FAOSTAT), surveys, reports, and personal communications; IRRI's RICESTAT database, which is based on primary data from requests and questionnaires and secondary data from statistical publications and international organizations including FAO, the International Labour Organization, the World Bank, etc.; and regional data from CIAT and WARDA.

We thank CABI, which has worked with us in producing a copublished hardcover version for distribution in developed countries.

We trust that the third edition of the Rice Almanac will continue to increase awareness of rice as the most important staple food in the world and of all that is involved in maintaining rice production.

Ronald P Cantrell

Ronald P. Cantrell
Director General, IRRI

Acknowledgments

The third edition of the Almanac builds on contributions to the first two editions by many people both within IRRI and around the world. Those who contributed to the third edition include

Contributors of chapters or sections: Lee Calvert and Luis Sanint (CIAT), Mahabub Hossain (IRRI), Aldo Ferrero (Italy), Tom Hargrove (USA), Ryoichi Ikeda (Japan), Marco Wopereis and Guy Manners [coordinator] (WARDA), and Nguu Nguyen (FAO).

Reviewers from IRRI: Gary Atlin, Alberto Barrion, Mark Bell, Roland Buresh, Mike Cohen, Hiroyuku Hibino, James Hill, Guy Kirk, Hei Leung, Duncan Macintosh, David Mackill, Graham McLaren, M.A. Hamid Miah, Josephine Narciso, Shaobing Peng, Mila Ramos, John Sheehy, Virendra Pal Singh, Boriboon Somrith, Mahyuddin Syam, Shengxiang Tang, To Phuc Tuong, Shiela Valencia, Sant Virmani, Len Wade, and Ren Wang.

External reviewers: B.N. Singh, Director General, Central Rice Research Institute (India), N.I. Bhuiyan, Aung Kyi, and U Tin Htut Oo, Director General, Department of Agricultural Planning (Myanmar); Leo Sebastian, Director General, PhilRice (Philippines); Somporn Isvilanonda, Kasetsart University (Thailand); and Bui Chi Buu, Director, Cuu Long Rice Research Institute (Vietnam).

IRRI production team: Grant Leceta (design and layout), Emmanuel Panisales and Ana Mae Wenceslao (figures and maps), and Juan Lazaro IV (cover).

The publishers wish to thank all these people—and those whose names we may have missed—for their time and effort to ensure the accuracy and usefulness of the Almanac.

The facts of rice

Production

Rice farming is the largest single use of land for producing food.

Rice is nearly all (90%) produced in Asia.

Rice production totaled 600 million tons in 2000.

Rice is the most important economic activity on Earth.

Thousands of varieties of rice are farmed.

Only 6–7% of all rice production is exported from its country of origin.

Rice fields cover 11% of Earth's entire arable land, or more than 500 million hectares.

Employment

Rice eaters and growers form the bulk of the world's poor.

Rice is the single most important source of employment and income for rural people.

Rice is grown on 250 million Asian farms, mostly smaller than one hectare.

Significance in human culture

Rice farming is about 10,000 years old.

Rice cultivation was once the basis of the social order and occupied a major place in Asia's religions and customs.

Rice is still sometimes used to pay debts, wages, and rent in some Asian rural areas.

Significance as food

Rice is the staple food for the largest number of people on Earth.

Rice is eaten by nearly half the world's population.

Rice is the single largest food source for the poor.

Rice is the source of one quarter of global per capita energy.

Rice is synonymous with food throughout Asia.

Rice is the most important food grain in most of the tropical areas of Latin America and the Caribbean, where it supplies more calories in people's diets than wheat, maize, cassava, or potatoes.

Toyota means bountiful rice field.

Honda means the main rice field.

Benefits of rice research

Research has provided 75% of the rice varieties now grown.

Research has increased potential yields from 4 to more than 10 tonnes per hectare per crop.

Research has been a major factor in more than doubling world rice production from 260 to 600 million tonnes over the past 40 years.

Research has provided rice plants that grow faster, enabling 2 or even 3 crops per year; plants that resist various pests and diseases, need less fertilizer, or thrive in saline water; and plants with enhanced levels of micronutrients.

Many more facts on rice production are contained in the Rice Facts on page 249.

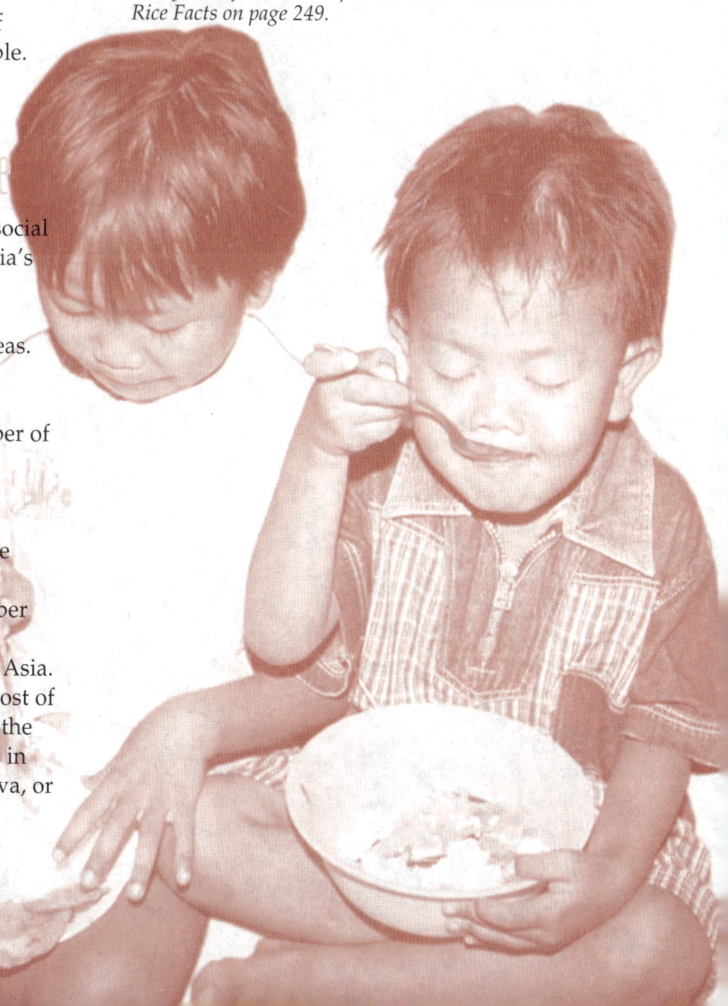

Importance of rice

Origin and diffusion

The origins of rice have long been debated. The plant is of such antiquity that the exact time and place of its first development will perhaps never be known. It is certain, however, that domestication of rice ranks as one of the most important developments in history. Rice has fed more people over a longer period than has any other crop.

Pottery shards bearing the imprint of both grains and husks of the cultivated rice species *Oryza sativa* were discovered at Non Nok Tha in the Korat area of Thailand. Plant remains from 10000 B.C. were discovered in Spirit Cave on the Thailand-Myanmar border.

In China, extensive archeological evidence points to the middle Yangtze and upper Huai rivers as the two earliest places of *O. sativa* rice cultivation in the country. Rice and farming implements dating back at least 8,000 years have been found. Cultivation spread down these rivers over the following 2,000 years.

Early spread of rice

From early, perhaps separate, beginnings in different parts of Asia, the process of diffusion has carried rice in all directions until today it is cultivated on every continent save Antarctica. In the early Neolithic era, rice was grown in forest clearings under a system of shifting cultivation. The crop was direct seeded, without standing water—conditions only slightly different from those to which wild rice was subject. A similar but independent pattern of the incorporation of wild rice into agricultural systems may well have taken place in one or more locations in Africa at approximately the same time.

Puddling the soil—turning it to mud—and transplanting seedlings were likely refined in China. Both operations became integral parts of rice farming and remain widely practiced to this day. Puddling breaks down the internal structure of soils, making them much less subject to water loss through percolation. In this respect, it can be thought of as a way to extend the utility of a limited water supply.

Banaue Rice Terraces, Philippines.

Transplanting is the planting of 1- to 6-wk-old seedlings in puddled soil with standing water. Under these conditions, the rice plants have an important head start over a wide range of competing weeds, which leads to higher yields. Transplanting, like puddling, provides farmers with the ability to better accommodate the rice crop to a finite and fickle water supply by shortening the field duration (since seedlings are grown separately and at higher density) and adjusting the planting calendar.

With the development of puddling and transplanting, rice became truly domesticated. In China, the history of rice in river valleys and low-lying areas is longer than its history as a dryland crop. In Southeast Asia, however, rice originally was produced under dryland conditions in the uplands, and only recently came to occupy the vast river deltas.

Migrant people from southern China or perhaps northern Vietnam carried the traditions of wetland rice cultivation to the Philippines during the second millennium B.C., and Deutero-Malays carried the practice to Indonesia about 1500 B.C. From China or Korea, the crop was introduced to Japan no later than 100 B.C.

Movement to western India and south to Sri Lanka was also accomplished very early. Rice was a major crop in Sri Lanka as early as 1000 B.C. The crop may well have been introduced to Greece and the neighboring areas of the Mediterranean by returning members of Alexander the Great's expedition to India around 344-324 B.C. From a center in Greece and Sicily, rice spread gradually throughout southern Europe and to a few locations in northern Africa.

Rice in the New World

As a result of Europe's great Age of Exploration, new lands to the west became available for exploitation. Rice cultivation was introduced to the New World by early European settlers. The Portuguese carried it to Brazil and the Spanish introduced its cultivation to several locations in Central and South America. The first record for North America dates from 1685, when the crop was produced on the coastal lowlands and islands of what is now South Carolina. The crop may well have been carried to that area by slaves brought from the African continent. Early in the 18th century, rice spread to what is now Louisiana, but not until the 20th century was it produced in California's Sacramento Valley. The introduction into California corresponded almost exactly with the timing of the first successful crop in Australia's New South Wales.

Genetic diversity

Two rice species are important cereals for human nutrition: *Oryza sativa,* grown worldwide, and *O. glaberrima,* grown in parts of West Africa. These two cultigens—species known only by cultivated plants—belong to a genus that includes about 20 other species.

Wild *Oryza* species are distributed throughout the tropics. They can be grouped into four complexes of closely related species (Table 1). Two species, however, seem to be different from others in the genus: the tetraploid *O. schlechteri* and the diploid *O. brachyantha.*

Species of the *O. ridleyi* complex inhabit lowland swamp forests and species of the *O. meyeriana* complex are found in upland hillside forests.

The *O. officinalis* complex consists of diploid and tetraploid species found throughout the tropics. All the species in this complex are perennial; some are rhizomatous and others form runners. They also differ in the habitats where they are found. Some occur in full sun, others in partial shade. Variation exists within these species as shown by the responses of different populations to pests and diseases.

The *O. sativa* complex consists of the wild and weedy relatives of the two rice cultigens as well as the cultigens themselves. The wild relatives of *O. glaberrima* in Africa consist of the perennial rhizomatous species *O. longistaminata,* which grows throughout Sub-Saharan Africa, and Madagascar, and the annual *O. barthii,* which extends from West Africa to East and Southern Central Africa. The annual and weedy relatives of *O. glaberrima* are found primarily in West Africa.

Among the wild relatives of *O. sativa,* the perennial *O. rufipogon* is widely distributed over South and Southeast Asia, southeast China, and Oceania; morphologically indistinguishable forms are found in South America, usually in deepwater swamps. A closely related annual wild form, *O. nivara,* is found in the Deccan Plateau and Indo-Gangetic Plain of India and in many parts of Southeast Asia. The habitats of *O. nivara* are ditches, water holes, and edges of ponds. Morphologically similar to (and sometimes in-

Table 1. Taxa in the genus *Oryza*: the complexes and genome groups to which they belong.

Complex/taxon	Genome group	Distribution
O. schlechteri	Tetraploid	Papua New Guinea
O. brachyantha	FF	Africa
***O. ridleyi* complex**		
O. longiglumis	Tetraploid	Papua New Guinea
O. ridleyi	Tetraploid	Southeast Asia
***O. meyeriana* complex**		
O. granulata	Diploid	South and Southeast Asia
O. meyeriana	Diploid	Southeast Asia
***O. officinalis* complex**		
O. officinalis	CC	Tropical Asia to Papua New Guinea
O. eichingeri	CC	East and West Africa
O. rhizomatis	CC	Sri Lanka
O. minuta	BBCC	Philippines, Papua New Guinea
O. punctata	BB, BBCC	Africa
O. latifolia	CCDD	Central and South America
O. alta	CCDD	Central and South America
O. grandiglumis	CCDD	South America
O. australiensis	EE	Australia
***O. sativa* complex**		
O. glaberrima (cultigen)	AA	West Africa
O. barthii	AA	Africa
O. longistaminata	AA	Africa
O. sativa (cultigen)	AA	Worldwide
O. nivara	AA	Tropical Asia
O. rufipogon	AA	Tropical Asia
O. meridionalis	AA	Tropical Australia
O. glumaepatula	AA	South America

distinguishable from) *O. nivara* are the very widely distributed weedy forms of *O. sativa* (*O. fatua*), which represent numerous different hybrids between *O. sativa* and its two wild relatives. Throughout South and Southeast Asia, these spontaneous forms are found in canals and ponds adjacent to rice fields and in the rice fields themselves.

The primary center of diversity for *O. glaberrima* is in the swampy basin of the upper Niger River. Two secondary centers are to the southwest near the Guinean coast.

Oryza glaberrima varieties can be divided into two ecotypes: deepwater and upland. In West Africa, *O. glaberrima* is a dominant crop grown in the flooded areas of the Niger and Sokoto River basins. It is broadcast on hoed fields. On shallowly flooded land, a rainfed wetland crop is either directly sown by broadcasting or dibbling, or transplanted. About 45% of the land planted to rice in Africa belongs to the upland (dryland) culture, largely under bush fallow or after the ground has been hoed. Some African farmers still use axes, hoes, and bush knives in land preparation. In hydromorphic soils, *O. glaberrima* behaves like a self-perpetuating weed. In wetland fields planted to *O. sativa*, *O. glaberrima* has become a weed.

Ecological diversification in *O. sativa*, which involved hybridization-differentiation-selection cycles, was enhanced when ancestral forms of the cultigen were carried by farmers and traders to higher latitudes, higher elevations, dryland sites, seasonal deepwater areas, and tidal swamps. Within broad geographic regions, two major ecogeographic races were differentiated as a result of isolation and selection: (1) indica, adapted to the tropics; and (2) japonica, adapted to the temperate regions and tropical uplands.

The combined forces of natural and human selection; diverse climates, seasons, and soils; and varied cultural practices (dryland preparation and direct seeding vs puddling of the soil and transplanting) led to the tremendous ecological diversity now found in Asian cultivars. Selections made to suit cultural preferences and socioreligious traditions added diversity to morphological features, especially grain size, shape, and color, and endosperm properties.

The complex groups of cultivars now known are categorized on the basis of hydrologic-edaphic-cultural-seasonal regimes as well as genetic differentiation (Fig. 1).

Within the last 2,000 years, dispersal and cultivation of the cultivars in new habitats have further accelerated the diversification process. Today, thousands of rice varieties are grown in more than 100 countries.

The full spectrum of germplasm in the genus *Oryza* consists of the following:

- Wild *Oryza* species, which occur throughout the tropics, and related genera, which occur worldwide in both temperate and tropical regions.

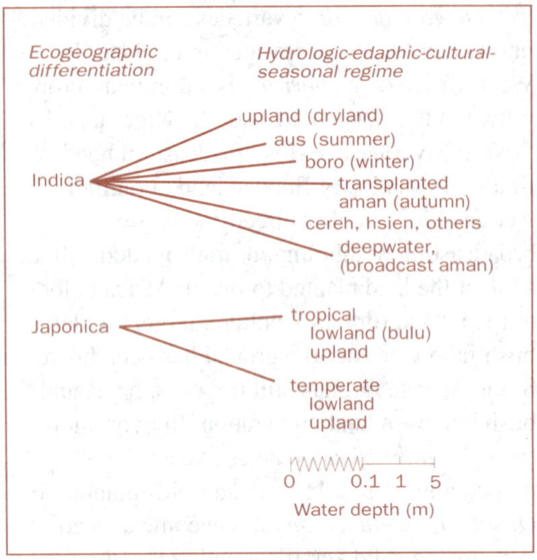

Fig. 1. Grouping of Asian rice cultivars by ecogeographic race, hydrologic-edaphic-cultural regime, and crop season. Cultivars grown in standing water belong to the lowland type.

- Natural hybrids between the cultigen and wild relatives, and primitive cultivars of the cultigen in areas of rice diversity.
- Commercial types, obsolete varieties, minor varieties, and special-purpose types in the centers of cultivation.
- Pure-line or inbred selections of farmers' varieties, elite varieties of hybrid origin, F_1 hybrids, breeding materials, mutants, polyploids, aneuploids, intergeneric and interspecific hybrids, composites, and cytoplasmic sources from breeding programs.

The diversity of Asian, African, and wild rice has given breeders a wealth of genetic material to draw on for breeding improved cultivars.

Rice-growing areas

Rice is produced in a wide range of locations and under a variety of climatic conditions, from the wettest areas in the world to the driest deserts. It is produced along Myanmar's Arakan Coast, where the growing season records an average of more than 5,100 mm of rainfall, and at Al Hasa Oasis in Saudi Arabia, where annual rainfall is less than 100 mm. Temperatures, too, vary greatly. In the Upper Sind in Pakistan, the rice season averages 33 °C; in Otaru, Japan, the mean temperature for the growing season is 17 °C. The crop is produced at sea level on coastal plains and in delta regions throughout Asia, and to a height of 2,600 m on the slopes of Nepal's Himalaya. Rice is also grown under an extremely broad range of solar radiation, ranging from 25% of potential during the main rice season in portions of Myanmar, Thailand, and India's Assam State to approximately 95% of potential in southern Egypt and Sudan.

Rice occupies an extraordinarily high portion of the total planted area in South, Southeast, and East Asia. This area is subject to an alternating wet and dry seasonal cycle and also contains many of the world's major rivers, each with its own vast delta. Here, enormous areas of flat, low-lying agricultural land are flooded annually during and immediately following the rainy season. Only two major food crops, rice and taro, adapt readily to production under these conditions of saturated soil and high temperatures.

The highest rice yields have traditionally been obtained from plantings in high-latitude areas that have long daylength and where intensive farming techniques are practiced, or in low-latitude desert areas that have very high solar energy levels. Southwestern Australia, Hokkaido in Japan, Spain, Italy, northern California, and the Nile Delta provide the best examples.

In portions of the rice world such as in South Asia, the crop is produced on miniscule plots using enormous amounts of human labor. At other locations, such as in Australia and the United States, it is raised on huge holdings with a maximum of technology and large expenditures of energy from fossil fuels. The contrasts in the geographic, economic, and social conditions under which rice is produced are truly remarkable.

Production

Among low- and middle-income countries, rice is by far the most important crop worldwide. In particular, rice is most closely associated with the South, Southeast, and East Asian nations extending from Pakistan to Japan. Here, the population pressure on limited land resources is high and a close balance is maintained between rice production and food needs. Within this area, rice is preeminent: it occupies more than one-third of total planted area in most countries and

Table 2. Rice production, area, and yield for countries producing more than 1 million t of rough rice, in order of decreasing rough rice production, 2000.

Country	Rough rice production (000 t)	Rice area (000 ha)	Arable land[a] (000 ha)	% of world rice		Rice yield (t/ha)
				Area	Production	
China	190,168	30,503	124,144	20	31.8	6.23
India	134,150	44,600	161,500	29	22.4	3.01
Indonesia	51,000	11,523	17,941	7	8.5	4.43
Bangladesh	35,821	10,700	7,992	7	6.0	3.35
Vietnam	32,554	7,655	5,700	5	5.4	4.25
Thailand	23,403	10,048	16,800	7	3.9	2.33
Myanmar[b]	20,125	6,211	9,548	4	3.4	3.24
Philippines	12,415	4,037	5,500	3	2.1	3.08
Japan	11,863	1,770	4,535	1	2.0	6.70
Brazil	11,168	3,672	53,200	2	1.9	3.04
USA	8,669	1,232	176,950	1	1.4	7.04
Korea, Rep. of	7,067	1,072	1,708	1	1.2	6.59
Pakistan	7,000	2,312	21,425	2	1.2	3.03
Egypt	5,997	660	2,834	0	1.0	9.09
Nepal	4,030	1,550	2,898	1	0.7	2.60
Cambodia	3,762	1,873	3,700	1	0.6	2.01
Nigeria	3,277	2,061	28,200	1	0.5	1.59
Sri Lanka	2,767	871	869	1	0.5	3.18
Iran	2,348	587	16,837	0	0.4	4.00
Madagascar	2,300	1,207	2,565	1	0.4	1.91
Lao PDR	2,155	690	800	0	0.4	3.12
Colombia	2,100	440	2,079	0	0.4	4.77
Malaysia	2,037	692	1,820	0	0.3	2.94
Korea, DPR	1,690	535	1,700	0	0.3	3.16
Peru	1,665	300	3,670	0	0.3	5.55
Argentina[b]	1,658	289	25,000	0	0.3	5.74
Ecuador	1,520	380	1,574	0	0.3	4.00
Australia	1,400	145	53,775	0	0.2	9.66
Italy	1,300	221	8,280	0	0.2	5.89
Uruguay	1,175	185	1,260	0	0.2	6.35
Côte d'Ivoire	1,162	750	2,950	0	0.2	1.55
World	598,852	153,766	1,380,239	100	100.0	3.89

[a]Arable land refers to land under temporary crops (double-cropped areas are counted only once), temporary meadows for mowing or pasture, land under market and kitchen gardens, and land temporarily fallow or lying idle. The figures are for 1998. [b]Data for rice area and production are for 1999. Source of basic data: FAO database, 2001.

one-fifth or more of planted area in the ecologically more diverse countries of China and India. Of 25 major rice-producing nations, 17 are located within this region (Table 2). The eight countries outside the region jointly produce less than 6% of the world's rice.

In the years since World War II, world rice area, yield, and production have changed considerably (Table 3). From 1948 to 1990, the area planted to rice increased by almost 71%, the mean yield obtained from that area went up by 110%, and total production more than tripled. During those four decades, rice became ever more important as a human food. World rice demand is predicted to increase at about 1% per year from 2001 to 2025, roughly equal to population growth in Asia during that period.

With the accelerating loss of productive rice land to rising sea levels, salinization, erosion, and human settlements, the problem becomes one of increasing yields under ever more trying circumstances. From 1965-67 through 1989-91, the improvements in productivity spawned by the Green Revolution spread rapidly. During those years, total rice production almost doubled. Most of this increase came from increased yields and increased cropping intensity, although some resulted from new land brought under cultivation or shifted into rice from other crops. Much of the yield increase could be traced to the introduction of the dwarfing gene and to the increased use of fertilizer, irrigation water, and other inputs. Further yield increases have been constrained by

Table 3. World rice area, yield, and production, various years.

Year	Arable land + permanent crops[a] (000 ha)	Rough rice Area (000 ha)	Rough rice Yield (t/ha)	Rough rice Production (000 t)
1948	1,232,000	86,700	1.68	145,400
1953	1,332,000	109,025	1.82	197,906
1958	1,390,000	117,017	1.92	224,093
1963	1,355,864	120,277	2.05	247,139
1968	1,377,311	129,449	2.23	288,714
1973	1,397,696	136,824	2.45	334,988
1978	1,423,870	143,638	2.68	385,106
1983	1,454,147	143,074	3.14	449,048
1988	1,497,719	146,252	3.34	488,292
1993	1,504,732	145,811	3.63	529,803
1998	1,380,239	152,002	3.81	578,786
1999	1,511,766	156,462	3.88	607,780
2000	1,511,766	153,766	3.89	598,852

[a]This column reports land in both temporary and permanent crops. Source of basic data: FAO production yearbook, 1952; FAOSTAT database.

diminishing returns and have been increasingly difficult to achieve.

Rice as human food

Rice, wheat, and maize are the three leading food crops in the world; together they directly supply more than 50% of all calories consumed by the entire human population. Wheat is the leader in area harvested each year with 214 million ha, followed by rice with 154 million ha and maize with 140 million ha (Table 4). Human consumption accounts for 85% of total production for rice, compared with 72% for wheat and 19% for maize.

Rice provides 21% of global human per capita energy and 15% of per capita protein. Although rice protein ranks high in nutritional quality among cereals, protein content is modest. Unmilled (brown) rice of 17,587 cultivars in the IRRI germplasm collection averages 9.5% protein content, ranging from 4.3% to 18.2%.

Environmental factors (soil fertility, wet or dry season, solar radiation, and temperature during grain development) and crop management (added N fertilizer, plant spacing) affect rice protein content. Breeding efforts to increase protein have been largely unsuccessful because of the considerable effects of environment and because of complex inheritance properties in the triploid endosperm tissue.

Rice also provides minerals, vitamins, and fiber, although all constituents except carbohydrates are reduced by milling. Milling removes roughly 80% of the thiamine from brown rice. A precook rinse or a boiling of milled rice results in additional loss of vitamins, especially B_1.

Average calorie intake per person in some developing countries—Cambodia and Madagascar—is as low as 2,000/day, while in some developed countries such as the U.S., it reaches almost twice that amount. The proportion that calories from rice make of the total calorie intake also varies greatly, from three-quarters in developing countries such as Bangladesh and Cambodia to only 2% in Turkey, Mexico, and the former USSR countries. The world average total calorie intake in 1999 was 2,808/person/day, of which rice made up 21%. The world average consumption of rice in 1999 was 58 kg, with the highest intake in some Asian countries; Myanmar has the highest annual consumption at 211 kg/person (Table 5).

Rice eaters and growers constitute the bulk of the world's poor: according to the UNDP *Human Development Report for 1997*, approximately 70% of the world's 1.3 billion poor people live in Asia, where rice is the staple food. To some extent, this reflects Asia's large population, but even in relative terms malnutrition appears to affect a substantially larger share of the population in South Asia than in Africa. For these people, rice is the most important commodity in their daily lives. In countries such as Bangladesh, Vietnam, and Myanmar, the average citizen consumes 150–200 kg annually, which accounts for two-thirds or more of caloric intake and approximately 60% of

Table 4. World food picture.

Human population (million)	6,055.1
Land use, 1998 (million ha)	
Total land area	13,048.4
Arable land	1,380.2
Permanent crops	131.5
Permanent pastures	3,426.5
Forest and woodlands (1994)	4,157.2
Other land	3,952.9

Food production

Crop	Area (million ha)	Production (million t)	Food (million t) 1999		Per capita/day (1999) Calories	Protein (g)
Rice (rough)	153.8	598.9	515.9	65% milling rate	577	11
Maize	139.7	590.8	113.0	80% for feed	156	4
Wheat	213.6	576.3	416.0	70% milling rate	537	16
Millet and sorghum	78.0	85.8	44.9	30% milling rate	62	2
Barley and rye, excluding beer	66.9	152.0	14.2	70% milling rate	17	0.4
Oats	12.8	26.0	2.9	65% milling rate	3	0.1
Potato	18.8	311.3	184.7	60% for feed	57	1
Sweet potatoes and yams	13.4	178.7	87.2	50% for feed	39	0.4
				Subtotal	1,448	34.9
				All foods	2,808	75

Table 5. Rice consumption, caloric intake, and percent of calories from rice, 1999.

Country	Milled rice consumption[a] (kg/capita/year)	Total[b] calories/ capita/ day	Rice calories/ capita/ day	% calories from rice
Myanmar	211	2,803	2,050	73
Lao PDR	171	2,152	1,516	70
Vietnam	170	2,564	1,676	65
Bangladesh	168	2,201	1,676	76
Cambodia	165	2,000	1,527	76
Indonesia	154	2,931	1,525	52
Thailand	101	2,411	1,004	42
Philippines	100	2,357	974	41
Madagascar	91	1,994	926	46
China	91	3,045	911	30
India	74	2,417	736	30
Japan	60	2,782	642	23
Egypt	41	3,323	426	13
Brazil	40	3,012	409	14
Nigeria	23	2,833	237	8
Pakistan	15	2,462	150	6
South Africa	12	2,805	119	4
USA	9	3,754	94	3
Turkey	7	3,469	65	2
Mexico	6	3,168	61	2
USSR (former area)	5	2,778	47	2
World	58	2,808	577	21

[a]Amount available for human consumption. [b]Data include all food available for human consumption. Source: FAO online database (FAO update 31 May 2001).

daily protein consumption. Even in relatively wealthier countries such as Thailand and Indonesia, rice still accounts for nearly 50% of calories and one-third or more of protein.

Rice is also the most important crop to millions of small farmers who grow it on millions of hectares throughout the region, and to the many landless workers who derive income from working on these farms. In the future, it is imperative that rice production continue to grow at least as rapidly as the population, if not faster. Rice research that develops new technologies for all farmers has a key role to play in meeting this need and contributing to global efforts directed at poverty alleviation.

Where rice is the main item of the diet, it is frequently the basic ingredient of every meal and is normally prepared by boiling or steaming. In Asia, bean curd, fish, vegetables, meat, and spices are added depending on local availability and economic situation. A small proportion of rice is consumed in the form of noodles, which serve as a bed for various, often highly spiced, specialties or as the bulk ingredient in soups.

Most rice is consumed in its polished state. When such rice constitutes a high proportion of food, dietary deficiencies may result. Despite the dramatic losses in food value resulting from milling, brown rice is unpopular because (1) it requires more fuel for cooking, (2) it may cause digestive disturbances, and (3) oil in the bran layer tends to turn rancid during storage even at moderate temperatures.

In contrast, parboiling rough rice before milling, as is common in India and Bangladesh, allows a portion of the vitamins and minerals in the bran to permeate the endosperm and be retained in the polished rice. This treatment also lowers protein loss during milling and increases whole-grain recovery.

Even though rice diets are often marginally deficient in protein, vitamins, and minerals, clinical manifestations of deficiency are not common among people whose diets are otherwise adequate in calories. The exception is when people do heavy labor and their higher calorie demand is met by an increase in rice without a corresponding increase in other foods such as legumes or fish. Under these conditions, there is danger of beriberi, which is related to a deficiency of thiamine or vitamin B_1.

Research is under way to fortify rice with micronutrients in areas where these are inadequate in the diet. Vitamin A is an important one—a severe lack causes irreversible blindness—and has now been incorporated in a variety known as Golden Rice. Another new variety is rich in iron and zinc, micronutrients often deficient in people consuming mainly rice. These fortified rice varieties are being tested in nutrition trials before they are grown commercially.

Rice and population

Agricultural population densities on Asia's rice-producing lands are among the highest in the world and continue to increase at a remarkable rate. Rapid population growth puts increasing pressure on the already strained food-producing resources.

The aggregate population of the less developed countries grew from 2.3 billion in 1965 to 4.4 billion in 1995. Asia accounted for 60% of the global population, about 92% of the world's rice production, and 90% of global rice consumption. Even with rice providing 35–80% of the total calories consumed in Asia and with a slowing of growth in total rice area, rice production more than kept up with demand in 2000. The largest producing countries—China, India, Indonesia, Bangladesh, Vietnam, and Thailand—together account for more than three-quarters of world rice production.

The world's annual rough rice production, however, will have to increase markedly over the next 30 years to keep up with population growth and income-induced demand for food. Although the global population expanded by only 40 million in 1950, about 80 million people are being added per year at the start of the 21st century (Table 6), most in the less developed countries where population growth is much faster than in industrialized countries.

From 1961 to the early 1990s, the proportion of rice traded on world markets ranged from 3.5% to 5% of total production. Since then, it has been gradually increasing to an average of 6%. However, most countries rely almost entirely on domestic production to feed their populations. With less reliable supplies in the past, the world rice trade was volatile, but now is no more unstable than that for wheat or maize. Nevertheless, rice is still regarded as a political commodity in some Asian countries.

Specialty uses of rice

Glutinous rice plays an important role in some cultures. In Lao PDR and northeast Thailand, for example, glutinous rice is the staple food. In other cultures, it is prepared in a sweetened form for snacks, desserts, or special foods for religious or ceremonial occasions. In a few areas, glutinous rice is pounded and roasted to be eaten as a breakfast cereal.

Table 6. World population, 1995-2000, with projection to 2050.

Year	Population (millions)	Annual growth rate (%)	Annual increase (millions)
1950	2,522	1.6	40
1960	3,022	1.8	55
1970	3,696	2.0	75
1980	4,440	1.9	82
1990	5,266	1.7	91
1995	5,666	1.5	84
2000	6,055	1.3	81
2010	6,795	1.2	79
2020	7,502	1.0	75
2025	7,824	0.8	66
2030	8,112	0.7	59
2040	8,577	0.6	48
2050	8,909	0.4	34

FAO Database, 2001.

Alcoholic beverages made from rice are found throughout the rice-producing world. The most common is a rice beer produced by boiling husked rice, inoculating the mix with a bit of yeast cake, and allowing the mixture to ferment for a short period.

The mash left at the bottom of the container is often prized. Among the Ifugao of the Philippines, the mash is frequently reserved for the village priest. Among the Kachins of Myanmar, it is the first food offered to a recently captured and hungry wild elephant. Kachins believe that the elephant will be loyal forever to the person who first provides such a meal.

Sake is widely consumed in Japan, as is wang-tsiu in China. These rice-based winelike beverages are served warm and featured at ceremonial feasts.

In some parts of the world, especially in North America and Europe, rice is developing a new market niche as a staple and as a gourmet food. This trend appears to be related to the arrival of large numbers of immigrants from Southeast Asia, who introduced aromatic rice to markets where it was previously unknown. It has been adopted by a food quality-conscious public over the past several years.

In much of Tanzania, rice is used for making bread; in the south, it is also used in ceremonies. In West Africa, rice bread, rice cake, and rice porridge are used for ceremonies such as funerals and weddings. Some "old" varieties (most likely *O. glaberrima*) are used in traditional religious rituals in West Africa, while certain parts of some varieties are used as medicines in the traditional treatment of illnesses.

An extensive list of other ways of using rice is given by FAO[1]:

- Milled rice is marketed precooked, canned, dried, and puffed for breakfast cereals as rice flour; extrusion-cooked foods; puddings and breads; cakes and crackers; noodles and rice paper; fermented foods and vinegars; rice starch; and syrups.
- Rice bran, which forms 5% to 8% of the grain weight, is used as livestock feed, a pickling medium, a medium for growing mushrooms, and as a growing medium for some enzymes, as well as for flours, concentrates, oils, and dietary fiber.
- Hulls and husks, about 20% of the grain weight, are used for fuel, bedding, and incubation material, and as a seedbed medium, as well as being sometimes incorporated in livestock feeds, concrete blocks, tiles, fiberboard, ceramics, cement, filters, charcoal briquettes, and cooking gas production.
- Rice straw, more or less equivalent in production weight to grain, is used as fuel for cooking, roofing material, livestock feed, fertilizer, and a medium for growing mushrooms.

[1]FAO Rice Information Vol. 2, January 2000. FAO, Rome.

The rice plant and its ecology

Morphology

Cultivated rice is generally considered a semiaquatic annual grass, although in the tropics it can survive as a perennial, producing new tillers from nodes after harvest (ratooning). At maturity, the rice plant has a main stem and several tillers. Each productive tiller bears a terminal flowering head or panicle. Plant height varies by variety and environmental conditions, ranging from approximately 0.4 m to more than 5 m in some floating rice. The morphology of rice is divided into the vegetative phase (including germination, seedling, and tillering stages) and the reproductive phase (including panicle initiation and heading stages).

Seeds

The rice grain, commonly called a seed, consists of the true fruit or brown rice (caryopsis) and the hull, which encloses the brown rice. Brown rice consists mainly of the embryo and endosperm. The surface contains several thin layers of differentiated tissues that enclose the embryo and endosperm (Fig. 1).

The palea, lemmas, and rachilla constitute the hull of indica rice. In japonica rice, however, the hull usually includes rudimentary glumes and perhaps a portion of the pedicel.

A single grain weighs about 10–45 mg at 0% moisture content. Grain length, width, and thickness vary widely among varieties. Hull weight averages about 20% of total grain weight.

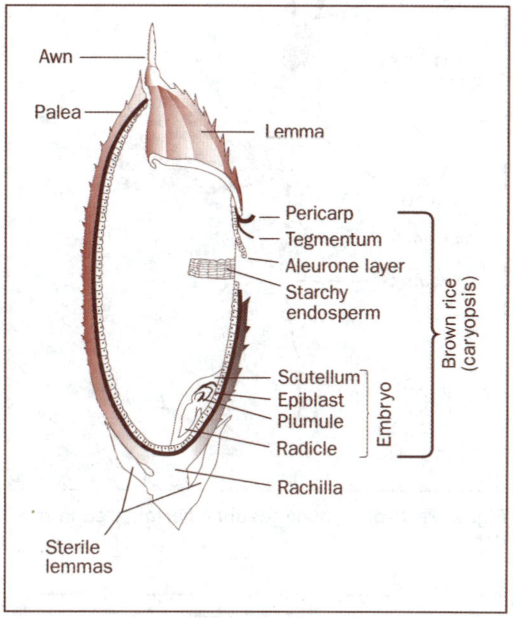

Fig. 1. Cross-section of the rice grain.

Seedlings

Germination and seedling development start when seed dormancy has been broken and the seed absorbs adequate water and is exposed to a temperature ranging from about 10 to 40 °C. The physiological definition of germination is usually the time when the radicle or coleoptile (embryonic shoot) emerges from the ruptured seed coat. Under aerated conditions, the seminal root is the first to emerge through the coleorhiza from the embryo, and this is followed by the coleoptile. Under anaerobic conditions, however, the coleoptile is the first to emerge, with the roots developing when the coleoptile has reached the aerated regions of the environment.

If the seed develops in the dark as when seeds are sown beneath the soil surface, a short stem (mesocotyl) develops, which lifts the crown of the plant to just below the soil surface (Fig. 2). After the coleoptile emerges, it splits and the primary leaf develops.

Tillering plants

Each stem of rice is made up of a series of nodes and internodes (Fig. 3). The internodes vary in length depending on variety and environmental conditions, but generally increase from the lower to the upper part of the stem. Each upper node bears a leaf and a bud, which can grow into a tiller. The number of nodes varies from 13 to 16, with only the upper 4 or 5 separated by long

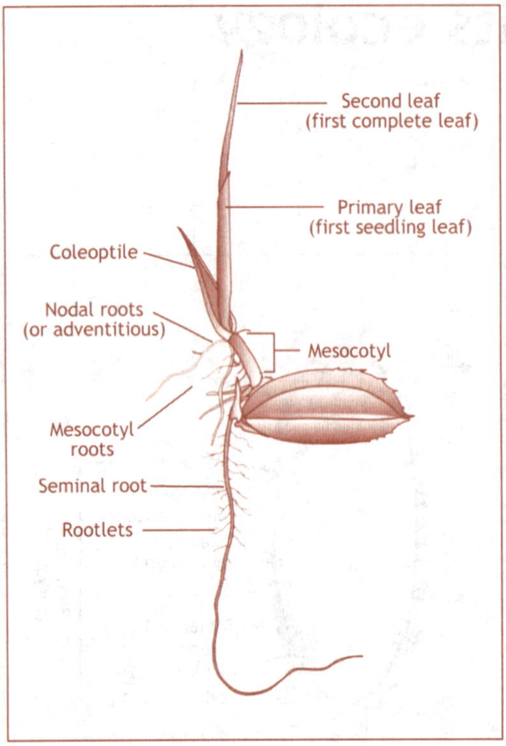

Fig. 2. Parts of a young seedling germinated in the dark.

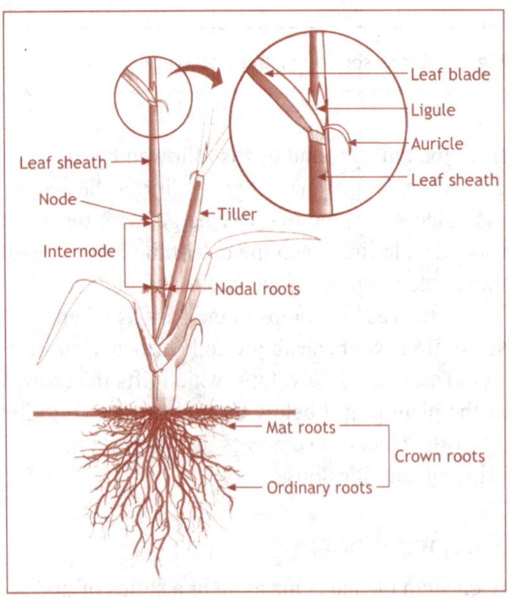

Fig. 3. Parts of the rice stem and tillers.

internodes. Under rapid increases in water level, some deepwater rice varieties can also increase the lower internode lengths by more than 30 cm each.

The leaf blade is attached at the node by the leaf sheath, which encircles the stem. Where the leaf blade and the leaf sheath meet is a pair of clawlike appendages, called the auricles. Coarse hairs cover the surface of the auricles. Immediately above the auricles is a thin, upright membrane called the ligule.

The tillering stage starts as soon as the seedling is self-supporting and generally finishes at panicle initiation. Tillering usually begins with the emergence of the first tiller when seedlings have five leaves. This first tiller develops between the main stem and the second leaf from the base of the plant. Subsequently, when the sixth leaf emerges, the second tiller develops between the main stem and the third leaf from the base.

Tillers growing from the main stem are called primary tillers. These may generate secondary tillers, which may in turn generate tertiary tillers. These are produced in a synchronous manner. Although the tillers remain attached to the plant, at later stages they are independent because they produce their own roots. Varieties and races of rice differ in tillering ability. Numerous environmental factors also affect tillering such as spacing, light, nutrient supply, and cultural practices.

The rice root system consists of two major types: crown roots (including mat roots) and nodal roots (Fig. 3). In fact, both these roots develop from nodes, but crown roots develop from nodes below the soil surface. Roots that develop from nodes above the soil surface usually are referred to as nodal roots. Nodal roots are often found in rice cultivars growing at water depths above 80 cm. Most rice varieties reach a maximum depth of 1 m or more in soft upland soils. In flooded soils, however, rice roots seldom exceed a depth of 40 cm. That is largely a consequence of limited O_2 diffusion through the gas spaces of roots (aerenchyma) to supply the growing root tips.

Panicle and spikelets

The major structures of the panicle are the base, axis, primary and secondary branches, pedicel, rudimentary glumes, and spikelets. The panicle axis extends from the panicle base to the apex; it has 8–10 nodes at 2- to 4-cm intervals, from which primary branches develop. Secondary branches develop from the primary branches. Pedicels develop from the nodes of the primary

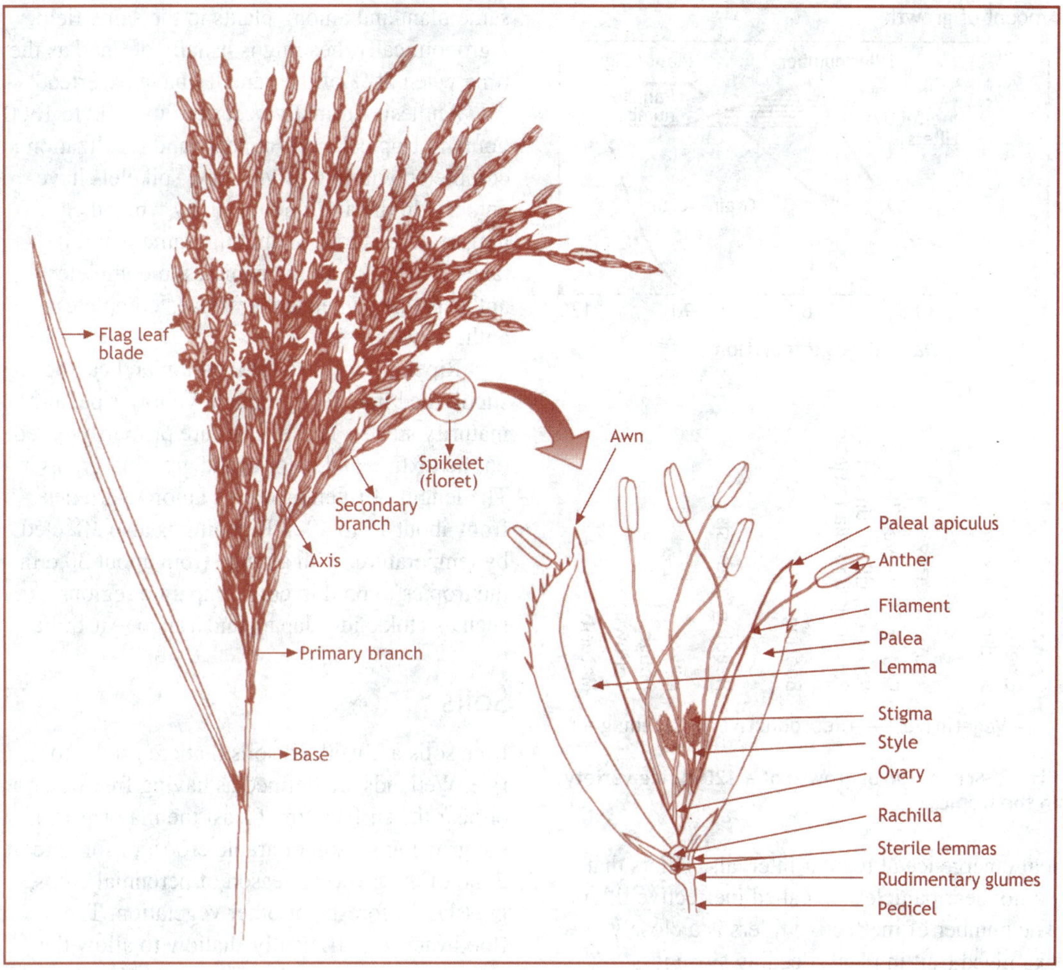

Fig. 4. Rice panicle and spikelets.

and secondary branches; the spikelets are positioned above them (Fig. 4).

Since rice has only one fully developed floret (flower) per spikelet, these terms are often used interchangeably. The flower is enclosed in the lemma and palea, which may be either awned or awnless. The flower consists of the pistil and stamens, and the components of the pistil are the stigmas, styles, and ovary.

Growth

The growth duration of the rice plant is 3–6 months, depending on the variety and the environment under which it is grown. During this time, rice completes two distinct growth phases: vegetative and reproductive. The vegetative phase is subdivided into germination, early seedling growth, and tillering; the reproductive phase is subdivided into the time before and after heading, that is, panicle exsertion. The time after heading is better known as the ripening period (Fig. 5).

Potential grain yield is primarily determined before heading. Ultimate yield, which is based on the amount of starch that fills the spikelets, is largely determined after heading. Hence, agronomically, it is convenient to regard the life history of rice in terms of three growth phases: vegetative, reproductive, and ripening. A 120-d variety, when planted in a tropical environment, spends about 60 d in the vegetative phase, 30 d in the reproductive phase, and 30 d in the ripening phase.

Vegetative phase

The vegetative phase is characterized by active tillering, a gradual increase in plant height, and

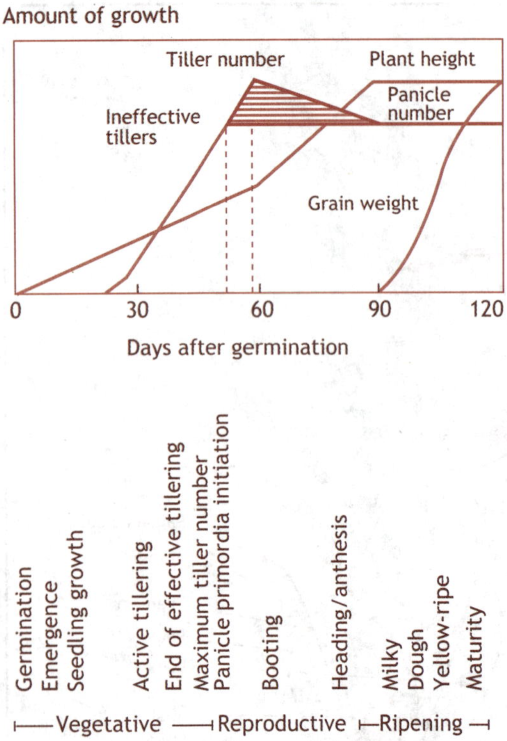

Fig. 5. Schematic of growth of a 120-d rice variety in the tropics.

leaf emergence at regular intervals. Tillers that do not bear panicles are called ineffective tillers. The number of ineffective tillers is a closely examined trait in plant breeding since it is undesirable in irrigated varieties, but sometimes an advantage in rainfed lowland varieties in which productive tillers or panicles may be lost because of unfavorable conditions.

Reproductive phase

The reproductive growth phase is characterized by culm elongation (which increases plant height), a decline in tiller number, emergence of the flag leaf (the last leaf), booting, heading, and flowering of the spikelets. Panicle initiation is the stage about 25 d before heading when the panicle has grown to about 1 mm long and can be recognized visually or under magnification following stem dissection.

Spikelet anthesis (or flowering) begins with panicle exsertion (heading) or on the following day. Consequently, heading is considered a synonym for anthesis in rice. It takes 10–14 d for a rice crop to complete heading because there is variation in panicle exsertion among tillers of the same plant and among plants in the same field. Agronomically, heading is usually defined as the time when 50% of the panicles have exserted.

Anthesis normally occurs from 1000 to 1300 hours in tropical environments and fertilization is completed within 6 h. Very few spikelets have anthesis in the afternoon, usually when the temperature is low. Within the same plant, it takes 7–10 d for all the panicles to complete anthesis; the spikelets themselves complete anthesis within 5 d.

Ripening follows fertilization and can be subdivided into milky, dough, yellow-ripe, and maturity stages. These terms are primarily based on the texture and color of the growing grains. The length of ripening varies among varieties from about 15 to 40 d. Ripening is also affected by temperature, with a range from about 30 d in the tropics to 65 d in cool temperate regions, such as Hokkaido, Japan, and Yanco, Australia.

Soils

Rice soils are wetland soils that are grown to rice. Wetlands are defined as having free water at or near the surface for at least the major part of the growing season of arable crops, or for at least 2 mo of the growing season of perennial crops, grasslands, forests, or other vegetation. The floodwater is sufficiently shallow to allow the growth of a crop or of natural vegetation rooted in the soil. Free surface water may occur naturally, or rainfall, runoff, or irrigation water may be retained by field bunds, puddled plow layers, or traffic pans.

Wetlands have at least one wet growing season, but may be dry, moist, or without surface water in other seasons. Wetland soils may therefore alternately support wetland and upland crops when cultivated.

The transition from wetlands to uplands is often gradual. It may fluctuate from year to year, depending on variations in precipitation, runoff, or irrigation. If water (both drainage and irrigation) can be fully controlled, farmers can choose to establish wetlands or uplands. But, in most wetlands, drainage capacities are insufficient to prevent soil submergence during the rainy season, particularly in the lowlands of the humid tropics.

The presence of "aquic" soil conditions is indicated by redoximorphic features, such as zones of accumulation and depletion of Fe and

Table 1. Typical profile of a flooded rice soil.

Horizon	Description
Ofw	A layer of standing water that becomes the habitat of bacteria, phytoplankton, macrophytes (submerged and floating weeds), zooplankton, and aquatic invertebrates and vertebrates. The chemical status of the floodwater depends on the water source, soil, nature and biomass of aquatic fauna and flora, cultural practices, and rice growth. The pH of the standing water is determined by the alkalinity of the water source, soil pH, algal activity, and fertilization. Because of the growth of algae and aquatic weeds, the pH and O_2 content undergo marked diurnal fluctuations. During daytime, the pH may increase to 11 and the standing water becomes oversaturated with O_2 because of photosynthesis of the aquatic biomass. Standing water stabilizes the soil water regime, moderates the soil temperature regime, prevents soil erosion, and enhances C and N supply.
Apox	The floodwater-soil interface that receives sufficient O_2 from the floodwater to maintain a pE + pH value above the range below which NH_4^+ is the most stable form of N. The thickness of the layer may range from several millimeters to several centimeters, depending on pedoturbation by soil fauna and the percolation rate of water.
Apg	The reduced puddled layer is characterized by the absence of free O_2 in the soil solution and a pE + pH value below the range at which Fe(III) is reduced.
Apx	A layer that has increased bulk density, high mechanical strength, and low permeability. It is frequently referred to as a plow or traffic pan.
B	The characteristics of the B horizon depend highly on water regime. In epiaquic moisture regimes, the horizon generally remains oxidized, and mottling occurs along cracks and in wide pores. In aquic moisture regimes, the whole horizon, or at least the interior of soil peds, remains reduced during most years.

Mn. Plowing and puddling often result in the development of a dense layer below the cultivated topsoil.

Three types of water saturation occur in rice soils: (1) endosaturation, in which the entire soil is saturated with water; (2) episaturation, in which upper soil layers are saturated but underlain by unsaturated subsoil layers; and (3) anthric saturation, a variant of episaturation with controlled flooding and puddled surface soil.

The properties of a typical soil profile of a flooded rice soil during the middle of a growing season are shown in Table 1.

Many rice-growing countries have developed classification systems that distinguish natural wetland soils from rice soils. The only soil classification systems applicable worldwide are found in the legend of the FAO-UNESCO Soil Map of the World and the U.S. Department of Agriculture Soil Taxonomy.

In the FAO-UNESCO Soil Map of the World, Gleysols, Fluvisols, Planosols, Plinthosols, and Histosols make up most of the wetland soils. Gleyic subunits of Arenosols, Andosols, Cambisols, Solonetz, Solonchaks, Chernozems, Pharozems, Greyzems, Luvisols, Podsoluvisols, Podsols, Lixisols, Acrisols, and Alisols are also mostly wetland soils. Although Vertisols, Nitosols, and Ferralsols have no gleyic subunits, these soils may be artificially flooded and sown to rice.

Soil taxonomy does not recognize wetland soils, but classifies soils with aquic conditions at the suborder level and soils with hydromorphism at the subgroup level. Most hydromorphic soils, which have an aquic moisture regime, are equivalent to wetland soils in soil taxonomy. Suborders with aquic conditions are found in the orders of Spodosols, Andosols, Oxisols, Vertisols, Ultisols, Mollisols, Alfisols, Inceptisols, and Entisols. Most Histosols are wetland soils per se. Practically all Aridisols are upland soils. Soils within the aquic subgroups showing hydromorphism generally are not wetland soils because signs of wetness are found only in subsoil horizons.

Research is now being conducted to grow rice on dry, but irrigated land. See "aerobic rice" on page 32.

Environments

The rice plant's environments have been divided into agroecological zones (AEZs), which are geographic mapping units developed by FAO. They are based on climatic conditions and land forms that determine relatively homogeneous crop growing environments. Characterization of AEZs permits a quantitative assessment of the biophysical resources upon which agriculture and forestry research depend.

The classification system distinguishes among tropical regions, subtropical regions with summer or winter rainfall, and temperate regions. These major regions are further subdivided into rainfed moisture zones, lengths of growing period, and thermal zones based on the temperature regime that prevails during the growing season.

The definitions of terms used in AEZ classifications are

tropics	regions with monthly mean temperature, corrected to sea level, of >18 °C for all months
subtropics	regions with monthly mean temperature, corrected to sea level, of <18 °C for one or more months
temperate	regions with monthly mean temperature, corrected to sea level, of <5 °C for one or more months
warm	daily mean temperature during the growing period of >20 °C
cool	daily mean temperature during the growing period in the range of 5–20 °C (including the moderately cool range of 15–20 °C)
warm/cool	daily mean temperature during part of the growing period´ <20 °C
arid	LGP[1] <75 d
semiarid	LGP 75–180 d
subhumid	LGP 180–270 d
humid	LGP >270 d

The distribution of AEZs in Asia is shown in Figure 6.

AEZs, population pressure, and food grain production

Population pressure on arable land is highest in the humid and subhumid subtropics and humid tropics but these AEZs also have favorable growing conditions for food grain crops. The production of food grain per hectare of arable land is about 5.7 times higher in the subhumid subtropics (southern and southwestern China and Taiwan) than in the semiarid tropics (southern and western India). The arid and semiarid tropics have the lowest production potential of the AEZs. The higher production potential in favorable AEZs, however, is offset by higher population pressure on land. So the difference between AEZs in food grain production per capita is small.

Analysis of land-use patterns in the AEZs of Asia indicates that rice is the dominant food crop in the humid subtropics, humid tropics, and subhumid tropics, where it makes up half to three-quarters of the area under food grains. It is a less important crop in the subhumid subtropics (36%) and semiarid tropics (19%), and is insignificant in the semiarid and cool subtropics.

Rice ecosystems

Rice ecosystems can be defined in various ways. This selection describes the four broad ecosystems used by IRRI—irrigated, rainfed lowland, upland, and flood-prone. In West Africa and Latin America and the Caribbean, some differently defined ecosystems are used (see "Rice around the world").

The most recent compilation of data, for the mid-1990s, on the extent of rice area in some ecosystems as defined by IRRI, is given in Table 2.

The irrigated rice ecosystem, which accounts for about half of the harvested rice area and contributes three-quarters of global rice production, is concentrated mostly in the humid and subhumid subtropics and humid tropics. This ecosystem is gradually losing land to urbanization and industrialization, and farm yields are approaching the ceiling of average yields obtained in experiment stations. It is characterized by high cropping intensity and intensive use of agrochemicals, with potential adverse effect on human health and sustainability of the natural resource base. The other, less favorable, rice ecosystems dominate the subhumid tropics (eastern India, Myanmar, Thailand) and large parts of the humid tropics (Bangladesh, Cambodia, Lao PDR). These are

[1]LGP = length of growing period: number of days during the year when available soil moisture supply (rainfed) is greater than half of potential evapotranspiration. Includes the period required to evapotranspire up to 100 mm of available soil moisture stored in the soil profile. Excludes any time interval when daily mean temperature is <5 °C.

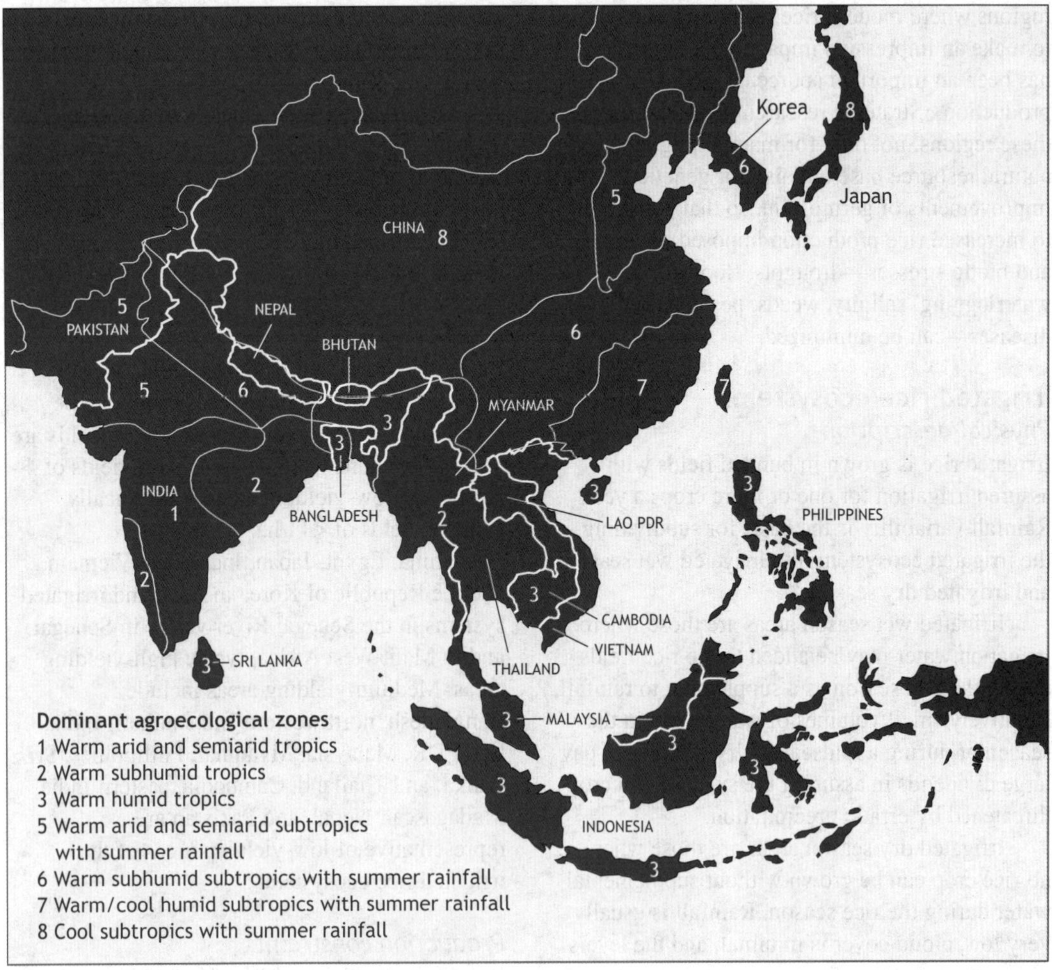

Fig. 6. Regional agroecological zones in Asia.

Table 2. Asian rice areas as of mid-1990s (000 ha).

State	Upland	Deepwater (flood-prone) (>100 cm)	Irrigated		Rainfed		Total rice area
			Wet	Dry	Shallow (0–30 cm)	Intermediate (30–100 cm)	
India	5,060	1,364	15,537	4,123	11,985	4,447	42,516
China	499	0	20,490	9,146	1,990	0	32,125
Indonesia	1,209	2	2,963	2,963	2,872	1,006	11,015
Bangladesh	697	1,220	351	2,267	3,271	2,873	10,679
Thailand	203	342	274	665	6,382	1,778	9,644
Vietnam	322	177	1,630	1,630	1,963	651	6,373
Myanmar	214	362	1,812	1,386	2,033	478	6,285
Philippines	165	0	1,175	1,029	911	341	3,621
Pakistan	0	0	2,125	0	0	0	2,125
Cambodia	24	152	140	165	1,069	349	1,899
Nepal	68	118	706	24	406	166	1,488
Korea, Rep. of	1	0	776	0	326	0	1,103
Sri Lanka	0	0	377	251	213	26	867
Korea, DPR	88	0	456	0	136	0	680
Malaysia	80	0	228	210	135	15	668
Lao PDR	219	0	33	11	348	0	611
Taiwan, China	0	0	133	133	0	0	266
Bhutan	4	0	5	0	16	1	26
Total	8,853	3,737	49,211	24,003	34,056	12,131	131,991

Source: Huke RE, Huke EH. 1997. Rice area by type of culture: South, Southeast, and East Asia. IRRI, Los Baños, Philippines.

regions where modern rice technologies have yet to make an impressive impact; area expansion has been an important source of growth in rice production. Strategic research is needed for these regions, not only for maintaining the natural resource base but also for genetic improvements of germplasm, so that constraints to increased rice production imposed by abiotic and biotic stresses—droughts, floods, waterlogging, salinity, weeds, pests, and diseases—can be minimized.

Irrigated rice ecosystem

Physical description

Irrigated rice is grown in bunded fields with assured irrigation for one or more crops a year. Rainfall variability is the basis for subdividing the irrigated ecosystem into irrigated wet season and irrigated dry season.

Irrigated wet season areas are those where irrigation water may be added to the rice fields during the wet season as a supplement to rainfall. Relatively small volumes of water early in the season or during a midseason dry period can pay large dividends in assuring the success of a crop threatened by erratic precipitation.

Irrigated dry season areas are those where no rice crop can be grown without supplemental water during the rice season. Rainfall is usually very low, cloud cover is minimal, and the levels of incoming solar radiation are markedly above those of the rainy season. In the dry season, evapotranspiration is high and water needs are considerably greater than during the wet season. In most areas, the combination of assured water throughout the season, high solar radiation, low pest incidence, and high input levels results in high yields.

Worldwide, about 79 million ha of rice are grown under irrigated conditions. Average yields vary from 3 to 9 t/ha. Because the risk of crop failure is lower than in the other ecosystems, farmers of irrigated land use more purchased inputs than in nonirrigated production. Even though irrigated rice represents only about one-half of the world's rice land, more than 75% of the world's rice supply is produced from irrigated rice. Improved rice cultivars that have been developed for irrigated rice are short in height and responsive to N fertilization. Equally important is their short duration, allowing two and sometimes three crops per year. Additionally, most improved cultivars have resistance to several insects and diseases and some tolerance for adverse soils.

East Asia, 93% of which is irrigated, accounts for about 43% of the world's irrigated rice area. In the tropics of South and Southeast Asia, only about 40% of the rice area is irrigated. In West Africa, only about 10.5% of the rice area is irrigated.

Productivity

Based on current national or regional yield averages, the irrigated rice ecosystem can be divided into high-yielding areas where yields are >5 t/ha, medium-yielding areas with yields of 4–5 t/ha, and low-yielding areas that typically achieve yields of <4 t/ha.

China, Egypt, Japan, Indonesia, Vietnam, and the Republic of Korea in Asia, and irrigated systems in the Senegal River valley in Senegal, and in Mali, West Africa, typify high-yielding areas. Medium-yielding areas include Bangladesh, northwestern and southern India, Lao PDR, Malaysia, Myanmar, Philippines, Sri Lanka, and Thailand. Cambodia, eastern India, Madagascar, Nepal, and Pakistan are representative of low-yielding areas of the irrigated rice ecosystem.

Production constraints

Beyond the basic yield-limiting factors, such as radiation, temperature, and availability of water, N, and P, and factors that relate to crop management, the main yield-limiting factors in the low- and medium-yielding areas are (1) poor management of production inputs; (2) losses from weeds, pests, and diseases; (3) inadequate land formation, leveling, and irrigation water; and (4) inadequate drainage, leading to the buildup of salinity and alkalinity. Some of the same problems limit yields in the high-yielding areas as well. The main problems encountered, however, are (1) an average on-farm yield plateau of about 6 t/ha in some countries, (2) yield instability because of pests, (3) difficulty of sustaining high yields under intensive cropping, and (4) environmental degradation because of excessive inputs.

Research accomplishments for the irrigated rice ecosystem have been significant. They include the development of semidwarf varieties with high yield potential (10–11 t/ha), short growth duration (around 100 d), greater yield stability due to genetic resistance to pests, and, to a lesser extent, tolerance for some environmental stresses such as problem soils.

Breeding programs have improved grain quality and increased grain size and milling recovery of the modern, high-yielding varieties. Increasing nutritional value by incorporating micronutrients such as iron and zinc or beta-carotene (vitamin A) are currently goals of the breeding program at IRRI.

The irrigated ecosystem benefited greatly from the development of associated management practices that allow the semidwarf varieties to express their yield potential. In partnership with national agricultural research and extension systems, international research on the irrigated rice ecosystem focuses on yield-limiting factors and technological advances to overcome them. Research is also conducted on the resilience of intensively cropped systems, including identifying factors that contribute to stagnating or even declining yields.

The objectives of IRRI's research are to extend the yield frontier through varietal improvement, more efficient use of inputs, and rehabilitation and maintenance of the physical, biological, and natural resource base. Technologies to increase the efficiency with which the rice crop uses inputs should improve rice lands, reduce dependence on pesticides, and increase the profitability of rice production.

WARDA emphasizes integrated rice management options and short-duration, high-yielding cultivars that are weed-competitive and N-use efficient to close the yield gap, that is, the difference between actual yields in farmers' fields and yield potentials of existing rice varieties.

Rainfed lowland rice ecosystem

Physical description

Rainfed lowland rice grows in bunded fields that are flooded for at least part of the cropping season to water depths that exceed 100 cm for no more than 10 consecutive days. Rainfed lowlands are characterized by a lack of water control, with floods and drought being potential problems. In favorable rainfed lowlands, small investments in water-control measures can increase water control considerably. In these cases, cropping systems are on a continuum between rainfed and irrigated ecosystems and available technologies of the irrigated sector need only minor adaptations. Considerable adaptations or different technologies are needed in the rainfed lowlands where risk of drought or flood is significant. About 34% of the world's total rice land or approximately 54 million ha are rainfed. Adverse climate, poor soils, and a lack of suitable modern technologies keep farmers from being able to increase productivity.

The rainfed lowland rice ecosystem can be divided into four subecosystems: (1) favorable rainfed lowland, (2) drought-prone, (3) submergence-prone, and 4) drought- and submergence-prone. Technologies for the irrigated rice sector can be applied in the rainfed lowland ecosystem only where there is no significant risk of drought or flood. However, the majority of lands in this ecosystem do face these risks, requiring different rice varieties and management strategies from those used in the irrigated ecosystem.

Another ecosystem that has significant potential for rainfed rice production is inland valleys, particularly in Sub-Saharan Africa. This is discussed in "International rice research and development" (pages 51-52).

Predominant cropping systems

The rainfed lowlands include areas in which farmers grow only one crop of rice, although in some areas farmers grow rice and a postrice crop, usually on a smaller area. The main post-rice crops are chickpea, mustard, linseed, rice, and mungbeans, followed to a lesser extent by vegetables, maize, wheat, and other legumes such as soybeans and lentils. The choice depends on water availability and season duration. Farmers often cultivate rainfed lowland rice at several toposequence levels such that on one farm some fields may be drought-prone, while others may be flood- and submergence-prone in the same season.

Productivity

The world's rainfed lowlands offer substantial potential for increasing rice production. Population pressure on arable land is low; the area planted is increasing, but yields remain low. Farmers have developed a range of practices that address variability across sites as well as heterogeneity within local ecosystems.

Production constraints

Uncertainty characterizes rice farming in rainfed lowlands. Crops suffer from droughts, floods, pests, weeds, and soil constraints. Since most rainfed lowlands depend on erratic rainfall, conditions are diverse and unpredictable. Understanding how farmers' practices help reduce risk and assure some production is essential to developing improved technologies for the rainfed lowlands.

Most rainfed lowland rice farmers are poor and must cope with unstable yields and financial risks. They adapt their cropping practices to the complex risks, potentials, and problems they face. They typically grow traditional, photoperiod-sensitive cultivars and invest their labor instead of purchasing inputs. Farmers bund the fields to store water. They weed, may redistribute seedlings to ensure good crop stands, and usually harvest by hand. Suitable modern varieties and associated production technologies have been limited.

Although new technology developed in the 1960s and 1970s focused on the irrigated sector, rainfed lowland rice farmers were not forgotten. Researchers have tried to produce new varieties and improved farming practices for nutrient management, crop establishment, on-farm water collection, and weed and pest control. These practices can potentially contribute to higher yields, especially in the favorable rainfed subecosystem.

Rainfed lowland rice farmers in less favorable areas use traditional varieties that do not respond well to higher fertilizer rates. In Bangladesh, eastern India, Indonesia, Philippines, and Thailand, however, adoption of new rice varieties is increasing as scientists are now developing new breeding lines with single or combined traits adapted to rainfed lowland stresses. Rice varieties that are bred for tolerance for submergence, late transplanting, and resistance to lodging allow farmers to apply fertilizers and use more intensive weed management practices. Decentralized regional breeding programs developed in northeastern Thailand and eastern India, focusing on drought- and submergence-tolerant improved lines, respectively, now have advanced lines under evaluation in farmers' fields.

Rainfed lowland rice varieties of the future will need to respond to improved management while retaining the tolerance of the traditional varieties for drought, floods, and soil stresses. Such varieties should perform well under favorable conditions and still equal the productivity of traditional cultivars under adverse conditions. Farmers would then be able to invest money and labor in potentially more productive land preparation and fertility management practices that will assure higher yields.

Upland rice ecosystem

Physical description

Upland rice is grown in Asia, Africa, and Latin America. Of about 150 million ha of world rice area in the mid-1990s, about 14 million ha were planted to upland rice: 8.9 million ha in Asia, 3.1 million ha in Latin America, and 1.8 million ha in Africa. Although upland rice constitutes a relatively small proportion of the total rice area, it is the dominant rice culture in Latin America and West Africa. In Asia, the area of the upland ecosystem is much larger than the area under rice, because rice is grown in rotation with many other crops.

Upland rice is grown in low-lying valley bottoms to undulating and steep sloping lands with high runoff and lateral water movement. In Southeast Asia, most upland rice is grown on rolling and mountainous land with slopes varying from 0% to more than 40%. In eastern India, upland rice is grown on several million hectares in permanently cultivated, level fields at the top of a toposequence ranging from rainfed lowland to upland. In West Africa, upland rice grows on hills in the humid zone and on flatland in the drought-prone and moist forest zones. Most upland rice in Brazil is on level to gently rolling (0–8% slope) land, much under mechanized cultivation. In northern and northeastern Brazil, some upland rice is grown on rolling topography under shifting cultivation.

Upland rice soils range from erodible, badly leached Alfisols in West Africa to fertile volcanic soils in some areas in Southeast Asia. Their texture, water-holding capacity (WHC), cation exchange capacity (CEC), nutrient status, and soil-related problems vary greatly. In Southeast Asia, many upland soils in high-rainfall areas where rice is grown are erodible, acidic, and highly P-fixing. Subsoil acidity and Al toxicity are common additional constraints, along with severe nutrient deficiency and poor WHC. In Brazil, where upland rice is a major crop, soils have extremely low CEC values, high P fixation, and high levels of exchangeable Al. Upland rice soils in most of Africa have low available WHC because of coarse texture, are often kaolinitic (i.e., low-activity clay soil with low CEC), and have severe nutrient deficiencies and Al and Mn toxicities.

Dry soil preparation and direct seeding in fields that are generally unbunded are common in upland rice culture. Surface water does not accumulate for any significant time during the growing season. In some countries such as in eastern India and Bangladesh, upland rice fields are frequently bunded to save scarce water.

Upland rice and food security

Upland rice is primarily grown as a subsistence crop. It is critical to the food security of impoverished communities that do not produce enough lowland rice to meet their needs. In eastern India and Bangladesh, upland rice is grown by farmers who also have lowland fields, but who face a "hungry month" before lowland crops can be harvested; the upland crop, which is harvested early, bridges the family through a period when food is extremely scarce. In other areas such as the mountains of northern Lao PDR, many farm families have no lowland holdings and are heavily dependent on upland rice for subsistence. A commercial cropping system with an upland rice component has emerged only in the Brazilian Cerrado region and in northern China (see below).

In Côte d'Ivoire, Guinea, Guinea Bissau, and Sierra Leone, upland rice is the only staple available between the maize and cassava harvests, or between sweet potato and cassava. Recently, interspecific hybrid rice (or NERICA) has been the main food available in the "hunger period" from late September to late November (see page 36).

Predominant cropping systems

Depending on the size of their farms and their resources, upland rice farmers use farming systems ranging from shifting to permanent cultivation. Shifting cultivation is common in Indonesia, Lao PDR, northern Thailand, and Vietnam in Asia, and in the forested areas of Latin America and West Africa. Farmers plant a rice crop alone or in association with other crops such as maize, yam, beans, cassava, or plantains. In many places, including Indonesia, the Philippines, Southwest China, and Brazil, upland rice may be intercropped with maize. In areas with sufficient rainfall, upland rice may be followed by a crop of maize, cowpea, beans, soybean, or sweet potato. In West Africa, one or more crops may be mixed with upland rice. Farmers use an area for 1–3 yr until soil fertility declines and weed and pest infestations increase. They then abandon the land and return to previously abandoned farmland or start cropping on other available fallow land. In traditional shifting cultivation systems, fallow periods were as long as 30 years. This lengthy fallow restored fertility and prevented weed seed buildup. Now, rotations in most upland areas practicing shifting cultivation have shortened to a 2- or 3-yr cycle because of increased population pressure. This has resulted in increased weed pressure and reduced soil fertility.

One variation of shifting rice cultivation is pioneer cultivation where fallow is replaced by perennial vegetation such as pasture or trees. Rice is intercropped with young fruit and forest trees for 2–3 yr (intercalary cultivation). As the trees grow, they shade more area and less rice is planted. After a few years, the rice crop is transferred to a new area. Pioneer cultivation is common in Brazil, but rare in Asia and Africa, although it is becoming popular in Indonesia under rubber, oil palm, and teak.

Permanent cultivation of upland rice is practiced in many Asian and Latin American countries. This is characterized by orderly intercropping, relay cropping, and sequential cropping with several crops. The largest area of permanent upland rice cultivation is in eastern India. In the Chhattisgarh Plateau area of Jharkand and Bihar, rainfall will support only a single upland crop per season. Rice may be grown continuously or rotated with legumes and pasture. Rice yields in continuous production tend to be very low (around 1 t/ha) because of nematode buildup and other biotic factors that are not well understood, but yields in rotation with legumes may exceed 2 t/ha. In eastern Uttar Pradesh, upland rice is usually double-cropped with gram, mustard, or other upland crops. Because the risk of crop damage or loss resulting from drought or pests is high, and because upland rice farmers are usually poor and have limited access to credit, most apply few inputs of animal manure, organic matter, chemical fertilizer, and pesticide to upland rice crops. This is particularly true in areas where farmers produce rice on both lowland and upland fields. In such areas, farmers invest more in their lowland fields, where greater response to inputs is usually obtained.

Upland rice is grown in an intensively managed commercial system in the Brazilian Cerrado, where it is rotated with soybean and pasture and often receives supplemental irrigation. Rice is grown no more often than one season in three because of the yield decline associated with continuous or short-rotation upland rice production. Commercial, high-input upland rice production with supplemental irrigation is also emerging in water-short areas of northern China. These intensively managed upland crops are now usually referred to as "aerobic rice" to distinguish them from traditional low-input rice crops. Areas planted to aerobic rice in Asia can be expected to increase because of water shortages in some areas currently growing irrigated lowland rice.

Productivity

Although accounting for 13% of total world rice area, upland rice contributes only 4% to total world rice production. Grain yields average about 1 t/ha in traditional low-input systems, but may reach about 2 t/ha in favorable upland environments such as those found in Mindanao, Philippines. In the Brazilian state of Mato Grosso, where upland rice is grown with supplemental irrigation and high inputs of N, yields averaged nearly 2.5 t/ha in 1998-99. Aerobic rice yields of more than 5 t/ha have been reported in China and Brazil.

Production constraints

Biological and physical constraints to upland rice yield are numerous. Weeds are the most severe biological constraint, followed by blast disease and brown spot. Weeds reduce upland rice grain yield and quality. Estimates of yield losses caused by weeds in upland rice range from 30% to 100%. Small farmers using traditional production systems cannot afford chemical weed control, but herbicides are used in commercial aerobic rice systems in Brazil and China. Stem borers and rice bugs are the predominant insect pests. Nematodes can cause yield losses of up to 30%. Rodents are reported in many countries as a major problem and birds are sometimes serious pests.

Amount and distribution of rainfall and poor soil fertility are the most serious physical constraints. In Asia, the monsoonal rainfall ranges from 1,500 mm to more than 3,500 mm. Rainfall is usually erratic during the monsoon season. In Africa, annual rainfall ranges from 1,200 to 2,000 mm where upland rice is grown. Rainfall is frequently erratic in semiarid areas of West Africa. Mean annual rainfall in Latin America ranges from 1,400 mm in central Brazil to 4,800 mm in Costa Rica. Dry spells frequently occur during the growing season in most of these areas, mainly in northern and central Brazil.

Upland rice soils are predominantly acidic (pH varies from 4 to 6) and depleted in major elements. In South and Southeast Asia, more than half of the upland rice is grown in infertile soils. In most upland soils, P rather than N is the most common limiting nutrient. Physiological disorders result from major and trace element imbalances. Major toxicities are caused by high Al and Mn content in strongly acidic soils.

The temperature of most regions of Asia, Latin America, and Africa is relatively favorable to upland rice except in some high-altitude areas in India, Indonesia, Myanmar, Nepal, and Thailand. Minimum and maximum temperatures average 20 and 30 °C for Latin America, 20 and 32 °C for South Asia, and 18 and 35 °C for Africa. The optimum temperature for maximum rice photosynthesis is 25–30 °C.

Flood-prone rice ecosystem

Physical description

The flood-prone ecosystem has several environments and incorporates a number of rice plant types. These rice plants must be adapted to conditions such as (1) deepwater: submergence in depths usually exceeding 100 cm and for durations ranging from >10 d to 5 mo, requiring the plant to elongate greatly to reach the surface; floating rice up to 5 m long is included here; (2) flash flood for periods of longer than 10 d; (3) salinity caused by tides in extensive low-lying coastal areas, where plants are subject to daily tidal submergence; the plants do not elongate greatly but tillering and tiller survival are at risk; and (4) problem soils, such as acid-sulfate and sodic soils, in which the problem is often excess water, but not necessarily prolonged submergence. Around 11 million ha of rice lands worldwide are affected by one or more of these conditions.

In these environments, rice yields are low and extremely variable because of problem soils and unpredictable combinations of drought and flood. Although average yields are only about 1.5 t/ha, these areas support more than 100 million people.

Crucial for survival and production of the rice crop are the age of plants at the start of inundation, the rate of water rise, and the duration of the flood. Many parts of the tidal, deepwater, and rainfed lowland rice areas are subject to sudden increases in water level, commonly called flash floods. These floods occur after heavy rain, local or remote, and may completely submerge the crop for several days. The water is often heavily laden with silt, which covers the leaves of the plants.

Deepwater rice and floating rice are mainly grown in fields on the floodplains and deltas of rivers such as the Ganges and Brahmaputra of India and Bangladesh, the Irrawaddy of Myanmar, the Mekong of Vietnam and Cambodia, the Chao Phraya of Thailand, and the Niger of West Africa. Flooding occurs in the later stages of plant growth and can last for several months.

Tidal wetland rice is cultivated during the wet season in scattered areas along the coastal plains of Bangladesh, India, Indonesia, Myanmar, Thailand, Vietnam, and West Africa. Tidal rice can tolerate submergence caused by flash floods or tidal fluctuations. Salt-tolerant tidal rice is grown in areas where there is salt water intrusion from the sea. The soils are of various ages and are complex and diverse, particularly in river floodplain areas. Many of these alluvial sediments have undergone little soil formation. Soil texture varies from medium to fine. It is usually more coarse on natural levees and finer in depressions and backswamps. Although many soils are fertile, problem soils such as peat, acid-sulfate, and saline soils occur, particularly near coasts.

Predominant cropping systems

The cropping systems vary depending on the time, depth, and duration of flooding. Most varieties of rice can survive complete submergence for only 3 to 4 d and some rainfed lowland rice can survive up to 10 d. In flood-prone areas, rice is the only food crop that can be grown during the rainy season (June to November in most of Asia). There are four major cultural conditions in the flood-prone environments for which rice varieties are selected: deepwater (up to 100 cm or more); very

Deepwater rice in Cambodia.

deepwater (100–500 cm); tidal; and dry-season irrigation.

Deepwater rice and floating rice are sown or transplanted before the floodwaters rise and flower near the time of maximum water depth. Deepwater rice varieties are adapted to maximum water depths of around 100 cm and most can elongate 2–3 cm/d when flooded.

Floating rice varieties are those that elongate very rapidly under submergence, sometimes up to 20 cm per day. They are adapted to rapidly rising water and very deeply flooded areas.

Tidal rice is cultivated in coastal zones during the wet season. It can tolerate submergence caused by tidal fluctuations or flash floods. It should not elongate stems when flooded since floodwaters recede within about 2 wk and the plants would fall over. Salt-tolerant tidal rice is needed where salt water intrudes from the sea. During the dry season, these lands are too dry or saline for cropping.

Irrigated rice is grown in flood-prone areas during the nonflood periods if irrigation is available. It is called boro rice in Bangladesh and India. Traditionally, boro rice was cultivated only in local land depressions, where there was sufficient residual water in the soil for a crop during the dry season. However, with improved irrigation, mainly from tube wells, boro rice is now cultivated on many floodplains. Although boro rice replaced some floating rice, there is now progress in integrating both crops. Floating rice has also been largely replaced by irrigated rice in southern Vietnam.

In Bangladesh, deepwater rice farmers have traditionally grown a wide range of crops after the floods by seeding or planting onto the straw of the deepwater rice crop. These crops include pulses, oilseeds, spices, potato, onion, garlic, wheat, and barley. Postflood and dry-season cropping are now widely practiced in Bangladesh, India, Myanmar, and Vietnam. Preflood cropping, although not widespread, includes mungbean, sesame, chili, and sorghum. These crops, if successful, are harvested before the main floods arrive. Sometimes jute or maize is planted along with deepwater rice or just before the deepwater rice crop is sown. In India, farmers in many deepwater rice areas have recently incorporated grain legumes (mainly mungbean), sesame, maize, and sorghum into their rotation. They grow these crops as a mixture with deepwater rice on stored soil water at the start of the wet season.

Productivity

Wet-season nonirrigated rice yields range from zero to 4.0 t/ha depending on the season, location, and rice type. Floating rice yields are usually low, from about 1.0 to 2.5 t/ha. Deepwater rice yields higher than floating rice, sometimes up to 4 t/ha, and generates high profits because of low production costs. Submergence-tolerant varieties are grown in more favorable areas and produce relatively high yields if flash floods are not severe. Yields of tidal rice vary widely and crop failure can occur in salt-affected areas.

Production constraints

Although fertilizer tends to increase nonirrigated rice yields, the response is often irregular because of environmental stresses. As a consequence, fertilizer use is limited and yields are also low compared with irrigated rice, which has controlled water supplies and high inputs of fertilizer and pesticides.

There are many opportunities for increasing the rice yields of the flood-prone ecosystem. The yield potential of deepwater rice could be improved by introducing an appropriate plant type. Yields of flash-flood and tidal areas could be increased by incorporating increased tolerance for submergence, salinity, and acidity. Varieties with tolerance for cold at the seedling stage and in the early vegetative stage could increase boro rice yield. However, yield increases for floating rice through varietal improvement are more difficult because survival is the first consideration.

Enemies and friends

Pests

Rice fields harbor a tremendous diversity of animals, plants, and microorganisms, some of which are harmful to the rice crop and many of which are beneficial. The goal of many scientists at IRRI and other institutions is to manage rice pests in ways that are safe, sustainable, and economical. Emphasis is placed on breeding rice varieties with resistance to insect pests and diseases and on minimizing the use of pesticides to promote natural biological control by beneficial insects, spiders, and microorganisms. The importance of biological control in rice was dramatically demonstrated in the 1970s, when the indiscriminate use of broad-spectrum insecticides devastated populations of beneficial insects and spiders and led to huge outbreaks of the brown planthopper, which had previously been a minor pest.

A rice pest is any organism that causes economic loss in rice production, including arthropods (insects and mites), pathogens (bacteria, fungi, and viruses), weeds, mollusks (snails), and vertebrates (rodents and birds). Some common pests are shown in Table 3. The damage they do ranges from severing stems or killing tissue to competing with the crop for nutrients and sunlight.

A recent survey found that rice pests in tropical Asia cause an average yield loss of 37%. Weeds are the most important source of loss: weeds taller than the rice plants caused 23% yield loss and weeds below the canopy 21%. The most damaging diseases are sheath blight, brown spot, and leaf blast, each causing 5–6% loss. The most important insect pests are stem borers, with damage at the reproductive stage causing 2.3% yield loss.

Weeds are an almost universal companion of rice in the tropics. In many situations, weed growth is prolific and weeds are a major constraint to crop yield. Weeding is a major production cost, with estimates of 50–150 person-days per hectare required for manual weeding, depending on the number of weedings and type of rice culture. For many farmers, weeding requires the greatest labor input during the agricultural cycle, and labor is often not available when weeds are most damaging to the crop. Upland rice more than any other crop shows the ravages of a lack of proper weeding. Sometimes, when the land is too weedy, the crop is abandoned.

The demands of transplanting and manual weeding and increasing shortages of labor have encouraged the move to direct seeding in irrigated and rainfed lowlands. Weeds become a major problem in these systems because rice and weeds emerge at the same time, and weed control by flooding is difficult in seeded rice. Herbicides are being used more to control weeds, and herbicide-resistant weeds and pollution are emerging problems in direct-seeded systems.

Table 3. Examples of organisms that may harm or compete with the rice crop.

Insect pests	
Stem borers	
African rice gall midge	*Orseolia oryzivora* (Harris & Gagne)
Yellow stem borer	*Scirpophaga incertulas* (Walker)
White stem borer	*Scirpophaga innotata* (Walker)
Striped stem borer	*Chilo suppressalis* (Walker)
Dark-headed rice borer	*Chilo polychrysus* (Meyrick)
Defoliators	
Rice leaffolders	*Cnaphalocrocis medinalis* (Guenée) and others
Rice caseworm	*Nymphula depunctalis* (Guenée)
Leafhoppers	
Green leafhopper	*Nephotettix virescens* (Distant)
	N. nigropictus (Stål)
	N. parvus Ishihara et Kawase
	N. cincticeps (Uhler)
Planthoppers	
Brown planthopper	*Nilaparvata lugens* (Stål)
Whitebacked planthopper	*Sogatella furcifera* Horvath
Rice bugs	
Malayan black rice bug	*Scotinophara coarctata* (Fabricius)
Rice grain bug	*Leptocorisa oratorius* (Fabricius)
Rodents	
Rice field rats	*Rattus argentiventer* (Rob. & Kloss)
	R. tanezumi (Temminck)
Diseases	
Viral diseases and their vectors	
Rice tungro	*Nephotettix virescens* (Distant)
	N. nigropictus (Stål)
Ragged stunt	*Nilaparvata lugens* (Stål)
Rice yellow mottle	*Chaetocnema pulla* Chapius
	Trichispa sericia (Guérin)
Bacterial diseases and their causal agents	
Bacterial blight	*Xanthomonas oryzae* pv. *oryzae* (Uyeda ex Ishiyama 1922) Swings et al 1990
Fungal diseases and their causal agents	
Blast	*Pyricularia oryzae* Cav.
Sheath blight	*Rhizoctonia solani* (*Thanatephorus cucumeris* [Frank] Donk)
Weeds	
Ageratum conyzoides L.	
Cyperus difformis L.	
Cyperus iria L.	
Echinochloa colona (L.) Link	
Echinochloa crus-galli (L.) P. Beauv.	
Fimbristylis miliacea (L.) Vahl	
Monochoria vaginalis (Burm. f.) Presl	

Insects attack all parts of the rice plant. Hundreds of species feed on rice, but only a few cause yield loss. The most important and widely distributed pest species are stem borers, leaffolders, planthoppers, and gall midge. Stem borers are chronic pests, found in every field in every season, but generally at low levels. Planthoppers and gall midge usually create localized outbreaks, causing high yield losses in relatively small areas. Biological control by natural enemies plays a critical role in the management of all insect pests. Resistant rice varieties are of importance in the control of planthoppers and gall midge. No strong sources of resistance to stem borers have been found in rice germplasm, although modern semidwarf rice varieties generally have lower levels of stem borer damage than the traditional varieties they replaced. Insecticides are used extensively against planthoppers in temperate areas of Asia, where mass immigration of planthoppers from tropical areas is a frequent problem.

Bacterial blight, blast, and sheath blight are the most important *diseases* of rice and have a worldwide distribution. Three insect-vectored viral diseases are also of importance: tungro in Asia, hoja blanca in South America, and rice yellow mottle in Africa. Bacterial blight and blast have been successfully controlled by resistant varieties for many years. However, the evolution of resistance-breaking strains of these pathogens has necessitated the continuing release of new resistant varieties. Strong sources of resistance for sheath blight have not been identified in rice germplasm. Sheath blight is a particularly important disease in intensive rice-growing conditions where high levels of nitrogen fertilizer are applied.

Integrated pest management

Modern approaches to crop protection rely on management rather than control or eradication. In this approach, a pest species is considered a pest only when it reaches numbers that can cause yield reduction. Natural factors—such as natural enemies—that prevent pest populations from increasing are emphasized. Indeed, most organisms inhabiting rice fields are never harmful. In addition, the use of rice cultivars that are resistant to major pest species is encouraged. These cultivars do not need prophylactic treatment to control the insects or diseases to which they are resistant. Using a combination of control tactics instead of relying on just one tactic, such as host-plant resistance or pesticides, and basing the decisions for control on sound economic grounds is called integrated pest management or IPM.

Pesticides should be used only as a last resort to bring abnormal pest densities down when crop loss is expected to exceed the cost of treatment. Pesticides are costly to farmers, can disrupt natural biological control, and are damaging to human health and the environment. In Thailand, for example, it is estimated that 40,000 rice farmers suffer from varying degrees of pesticide poisoning every year. In the Philippines, there is a similar high level of poisoning and the overall cost of crop loss from pests is less than the farmers' resulting health costs. The total extra costs of pesticide use in Thailand, including health, monitoring, research, regulation, and extension, amount to $128 million per year.

Environmental effects are also serious in Thailand, where pesticide residues are found in nearly all samples of soils, river sediment, fish, and shellfish. One survey found organophosphorus insecticide residues in three-quarters of tangerine samples and one-third of all vegetable samples.

Farmer education is a central feature of IPM, and various approaches have been employed. Many countries have implemented "farmer field schools," which entail season-long weekly meetings in which farmers learn about the value of natural enemies and other aspects of growing a healthy crop. Another approach is the mass communication of "simple rules" to farmers. A campaign to encourage farmers to experiment in eliminating early season insecticide applications has been very successful in Vietnam and is now being attempted in Thailand. The campaign used billboards, cartoon characters, information handouts, and humorous radio programs; it reduced rice farmers' insecticide use by an estimated 70% and the proportion of farmers believing that insecticides bring higher yields fell from more than 80% to just 13%.

Landscape manipulation is a new approach to IPM. The concept of using alterations to the rice field and its surrounding areas involves the roles of bordering vegetation and levees, and

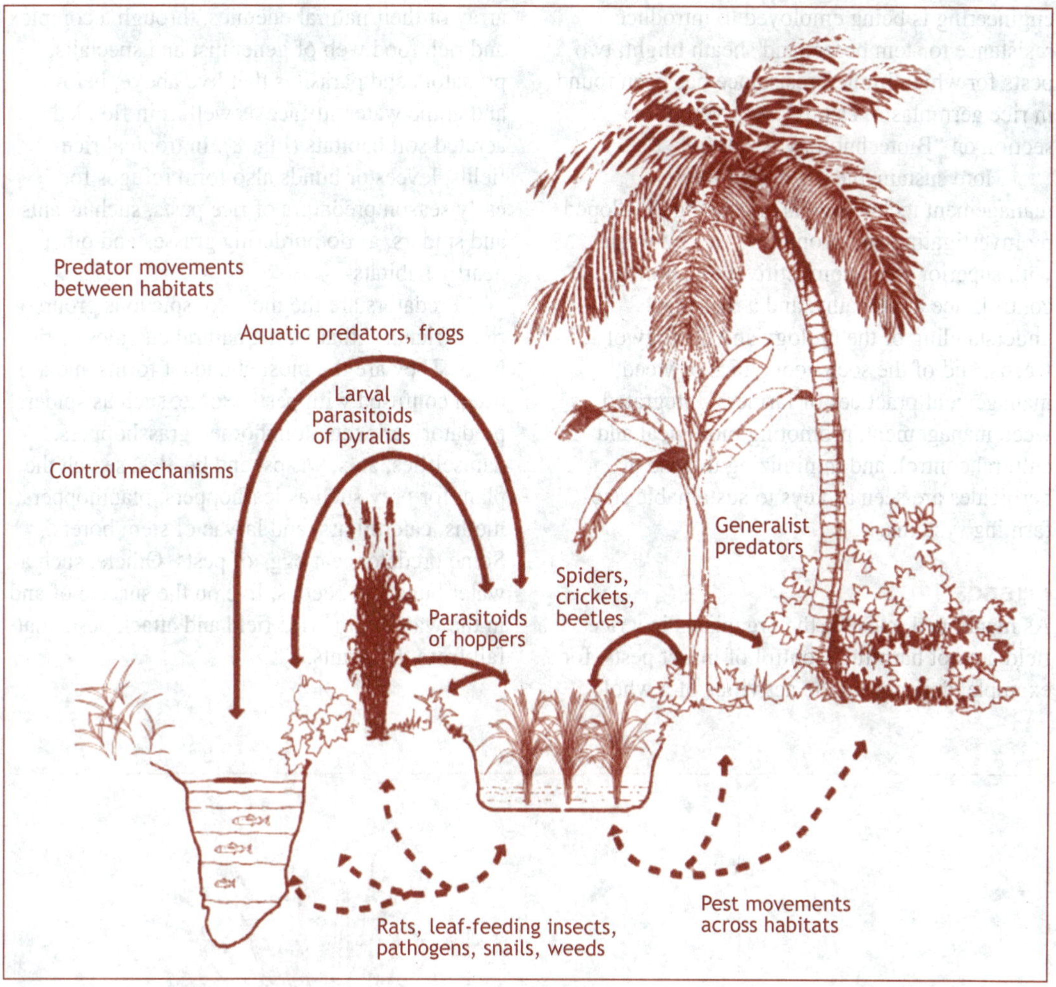

Fig. 7. Landscape features and organism associations of the rice ecosystems.

patterns of distribution of pests and their natural enemies in rice and nonrice habitats, within rice fields and between the edges and interiors of rice fields. Some of these associations are shown in Figure 7.

Another biodiversity-based approach, varietal mixtures, has been implemented to manage blast disease in Yunnan Province, China. Blast was causing a great yield loss in traditional glutinous rice varieties and farmers were spraying fungicides up to seven times on one crop. IRRI scientists introduced the concept of varietal diversification and worked with scientists and farmers in China to practice the interplanting of glutinous varieties with blast-resistant semidwarf varieties. This added diversity of varieties prevented the fungus from achieving the population explosions that had previously occurred in the monoculture fields of the glutinous varieties. The result has been dramatic, in terms of both increased yield and reduced use of fungicide. The interplanting methodology is spreading like wildfire across Yunnan Province and has great potential for application in other areas of Asia.

New biotechnology-based approaches are being applied to produce new insect- and disease-resistant varieties. DNA marker-assisted selection has been used to increase the efficiency of breeding for pest resistance and to enable the "pyramiding" of multiple genes for resistance to a single insect or disease. Wide hybridization techniques enable resistance genes to be introduced from wild rice species. Genetic engineering is being employed to introduce resistance to stem borers and sheath blight, two pests for which limited resistance has been found in rice germplasm. Details are given in the section on "Biotechnology issues".

More sustainable integrated weed management technology is also being developed by investigating and promoting rice cultivars with superior weed competitiveness, biological control, rice allelopathy, and a thorough understanding of the biology and ecology of weeds, and of the socioeconomics of weed management practices of farmers. Integrated weed management, promoting biological and cultural control, and minimizing the use of herbicides are seen as keys to sustainable rice-farming systems.

Friends

As mentioned, most of the organisms in a rice field are not harmful. Control of insect pests, for example, depends on the activities of a whole array of their natural enemies, through a complex and rich food web of generalist and specialist predators and parasites that live above, below, and at the water surface as well as in flooded and aerated soil habitats (Fig. 8). In tropical rice fields, levees or bunds also form refuges for early season predators of rice pests, such as ants and spiders, as do bordering grasses and other nearby habitats.

Predators are the most conspicuous group of rice "friends," that is, the natural enemies of rice pests. They are the most abundant forms and are often confused with pests. Some, such as spiders, predatory crickets, longhorned grasshoppers, damselflies, ants, wasps, and beetles, search the plant for prey such as leafhoppers, planthoppers, moths, caterpillars, and larvae of stem borers. Some predators eat eggs of pests. Others, such as water bugs and beetles, live on the surface of and in the water in the rice field and attack pests that fall from the plants.

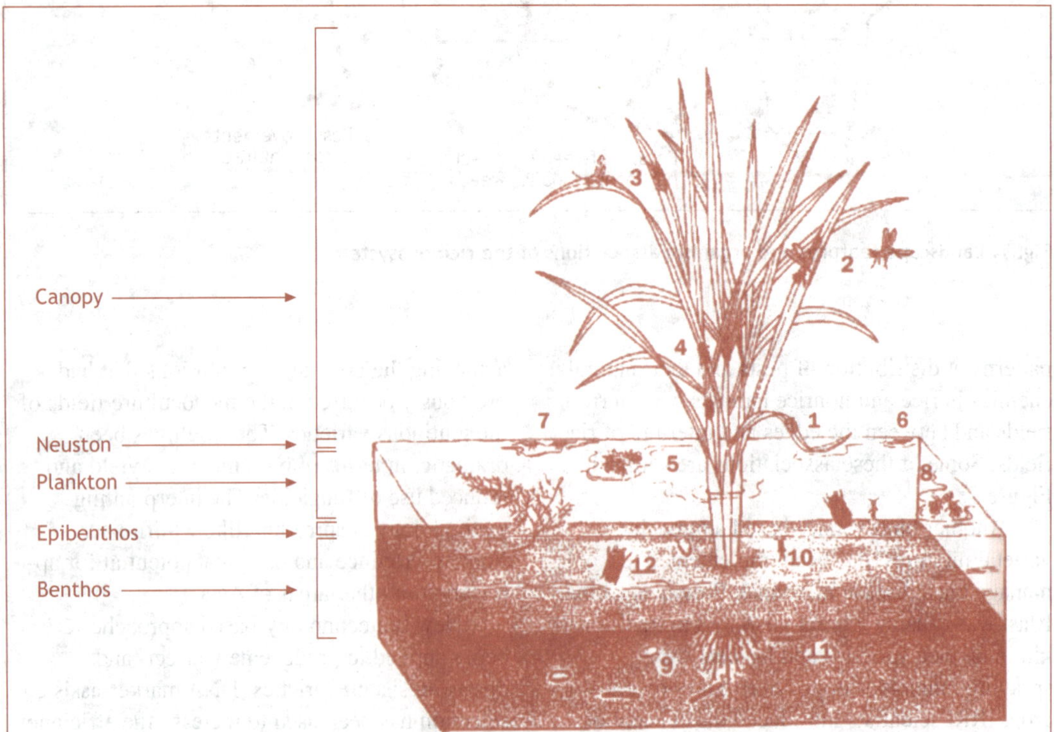

Fig. 8. Habitat zones and dominant taxa of an irrigated rice farmer's rice field in Calauan, Laguna, Philippines. Canopy taxa are *Stilbus* sp. adults (1), Chironomidae adults (2), *Cardiochiles philippinensis* adults (3), and *Egadroma* sp. adults (4). Neustonic (water surface) taxa are *Microvelia atrolineata* adults (5), *Ephydra* sp. immatures (6), Entomobryidae adults (7), and *M. atrolineata* immatures (8). Planktonic and benthic taxa are *Heterocypris luzonensis* (9), *Eucyclops serrulatus* adults (10), Chironomidae immatures (11), and Hydrophilidae adults (12).

Wolf spiders feed on a range of insect pests.

Spiders are especially numerous and diverse partners in rice fields. In South and Southeast Asia, at least 340 species are known from rice lands. Parasitoids are generally more host-specific than predators. Although parasitoids are easy to overlook, their effect on pests can be extremely important. Parasitoids lay their eggs in, on, or near the host. When the egg hatches and the young parasitoid develops, the host eventually stops feeding and dies. Various pathogenic microorganisms (fungi, bacteria, and viruses) infect and kill insect pests of rice. Other species of fungi and bacteria attack weeds or are important in the biological control of rice diseases.

International issues

Past scientific research on various aspects of rice cultivation and on the rice plant itself has been responsible for remarkable gains in the productivity of rice-farming systems. However, rice production must increase if it is to keep pace with population growth. The scope for expansion of rice-growing areas is limited. Most of the increase will have to come from improved plants and from production methods that are more efficient yet more environmentally sustainable.

Here, we look at several major issues: dealing with a future in which less water is available, ways to increase the yield potential of the rice plant, steps toward making rice more nutritious, forecasting and dealing with the effects on rice plants and ecosystems of global climate change, reducing greenhouse gas emissions from rice fields, the new research area of functional genomics, biotechnology, and integrating knowledge through crop modeling.

The looming water crisis

The amount of water needed to produce one ton of rice would fill between two and three Olympic-sized swimming pools.

Nearly 90% of fresh water diverted for human use in Asia goes to agriculture and, of this, more than 50% is used to irrigate rice. China's Yellow River, which flows for 4,600 km through some of Asia's richest farmland, has run dry nearly every year since 1972. In South Asia, the Ganges and Indus rivers have almost no outflow to the sea in the dry season.

Too much water or too little

Asia's water problems are caused partly by the uneven distribution of water. On the one hand, about half of China receives less than 400 mm of rainfall a year, and extensive areas of northwestern, central, and southern Asia are drought-prone. On the other hand, the Ifugao rice

Precious irrigation water in Bangladesh.

terraces of the northern Philippines are situated in one of the wettest rice-growing regions of the world, with an average annual rainfall of 3,530 mm.

As a further complication, when rain comes in Asia it usually arrives in torrents over a short period, during a single monsoon that lasts from 4 to 6 mo. The rest of the year is almost dry. As a result, much of the runoff simply flows into the ocean as waste, at the same time eroding the uplands, sometimes catastrophically. Furthermore, the monsoon is often erratic, so that, in many countries, floods and seasonal water shortages occur concurrently.

Environmental costs of increased rice production

Water was a critical input to the Green Revolution, through irrigation, flood control, and drainage, and it has contributed most to the growth in rice production for the past 30 years. But this expansion has occurred at a cost to the environment: a proportion of the chemicals applied as fertilizer and for pest and weed control pollutes rivers and lakes through runoff or groundwater through leaching.

In some upland areas, intensive agricultural practices, coupled with deforestation, have resulted in high rates of soil erosion and degradation of both land and water resources in lowlands below. The effects can reach as far as coastal waters, with a consequent effect on riverine and marine life.

Another problem involves long-standing surface water, which causes waterlogging, makes the land unproductive, and leaves soils salty as the water evaporates. In India, about 6 million ha of irrigated land are known to be affected by waterlogging. Nearly 10% of Pakistan's irrigated 13.5 million ha is estimated to be affected by salinity, and northwestern India and northeastern China are similarly degraded.

Overexploitation of tube and shallow wells also presents problems. This is the case in large areas of India, Pakistan, and Bangladesh. The practice causes shortages of drinking water and pollution when aquifers are recharged with irrigation water contaminated with chemicals.

Capital costs of irrigation systems have recently soared. In Sri Lanka, it cost almost three times as much per hectare of land to set up an irrigation system in the 1990s as it did in the 1960s, twice as much in India and Indonesia, and nearly 50% more in the Philippines and Thailand. At the same time, market prices for rice have plummeted by nearly 40% over the past 30 years, while political pressure from environmental groups against large-scale projects is mounting.

What this will mean for future rice production is that it will depend heavily on the development of water-efficient measures producing more rice per unit of water input. The trend now is to develop management policies for the efficient operation of irrigation systems; technologies that reduce water consumption; changes in the rice plant itself and the ways in which it is grown, so as to use water more efficiently; and to provide economic incentives to farmers to reduce water losses.

Aerobic rice

One way to save water is to make a rice plant that needs less of it. IRRI scientists are working on creating a high-yielding tropical rice plant that grows on dry but irrigated land instead of in flooded paddies, calling it "aerobic rice." An Aerobic Rice Working Group has been formed, consisting of plant breeders, physiologists, and water and soil scientists, to address the many difficulties of taking the plant out of its natural environment and growing it as a dryland crop like maize. Some upland rice varieties already withstand drought but they are low-yielding and do not respond to fertilizers.

A first step has been testing aerobic varieties, already used commercially in Brazil and experimentally in China, in tropical conditions. In Brazil, the aerobic rice crops are rotated with other crops. However, when the plants are used as a single crop, as on most Asian farms, yields fall dramatically after the first few seasons, a phenomenon called "yield collapse." This is the major problem that the research is now addressing.

Dry seeding—which can avoid the waste of 400–600 mm of rainfall—is assuming an important role in rice production in rainfed areas. The early harvest of dry-seeded rice can allow planting of a second crop, which makes use of rainwater that arrives later in the season. This practice also reduces risk of drought where the rainy season is short, and because dry-seeded cultivars can generate more roots.

Aerobic rice cropping is not an option in semiarid to arid climates, nor on saline and acid-sulfate soils. Removal of a permanent water layer and thereby reduced percolation could result in soil degradation such as salinization, alkalinization, and acidification. Water-saving efforts under these conditions concentrate on the replacement of long- and medium-duration varieties by short-duration varieties and on good water management (e.g., irrigation stops 2 weeks before maturity).

In the Sahel, short duration has been a target for introduction and breeding for some time. In the mid-1990s, cultivar Sahel 108 was introduced and quickly proved popular, occupying one-third of the rice area of the Senegal side of the Senegal River Valley in the wet season and two-thirds in the dry season, and up to a third of the area on the Mauritanian side of the river in 1999. It thus saves at least 11 million m^3 of water use by plants; at 40% irrigation efficiency, this amounts to 28 million m^3, or $400,000 in saved fuel costs for pumping.

The water-weed connection

One reason farmers keep rice fields continuously flooded is to keep down weeds, which compete less well with rice under such conditions. But if flooding is reduced, other ways of controlling weeds will be necessary.

Scientists are now examining the place of weeds in the entire rice-growing system, with the aim not simply of attacking and killing them by one means or another, but of managing them in a way that will not interfere unduly with rice production. This involves studying the interaction of such components as water, tillage practices, and weeds, and incorporating in the rice plant characteristics that increase competitiveness with weeds, such as early vigor. It also involves investigating naturally occurring pathogens of weeds and how they might be used to suppress weed growth.

By such means, scientists at CIAT, IRRI, and WARDA are developing environmentally sound integrated weed management systems that they hope will prevent the kind of excessive reliance on chemical herbicides that happened with insecticides a decade or so ago. Meanwhile, herbicides remain an important component of integrated weed management and their use is rising rapidly. At the same time, weed resistance to them is increasing. Up to 15 major weed types now show such resistance and strategies are needed that minimize herbicide use and promote the development of further resistance.

A new research initiative

So important is the water problem worldwide that the Consultative Group on International Agricultural Research (CGIAR)—a consortium of donors supporting 16 international research centers including CIAT, IRRI, and WARDA—has mounted a water conservation campaign. The International Water Management Institute and several other CGIAR and national research centers are investigating the effects of on-farm water management on system performance through a network called SWIM—Systemwide Initiative on Water Management. The objective is to enhance the productivity of water in an environment of growing scarcity and competition. The result is expected to include not only ways to increase water productivity but also the development of methods to measure water productivity that will be widely adopted by researchers and the building of research capacity in water management. The main focus of SWIM is a comprehensive assessment of the benefits, costs, and future directions of irrigated agriculture.

Adding protein, vitamin A, and other micronutrients

Vitamin A is one of the most important nutrients in maintaining life and health. Consequently, dietary lack or deficiency of this vitamin leads to severe clinical symptoms. Africa, Asia, and Latin America and the Caribbean have areas where this is a severe problem. Overall, around half a million children become irreversibly blind each year as a result of vitamin-A deficiency. A much larger number of children develop other eye problems—5 million children per year in Southeast Asia alone. Worldwide, an estimated 124 million children are deficient in vitamin A to some extent, and improved intake would prevent 1–2 million deaths each year of 1- to 4-year-old children.

IRRI is joining an international research endeavor to find out whether rice consumption can help prevent this problem. The research has been made possible by the development in European laboratories of Golden Rice, a genetically modified rice variety that contains precursors of vitamin A—beta-carotene and other carotenoids—in its seeds. Rice plants already have the pathway for producing beta-carotene in their vegetative tissues; by genetic engineering, genes for beta-carotene driven by endosperm-specific promoters express beta-carotene (provitamin A) in rice seeds.

National institutes and IRRI are working together in finding suitable, well-known local varieties in which to transfer the new pathway. The resulting new lines will be released to national agricultural research centers, which will undertake their own research on the safety and effectiveness of these vitamin-enriched varieties in combating vitamin-A deficiency. Each country will make its own evaluation.

The interspecific varieties (NERICAs) recently developed by WARDA for West African upland ecologies have a higher protein content than either of their parent varieties. They therefore have the potential to improve the nutritional status of the subsistence-oriented farming families who are their target beneficiaries.

Another new rice variety, one that is particularly rich in iron and zinc, is awaiting a large-scale nutrition trial in the Philippines. This variety is not transgenic and was developed by exploiting natural variation in rice germplasm. Iron and zinc are usually deficient in people eating a rice-heavy diet. Lack of iron causes widespread anemia in some countries. Zinc enhances the body's capacity to absorb iron and combats diarrhea and cholesterol accumulation in blood vessels. The new variety, developed by IRRI, was tested by a group of 27 religious sisters in Manila, who ate the rice exclusively for six months, resulting in higher iron levels in their blood. However, a larger sample is needed to verify the results and a new trial involving 300 persons is beginning, under the supervision of two U.S. universities. The trials are organized by the International Food Policy Research Institute in Washington, D.C.

Potentially, these two new rice varieties can help prevent some of the most widespread micronutrient deficiencies in Asia.

Increasing yield potential

Since the late 1980s, scientists from IRRI and other institutions have been developing a new rice plant architecture, one that would increase yield potential in the tropics from 10 to 12 t/ha per crop: a fast-growing plant with a deep root system, dark green erect leaves, and 200 to 250 grains per panicle. It has been a massive task; for example, a grain-filling problem took three years of breeding effort to solve. When the number of tillers had to be increased to deliver a bigger harvest, the program adjusted to the new goal. Work on the plant is nearing fruition and in temperate China it is showing yields of 13 t/ha. Lines have been developed with resistance to various pests and diseases. Breeders in different countries are selecting lines best suited for local conditions and multiplying seeds for distribution. It may be four more years before they are available to farmers.

Meanwhile, global warming is creating new problems for rice farming and plants are needed that not only are higher yielding but also will thrive in higher temperatures, with more carbon dioxide and pollutants in the air, and where extremes of weather are commonplace. The plants will have to use less water and use nitrogenous nutrients very efficiently. This is an even more challenging task. There are biophysical limits to how much grain rice plants can produce, imposed by the photosynthetic pathway they use to convert sunlight into organic matter. Only by changing that pathway can the limit be overcome.

C_4 rice

In an attempt to improve yields, scientists from several research centers in both developed and developing countries are beginning work to modify the photosynthetic pathway of rice plants from a C_3 to a C_4 system such as is found in maize, a more efficient user of sunlight. Most, if not all, of the genes required for a C_4 system actually exist in the rice plant, but they are expressed differently. Importantly, C_4-type plants operate well at high temperatures, are extremely water-efficient, and require less nitrogen, and it might be these characteristics that will be most important in a world of climate change. It is something of an enigma that rice does not already operate on a C_4 system, given the advantages of that system. As carbon dioxide levels increase in the future, the yield gap between a C_3 plant and a C_4 plant will close, but the difference in water-use efficiencies will widen. Nonetheless, this is unlikely to happen before the end of the 21st century. Improving rice

International Rice Information System

The International Rice Information System (IRIS) is a database for managing and integrating information on genetic resources, phenotypic and molecular characterization, crop improvement, and crop management. Although major international initiatives for germplasm collection and conservation followed the Green Revolution, much collected material is still not used because information about it is difficult to access and is not integrated with other data sources. As a result, its potential impact on agriculture has not yet been realized. The sheer volume of unmanaged genealogical information produced by plant scientists precluded its wide use until now.

IRIS is the implementation for rice of the International Crop Information System (ICIS), which is being developed by scientists of several CGIAR centers, national agricultural research and extension systems, and research institutes to address this constraint for different crops.

The core of ICIS is a genealogy management system that facilitates the unique identification of germplasm and manages synonyms and homonyms. It allows users to trace cultivars back to their ancestral landraces or to lines of unknown pedigree; genealogies are automatically updated as additional information on ancestors is discovered. This information can be accessed not only as text but also as family trees—genealogical dendrograms. The genealogy management system links to a data management system, which integrates data from different scientific disciplines and will provide new insights into crop adaptation. Geographical information about germplasm evaluation sites is associated with latitude and longitude and can be analyzed with geographic information systems.

For more information on IRIS, see www.iris.irri.org.

photosynthesis may be the only way to guarantee feeding the growing population of Asia over the next 50 years.

Hybrid rice

Hybrid vigor or heterosis is a universal phenomenon that occurs in all biological systems. In plant breeding, hybrid vigor denotes the expression of more vigor than in the better parent and the existing commercial variety.

Commercial exploitation of heterosis to increase rice yields has been successfully demonstrated in China, where nearly 18 million ha of a total of 33 million ha of harvested rice land were planted to F_1 hybrids by 1992. Rice hybrids yield about 20% higher than do inbred rice varieties.

Research in other countries also indicates that heterosis in rice can increase yields by 15–20% over those of the best available semidwarf inbreds under irrigated conditions. Commercial hybrids have been identified and/or developed and grown on several hundred thousand hectares in Vietnam, India, and the Philippines. Several other countries are still searching for suitable hybrids for commercial cultivation.

Cytoplasmic male sterility is the most effective genetic tool for developing hybrid rice. Photoperiod-sensitive and thermosensitive male sterility systems have shown promise and have resulted in some two-line commercial rice hybrids in China. IRRI and some national institutions are still developing these male sterility systems. The use of chemically induced male sterility has not been efficient enough for deployment on a large scale.

Well-developed hybrid seed production practices give average yields of 2 t/ha of seed. Seed yields of 1–2 t/ha have been obtained by commercial seed growers in India, Vietnam, and the Philippines.

Tropical hybrid rice

The search for better tropical hybrids has continued at IRRI for the past 20 years. The 15 commercially usable, male-sterile parental lines bred to date have been designed for irrigated cultivation. Experimental hybrids for rainfed lowland systems have now been produced and are being tested in India, the Philippines, and Thailand. At the same time, new farm management packages for the hybrids are being developed. Private industry is showing considerable interest in tropical hybrids to become the provider of seed for each crop. Farmers are wary of using hybrids: having to buy seed for each crop means that their costs increase by about $50 per hectare. But the increased yields mean extra income of $135 to $200 per hectare per crop.

Hybrid breeding has also begun with IRRI's new plant type and could give yields of up to 15 t/ha. Golden Rice hybrids are another possibility for the future. An International Hybrid Rice Network extends across Asia, involving national research institutions in Bangladesh, China, India, Indonesia, the Philippines, Sri Lanka, and Vietnam.

Researchers in China have used NERICA lines (see section on "Interspecific rice" below) as fertility restorers for cytoplasmic male-sterile lines. Crosses proved that restoration was controlled by a single dominant gene from the NERICA lines.

Interspecific rice ("New Rice for Africa," or NERICA)

The potential of the cultivated rice species in Africa (*Oryza glaberrima*) has been "unlocked" through the successful crossing of this species with *O. sativa* and the development of true-breeding "interspecific hybrid progenies" (the NERICA varieties). The yield potential of these varieties is enhanced by the combination of the African species' adaptation to the West African environment with yield attributes from *O. sativa*. In addition, secondary branching on the panicles (from *O. sativa*) combined with transgressive segregation (a form of heterosis) gives NERICAs more than 400 grains per panicle, compared with about 250 in *O. sativa*. These new upland varieties also combine noninput dependence with input responsiveness—yielding more grain as farmers earn more to invest in their crop.

Functional genomics

The rice plant has 12 chromosomes, the tiny strands of DNA within each cell that hold its genetic information. Along the chromosomes, about 50,000 genes make up the genome. Scientists have been working for several years on rice gene sequencing: pinpointing each gene and deciphering DNA sequence structure, variation, and function. The study is called genomics. The entire sequence of genes along the rice chromosomes is being elucidated by various groups. Syngenta, a multinational agribusiness corporation, and the Beijing Genomics Institute published their sequencing of the rice genome in April 2002. The International Rice Genome Sequencing Project led by Japan expects to complete its task by the end of 2002. The international project, largely supported by governments, is committed to providing all sequence information to the public.

The rice genome represents an enormous pool of information for rice improvement through marker-aided selection or genetic transformation. However, a full application of this wealth of information will not be possible until the biological functions encoded by the sequenced DNA are understood. Functional genomics is the aspect of discovering what the genes do: how they function, how their functions combine with those of other genes, and for what purpose. Thus, functional genomics is expected to become the engine that drives discovery of traits and helps solve presently intractable problems in crop production.

IRRI is in a unique position to contribute to this study, backed by the vast collection of rice germplasm that it holds in trust.

One approach to the task is to delete a particular gene from the plant using chemicals or irradiation, then examine the plant for missing characteristics as it grows. IRRI already has a collection of more than 18,000 of these "deletion mutants" and the number is growing rapidly. The Institute is also developing a large collection of "introgression lines," plants that carry a wide range of chromosome segments implanted from commercially used varieties and wild rice. These will be used in the discovery of the functional diversity of the genes, and to understand the overall genetic, biochemical, and physiological systems in the plant. The mutants and introgression lines can be supplied to other institutions to assist them in the challenging work of assigning functions to the rice genes.

So far, the functional genomics team at IRRI has identified several genes giving the plants enhanced resistance to various types of pathogens that cause diseases. The team has also produced plants containing small chromosome segments from wild rice that confer resistance to several diseases and pests. In fact, more than 100 genes that can help the plants defend themselves against pathogens have been found and are already being used to select better disease-resistant varieties. The scientists have also found introgression lines and mutants that exhibit variations in growth and yield under water stress. Such genetic variation is the prerequisite for selecting better performing germplasm in soil with too much or too little water. Scientists also studied the drought response of the plants in different water conditions and identified a variety of proteins produced by the plant in response to drought and salinity stress.

A new tool under development for the work is the "microarray," in which about 20,000 genes can be displayed at once on a "chip," which can then be used as a sensor to detect genetic messages that are turned on or off when plants are exposed to stress. Thus, the expression of the genes can be recorded and analyzed to give a total picture of how the plant behaves in different conditions. In this way, scientists will be able to identify hundreds or perhaps thousands of genes that may combine and interact to achieve a particular function, such as tolerating drought, resisting disease, or producing more nutritious grains. This technology speeds up the discovery of gene functions enormously, doing in weeks what would have previously taken years.

Details of the methods used in functional genomics are given in the section on "Biotechnology issues" later.

Functional genomics requires diverse genetic resources, expertise in evaluating and identifying important traits, and an extensive collaborative network to evaluate newly found traits in different environments. Broad participation by the rice research community is needed to realize the potential offered by genomics.

To meet this need, the International Rice Functional Genomics Working Group has been formed to

- build a research community with a shared vision on rice functional genomics,
- create a common resource platform to broaden access to new knowledge and tools in functional genomics, and
- accelerate the application of functional genomics to rice improvement.

Three activities are being given high priority by this rice research group:

1. Creating a vehicle to communicate information related to functional genomics.
2. Promoting the sharing of genetic stocks.
3. Facilitating the sharing of resources for microarray.

So far, 13 research groups have joined the effort and national research centers in particular are sought to become involved. More information can be found at www.irri.org/genomics.

Of paramount importance is protecting the interests of rice farmers and consumers by ensuring that all information is available to the public and not locked up in private companies or even universities and public institutions, exercising their "intellectual property rights." As part of the effort to broaden access to the information, IRRI has launched a database on the Internet that describes the biological characteristics of the collection of deletion mutants; a similar database has been developed for information on stress-response genes. These will be linked to genome sequence databases to improve information exchange.

Ultimately, plant breeders may be able to find in a database the genes they need to achieve particular traits, select the genes from the genome map, and mix them as required to produce the desired plant type.

Global climate change and rice

Climate change is not a new phenomenon. Change has been a consistent feature of Earth's climate. Periods of relatively cool temperatures caused the Ice Age. For the past 10,000 years, however, Earth has experienced the longest period of consistently warm temperatures since the beginning of life. That warm period almost exactly matches the period over which modern agriculture has evolved.

For the first time in history, climate appears to be changing as a direct result of human activity. People have released chlorofluorocarbons into the atmosphere, thereby degrading stratospheric ozone and increasing biologically harmful ultraviolet (UV) radiation that reaches Earth's surface. Through mining and combustion of fossil fuels, deforestation, maintenance of livestock herds, and even through rice cultivation, people have released enormous quantities of carbon dioxide (CO_2), methane (CH_4), and other "greenhouse" gases into the atmosphere.

Samples from ice cores show that past fluctuations in global temperatures were strongly correlated with concentrations of atmospheric CO_2. Simulation models of global atmospheric circulation predict that greenhouse gases will cause a 2–8 °C global temperature rise before the end of the 21st century.

UV radiation effects

UV-B radiation damages leaf tissues in rice seedlings. Leaves become stunted, stomata collapse, and photosynthesis decreases. Some rice varieties appear to be better able than others to withstand the adverse effects of UV radiation. Leaves of tolerant varieties contain phenolic compounds, which are natural chemicals that filter out harmful UV-B radiation before it can damage sensitive tissues. Research is now in progress to predict possible regional losses in rice productivity if UV-B radiation continues to increase, and whether plant breeders can prevent those yield losses by developing new varieties that tolerate UV radiation.

In addition to its adverse direct effects on rice plants, UV-B may change the susceptibility to and/or tolerance for disease. Although there is no evidence yet that UV-B affects susceptibility to blast, it appears that the tolerance for blast decreases. In other words, UV-B does not increase disease frequency, but enhances the effects of disease on plant growth.

Global warming

Although increasing atmospheric CO_2 stimulates plant growth, the beneficial effects on rice growth have been observed for levels up to only 500 ppm. Some plant species respond positively to CO_2 levels up to 1,000 ppm. Experts predict that atmospheric CO_2 will surpass 650 ppm before the end of the 21st century. Furthermore, the benefits of increased CO_2 would be lost if temperatures also rise. That is because increased temperature shortens the period over which rice grows. Research is being conducted to identify means by which rice plants may better benefit from increases in atmospheric CO_2 while minimizing the adverse effects of warmer temperatures (Fig. 1).

Emissions of greenhouse gases from rice fields

Methane (CH_4) is second in importance to CO_2 as a greenhouse gas. CH_4 concentration in the atmosphere has more than doubled during the last 200 years. Some of this CH_4 is produced by rice fields (Fig. 2). To reduce the burden and harmful effects of CH_4 in the atmosphere, emissions from all anthropogenic sources have to be mitigated.

Methane is produced in the anaerobic conditions associated with submerged soils. Much of it escapes from the soil to the atmosphere via gas spaces in rice roots. The remainder bubbles up from the soil or diffuses slowly through the soil and overlying floodwater.

The potential for CH_4 emissions from rice fields has long been noted, but comprehensive measurements of CH_4 fluxes in rice fields have been reported only since the early 1990s. Water regime, organic matter management, temperature, and soil properties as well as rice plants are the major factors determining the production and flux of CH_4 in rice fields. Irrigated rice areas are the major source of CH_4 emissions from rice fields. The assured water supply and control, intensive soil preparation, and resultant improved growth of rice favor CH_4 production and emissions.

With financial support from the U.S. Environmental Protection Agency, IRRI undertook baseline research on CH_4 fluxes in rice fields in collaboration with the Fraunhofer Institute for Atmospheric Environmental Research, Germany, and the Wetland Biogeochemistry Institute of Louisiana State University, USA. Other collaborating institutes were the Wageningen Agricultural University, The Netherlands; the Laboratory for Microbiology, French Institute of Scientific Research for Development Cooperation; the Université de Provence, France; and the University of Georgia, USA. IRRI also coordinated an interregional research program on CH_4 emissions from rice fields funded by the Global Environmental Facility of the United Nations Development Programme. This activity comprised collaborative CH_4 research on irrigated, rainfed, and deepwater rice in China, India, Indonesia, the Philippines, and Thailand.

The studies, which took almost a decade of work, concluded that CH_4 emissions from rice fields are much smaller than originally thought, contributing on the order of only 10% of total global CH_4 emissions. High methane emissions are associated with specific rice management practices, and management practices can be modified to reduce emissions without reducing yields.

One result was the finding that CH_4 production from Indian rice production, originally estimated to be some 38 million t per

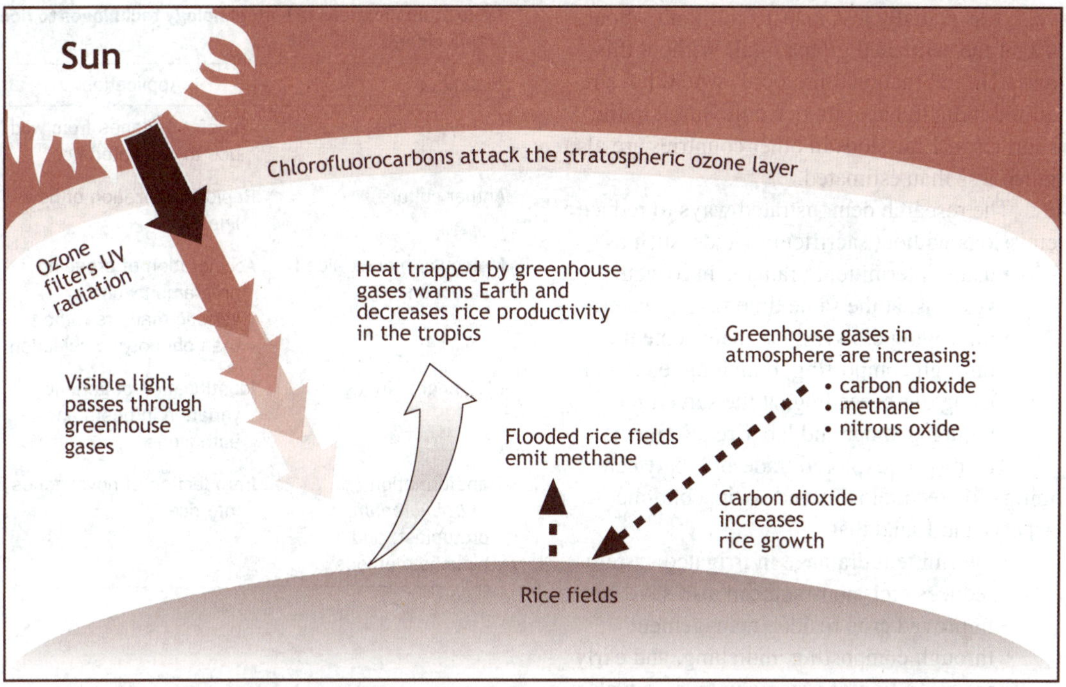

Fig. 1. How global climate change affects tropical rice. Greenhouse gases trap heat and warm Earth's surface. Higher temperatures reduce the productivity of tropical rice. The two most important greenhouse gases for rice are CO_2, which may enhance rice growth, and CH_4, which is emitted by flooded rice fields. Degradation of the atmospheric ozone layer allows more biologically destructive UV-B radiation to reach Earth's surface. The effects of UV-B on rice are not known.

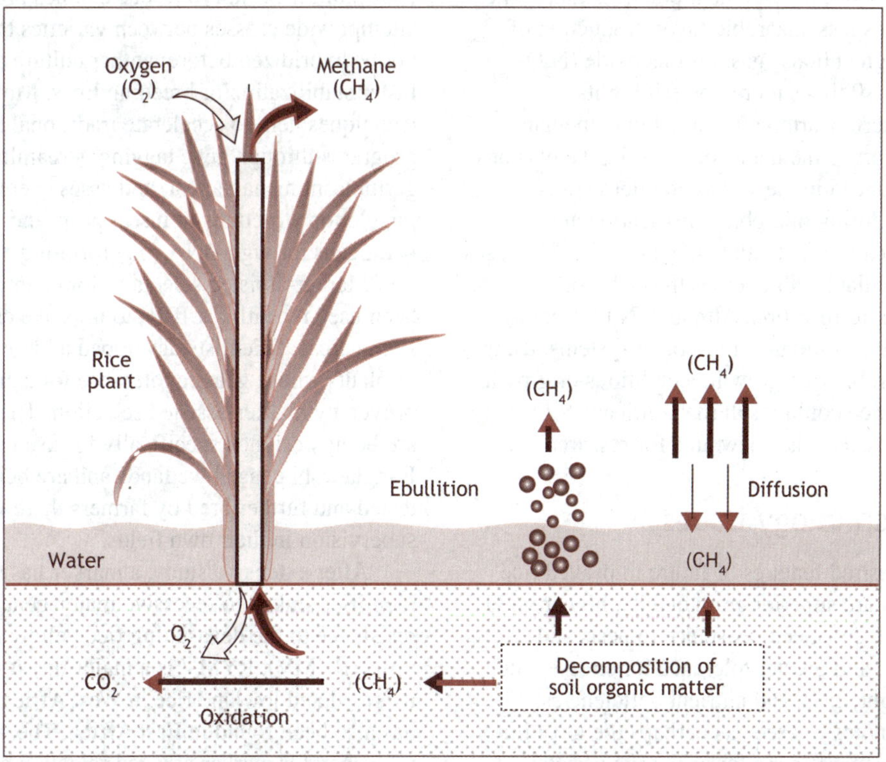

Fig. 2. Methane gas escapes into the atmosphere from flooded rice fields in three ways. Up to 80% travels through the plant from the roots. Ebullition, in which gas bubbles up to the water surface, and diffusion each contribute smaller amounts.

year, was actually less than 19 million t, about 4% of the estimated global total. Without this research, international protocols would have required India to halve its rice cultivation in the short term! Emissions in other countries are also much less than estimated.

The research demonstrated ways to reduce emissions without sacrificing yields, such as

- using intermittent drainage in irrigated systems, at the same time saving water;
- improving crop residue management through composting, mulching, etc.; and
- using direct seeding, at the same time reducing water and labor requirements.

The project explored trade-offs between mitigation technologies and socioeconomic aspects and found that

- intermittent drainage in irrigated systems reduces emissions and can also save water;
- improved crop residue management through composting, mulching, and early incorporation of organic manure can also reduce emissions; and
- direct seeding results in less labor and water input and at the same time reduces methane emissions.

However, management practices that make conditions less anaerobic favor production of another greenhouse gas, nitrous oxide (N_2O), which is 10 times more powerful in its atmospheric warming potential than methane. Agriculture is the main source of N_2O emissions; it is produced in the soil as an intermediate product during microbial nitrification and denitrification. Potential N_2O emissions increases when available N increases through, for example, fertilization. Although N_2O is not a problem in continuously flooded systems, using more aerobic rice-growing conditions or growing aerobic rice could result in significant N_2O emissions. This is a new area for research.

Biotechnology issues

Some desired features of future improved rice varieties are superior grain quality, higher yield potential, enhanced resistance to pests and diseases, and greater tolerance for stresses such as drought, cold, and nutrient deficiencies. Biotechnology is seen as perhaps the most important new resource for achieving varietal improvement (Table 1).

Table 1. Applications of biotechnology techniques to rice improvement.

Technique	Application
Embryo rescue	Transfer of genes from wild rice to cultivated rice
Anther culture	Rapid stabilization of new lines
Molecular marker-aided selection	Acceleration of breeding programs by use of genetic markers rather than phenotypic selection
DNA fingerprinting	Identification of genetic variation in pests and pathogens
Transformation (*Agrobacterium*, protoplast, and biolistic methods)	Introduction of novel genes into rice

Rice biotechnology techniques encompass plant tissue culture and molecular biology. Tissue culture techniques such as embryo rescue and anther culture have made important contributions. Embryo rescue enables breeders to attempt wide crosses between varieties that could not be hybridized before; anther culture allows faster stabilization of breeding lines. Molecular techniques help to accelerate traditional breeding programs through gene tagging, streamlined germplasm management, and assessment of population structures in insect pests and pathogens through DNA fingerprinting.

Marker-assisted selection, for example, has been used recently at IRRI to improve crops in saline areas. Scientists developed a large number of plants whose genetic tolerance for salinity was proven by marker-assisted selection. The plants are being designed specifically for use in Bangladeshi coastal wetlands and are being tested and further bred by farmers there under supervision in their own fields.

After extensive study, a marker has been identified linked to recessive major-gene resistance to the devastating rice yellow mottle virus (RYMV). RYMV is a major threat to irrigated and lowland rice in West Africa, and has also been found in East Africa. The way is now open for marker-assisted selection of resistant varieties.

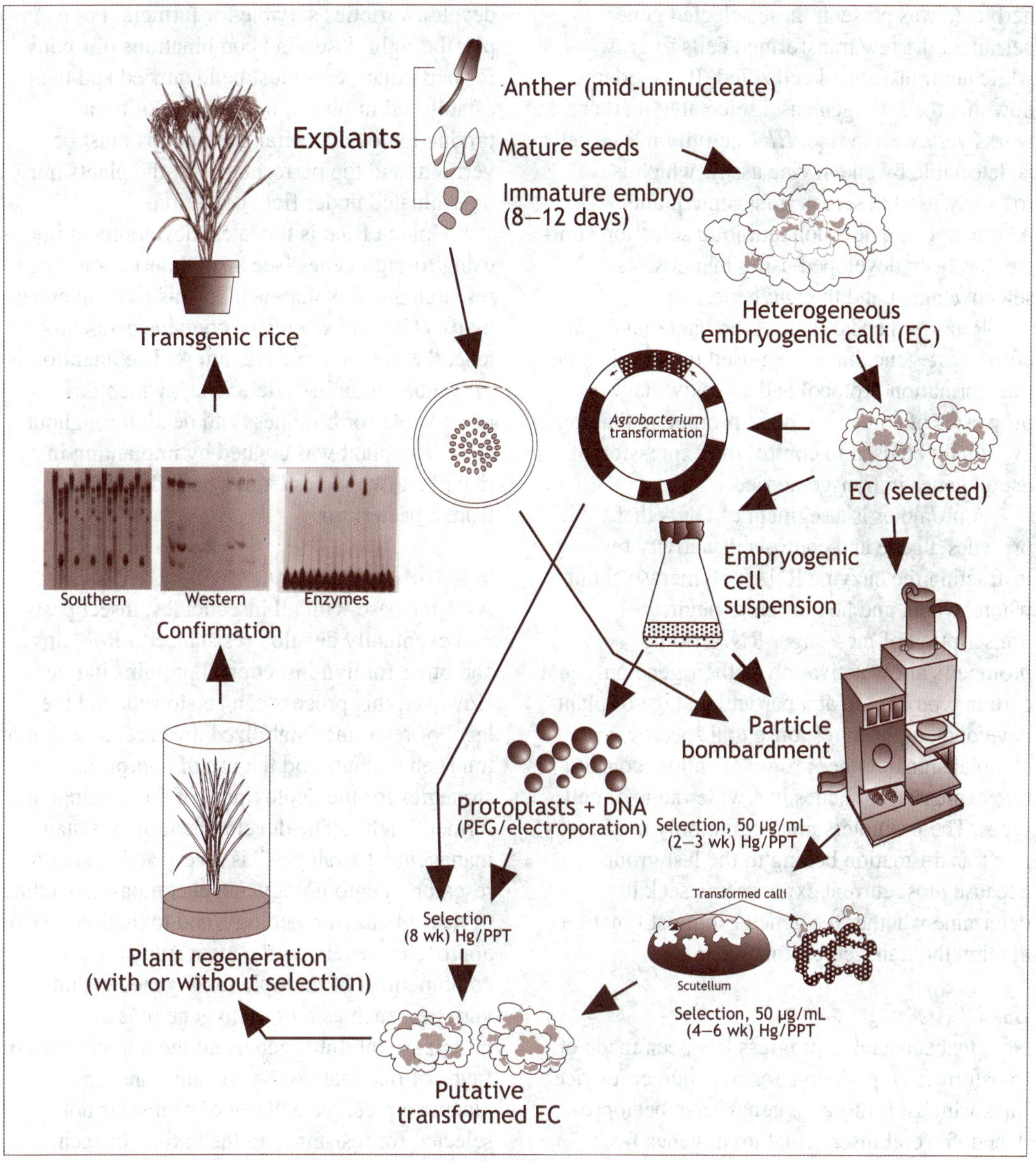

Fig. 3. A. schematic protocol for production of fertile transgenic rice plants using biolistic, protoplast, and *Agrobacterium* systems.

But biotechnology's most novel contribution will probably be in adding alien genes to the rice gene pool through genetic engineering. Genetic engineering also allows the reintroduction of rice genes that have been extracted and modified to give altered properties. Such gene transfers are impossible with conventional breeding methods. Genetic engineering also allows introducing one or two well-characterized genes at a time. There is no need for the extensive backcrossing done in conventional hybridization to remove undesirable genes.

Three methods have been used successfully to transfer genes into rice: protoplast transformation, biolistics or particle bombardment, and *Agrobacterium*-mediated gene transfer (see Fig. 3).

Marker genes and promoters

The first foreign genes expressed in rice were bacterial genes conferring antibiotic (*HPT*, hygromycin phosphotransferase) or herbicide resistance (*ppt,* phosphinotricin acetyl transferase). When the appropriate antibiotic or herbicide was present, these selected genes permitted the few transformed cells to grow while nontransformed cells died. It is evident now that the *HPT* gene as a selectable marker works very well in rice. *HPT* activity in rice cells is detectable by an enzyme assay, which is routinely used to screen transgenic plants. Alternatively, a new nonantibiotic selection strategy has been developed using mannose as a selective agent and the *pmi* gene.

Reporter genes, such as the bacterial *ß-glucuronidase* gene, have been used to optimize the transformation protocol and to study the properties of some new plant promoters that may eventually be used to control the expression of useful genes in transgenic rice.

A promoter is a segment of DNA that precedes a gene and controls its activity by instructing the enzyme RNA polymerase about where, when, and how often to begin transcription of messenger RNA. Some promoters allow activation of their gene only in a certain type of cell, at a particular stage of plant development, or in response to a specific external signal. Other promoters allow constitutive expression of genes in a wide range of cell types. The promoters most commonly used for rice transformation belong to the last group because most current experiments seek to determine whether the gene of interest functions at all in the transgenic situation.

Useful foreign genes

Now that substantial progress has been made on transformation protocols for rice, numerous rice lines with useful foreign genes have been produced. Several insecticidal toxin genes from *Bacillus thuringiensis* (*Bt*) have been transferred to rice, including *cry1Ab*, *cry1Ac*, and *cry2A*. The process is shown in Figure 3. Plants containing *Bt* genes have been evaluated in greenhouses in numerous countries and have been field-tested in China. They have shown substantial resistance to caterpillar pests such as stem borers and leaffolders. Genes encoding proteinase inhibitors from cowpea, soybean, and potato have also been transferred to rice for enhanced resistance to caterpillars. Rice has also been transformed with a lectin gene from the snowdrop plant that encodes a protein toxic to the brown planthopper and green leafhopper. Work is in progress with all these genes to develop varieties suitable for farmers. For example, the right doses and combinations of toxins for particular pests must be identified and then established in plants, the stability of toxin production over several generations must be verified, and the performance of the plants must be evaluated under field conditions.

Golden Rice is the latest development in using foreign genes (see also "International rice research and development"). This rice, invented by two German scientists, contains beta-carotene, the precursor of vitamin A. The intention is to combat vitamin-A deficiency, which is responsible for blindness and death throughout Asia. The plant was created by implanting in a rice plant two genes from a daffodil and one from a bacterium.

Sustainable use of transgenic plants

As is the case with all insecticides, insect pests will eventually develop resistance to *Bt* toxins and other foreign insecticidal proteins in rice. However, this process can be slowed, and the level of resistance stabilized, by careful design of transgenic plants and the use of appropriate strategies for the deployment of these plants in farmers' fields. The development of "resistance management strategies" is a very active area of research in entomology and encompasses studies of insect behavior, ecology, and toxicology. Two approaches receiving the most study are combinations of multiple toxin genes within varieties, such as a *Bt* toxin gene plus a proteinase inhibitor gene, and the maintenance of fields of rice that do not contain transgenic plants, to preserve a "refuge" of insects not selected for resistance to the toxins. In each generation, insects from the refuge fields would mate with resistant insects that survived in the transgenic fields, maintaining the number of resistant offspring at a low level.

Biosafety

Genetic engineering is a relatively new technology and most countries closely regulate the testing and commercial release of transgenic organisms. The Philippines was the first country in Southeast Asia to have biosafety guidelines. The National Committee on Biosafety of the Philippines (NCBP), established in 1990, develops policies to regulate the use of transgenic plants in the country, including those being developed at IRRI. Similar biosafety committees have been established in other rice-growing countries in Asia.

One of the principal biosafety issues is the assessment of the movement of foreign genes by pollen dispersal from transgenic rice to other rice varieties and wild rice. In 1995, IRRI began operating a greenhouse designed to prevent pollen dispersal from transgenic plants in the early stages of evaluation. This 320-m^2 greenhouse enables researchers to grow and multiply transgenic plants and to evaluate them for resistance to insects, disease, and other stresses. A recent study using transgenic IR72 with *Xa21* in a screenhouse revealed that there was no functional transfer of transgenes from transgenic IR72 to nontransgenic control plants, a finding that ensures the potential use of this technology in improving rice germplasm.

Crop modeling to integrate knowledge

Crop modeling enables researchers to integrate knowledge from different disciplines in a quantitative way. That, in turn, helps researchers to understand the underlying processes that determine the behavior of complex agricultural systems. Mathematical models are representations of systems made from mathematical equations. Integrating and solving the equations enable a numerical description of the system to be produced. During the first phase of a modeling exercise, the modeler seeks to give names, magnitudes, and units to the component parts of the problem. In the second phase of modeling a problem, the processes are described as mathematical functions. In the final phase, "what-if" questions can be asked about the functioning of a system and numerical answers provided. Mathematical models that contain no clear logical link with the basic processes governing the relationship between the system inputs and outputs are unlikely to contribute much of significance to any debate concerning strategic decisions in relation to research management.

Models at different levels of detail are developed to meet different objectives, ranging from a thorough understanding of an existing system to the prediction of crop production in untested conditions. Four types of crop production systems can be distinguished:

1. Potential production, where production is determined by solar radiation, temperature, and crop and varietal characteristics.

Transgenic greenhouse at IRRI.

2. Water-limited production.
3. Water- and N-limited production.
4. Water-, N-, and other nutrient-limited production.

Going from type one to type four, production generally decreases and the variables that determine system behavior increase. At all levels, growth-reducing factors such as insects, pathogens, and weeds can be introduced. Models for all production levels can be developed. Models at the first level are further developed than models at the others.

Well-developed models that simulate the growth of a crop in relation to its dynamic environment can be used to help prioritize research. Crop modeling combined with geographic information systems (GIS) analysis enables researchers to distinguish agroecological zones and to rank quantitatively the technical constraints to agricultural production within them. These models allow the impact of new technology on agricultural production to be assessed before the technology is introduced. The GIS database can link the models directly with socioeconomic aspects.

Crop simulation models have many uses. Models can be used as a research tool and to support problem solving, risk assessment, and decision making. They can guide researchers in prioritizing their research and in integrating quantitative knowledge from different disciplines. Also, models can be used as a framework for training. Further, models can be used to extrapolate research findings over broad regions and extended time, since the models account for crop-environment interactions. Using long-term weather data, yield probabilities can be simulated.

Crop models are particularly useful in the rainfed lowland rice ecosystem, which is characterized by high temporal variability and spatial heterogeneity of the environment. A limited number of field experiments cannot provide a reliable basis for management strategies under the myriad conditions that exist. Simulation models can replace expensive and time-consuming experiments because of their ability to generalize experimental findings and help interpret the results of a few selected experiments.

An aspect that is beginning to gain more importance is the use of models to set breeding goals. The physiological attributes that contribute significantly to crop production in a given environment lend themselves to definition by crop modeling. Models can serve to bridge between functional genomics, ecophysiology, and agronomy. Ecophysiological models are used to analyze the influence of different plant parameters (e.g., plant architecture, nutrient status, partitioning) on desired agronomic traits such as yield, weed competitiveness, or drought tolerance. Functional genomics supplies genetic markers for these traits, which then assist breeders during the selection process.

Modeling can also improve crop management. In West Africa, for example, a framework was developed combining the use of simulation models, field data, and long-term weather data to design site-specific crop management options. This helps farmers to improve the timing of seeding, transplanting, irrigation, and fertilizer application, as well as to determine type and dose of fertilizer in a range of biophysical and socioeconomic environments. One component, a decision tool called RIDEV, is already widely used by extension agencies in Senegal and Mauritania.

Modeling is especially useful in yield gap analysis, a method for identifying constraints to agricultural production in different agroclimatic zones. From yield gap analysis, constraints that can be reduced can be identified. Researchers then concentrate on ameliorating those factors that contribute to the gap between farm yield, potential farm yield, and potential experiment station yield (Fig. 4).

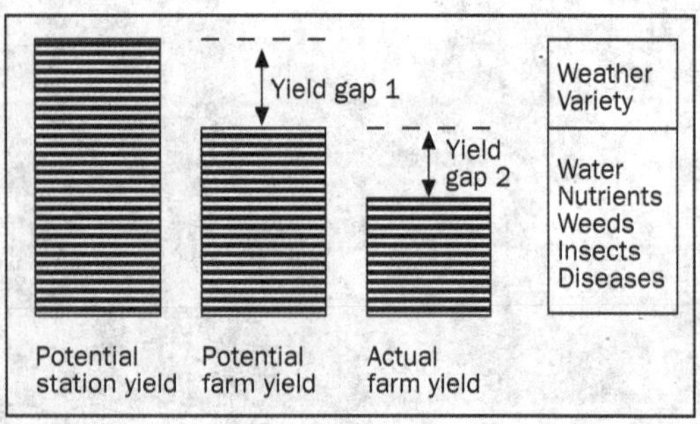

Fig. 4. Modeling applied to yield gap analysis helps to identify constraints that can be reduced, thus contributing to higher yields in farmers' fields.

International rice research and development

Research and development on rice are carried out by many institutions in both developed and developing rice-growing countries. Internationally, three research centers belonging to the Consultative Group on International Agricultural Research (CGIAR) have mandates to pursue research at the regional or global level: the International Rice Research Institute based in the Philippines, which is the foremost global rice research center; the West Africa Rice Development Association in Côte d'Ivoire, responsible for rice in that region; and the Centro Internacional de Agricultura Tropical in Colombia, which undertakes rice research relevant to Latin America and the Caribbean (as well as global research on some other crop commodities).

These international research centers are working on some or all of the critical issues mentioned in the previous chapter. They do not work in isolation, but collaborate with one another and with other, national agencies through research consortia, different types of networking, and joint ventures. Rice research issues can be resolved more easily and in relatively less time through such collaboration. Working with relevant partners also has many synergistic benefits: speeding the transfer of information and advanced research methodologies, shortening the time needed to solve problems, enabling scientific collaboration across political borders and economic barriers, and stretching scarce research resources. Similarly, the Food and Agriculture Organization of the United Nations (FAO) assists countries in rice sector development by organizing research, development, and extension activities, particularly through networking. Descriptions of the rice-related activities and collabora-

Recent advances in molecular biology have aided rice researchers immensely.

tive mechanisms of the three international research centers and FAO are given here.

In the following sections, brief descriptions of the research activities of the three CGIAR centers involved in international rice research are given. Progress in and results of their research are documented in a variety of scientific and popular publications. Each center generally produces a corporate annual report, a volume of research highlights, and a list of publications that are available from the center on request.

International Rice Research Institute (IRRI)

Mandate and structure

IRRI's mandate is to improve the well-being of present and future generations of rice farmers and consumers, particularly those with low incomes, by generating and disseminating rice-related knowledge and technology of short- and long-term environmental, social, and economic benefit, and to help enhance national rice research systems.

IRRI was established in 1960 by the Ford and Rockefeller Foundations with the support of the Government of the Philippines, and consists of laboratories and training facilities on a 252-ha experimental farm in the Philippines. It has more than 50 international staff there plus outposted scientists working with national programs in various countries in Asia.

Research programs

IRRI's research programs in the past were built around the different rice-growing ecosystems. In 2000, a new program structure was developed, based on problem-focused tasks and emerging issues in the main rice production systems. This structure allows more efficient allocation of resources and fast tracking of impact. The new programs, outlined below, take full advantage of the advancement of science to address emerging development concerns and provide stronger linkage of IRRI's research with its outreaching staff and activities as well as with national research institutions.

Genetic resources conservation, evaluation, and gene discovery

IRRI's work on the collection, conservation, characterization, documentation, and exchange of germplasm for research on genetic enhancement and sustaining biodiversity is housed in this program, which has two elements:

- germplasm conservation, characterization, documentation, and exchange and
- functional genomics.

The first entails maintenance of IRRI's efforts to collect and conserve the genetic resources of rice, now held in trust in the International Rice Genebank; and strengthening of efforts to characterize and evaluate the conserved germplasm, explore important traits, and describe allelic diversity using molecular techniques. The second, on functional genomics, aims to understand the biological functions encoded in rice genes, taking advantage of the investment made in the private sector in sequencing of the rice genome. IRRI remains committed to ensuring public access to rice genetic information. IRRI will develop additional genetic databases and establish a bioinformatics system (i.e., integration of data from DNA sequences, phenotypes, and functional diversity of rice genes in the germplasm pool) to assist national institutions in the discovery of new genes and traits.

Enhancing productivity and sustainability of favorable environments

The major factor in poverty alleviation in recent years has been the reduction in the unit cost of production and the downward trend in real prices of food. Improved technologies were adopted fastest in the favorable irrigated environment, which accounts for nearly 45% of the rice land and more than 70% of total rice production. Because of its importance, we must continue to focus on this favorable environment as the major source of rice supply to meet the growing demand from the expanding urban population and the rural landless. The challenges are how to sustain the high yields already achieved in this ecosystem and whether modern science can be used to explore possibilities of a further shift in yield potential. The options for extending the area under high-yielding modern rice varieties by developing irrigation infrastructure will no longer be available for many countries because of the looming water crisis. Farmers need assistance from rice scientists on how to grow rice with less water and how to operate irrigation systems more efficiently. Technological options must be developed to maintain soil fertility and control pest pressure.

A consortium of national research centers throughout Asia and institutions in Australia, Europe, and North America are active in all the major irrigated rice countries and will continue to address important problems of sustainability and productivity by forming multi-institutional research partnerships. These partnerships should result in the development and dissemination of new technologies for irrigated rice production.

The program has four elements:

- genetic enhancement for yield, grain quality, and stress resistance;
- managing resources under intensive rice-based systems;
- enhancing water productivity in rice-based systems; and
- the Irrigated Rice Research Consortium.

Improving productivity and livelihood for fragile environments

The fragile environments—the infertile uplands, rainfed lowlands subject to frequent droughts and submergence, and the deepwater and coastal areas that suffer from flooding, strong winds, salinity, and other soil-related problems—have had limited benefits from rice science and technological developments. The available modern varieties do not adapt well to these ecosystems and farmers get low and unstable yields and limited gains in profits when they adopt them. As a result, the traditional varieties are still widely grown and the average rice yield has remained low at 1.5 to 2.5 t/ha. These ecosystems, characterized by low farm incomes and high incidence of poverty, account for about 55% of the rice lands. High-yielding modern varieties that are tolerant of droughts, submergence, and problem soils are needed, along with efficient crop management practices, to induce farmers to adopt them and to make a direct assault on poverty.

Recent advances in molecular biology—in tagging and characterization of genes and their transfer to other species—have greatly improved the probability of success in this area. Since the environments are diverse and their domains vary across countries, the research must be done in partnership with national institutions, drawing on local scientific expertise and farmers' indigenous knowledge.

Two consortia of research institutions are in place to address the problems of increasing and stabilizing productivity in the rainfed lowlands and uplands, respectively. In these consortia, groups of national research centers are working with IRRI on prioritizing regional research needs, undertaking interdisciplinary research on productivity, sustainability, and diversity of rice-based cropping systems, and exchanging and evaluating rice germplasm and technology.

The program's three elements are

- genetic enhancement for improving productivity and human health in fragile environments,
- natural resource management for rainfed lowland and upland rice ecosystems, and
- rice research consortia for fragile environments.

Strengthening linkages between research and development

Demand has been growing for a more interactive process in research planning, combining top-down and bottom-up approaches for a better assessment of the technology needs of farmers, the priorities of rice research, the probability of research success in addressing emerging problems, and feedback on farmers' criteria for evaluating scientific knowledge in the context of their traditional knowledge. Such an approach may help improve the efficiency of research, reducing the risk that technologies and scientific output will remain on the shelf or be used only for academic purposes.

The new program in this area will incorporate ongoing socioeconomic research on understanding rural livelihoods, assessing technology needs of farmers, and validating technologies through farmer-participatory experiments in three projects:

- understanding rural livelihood systems for research prioritization and impact assessment,
- enhancing ecological sustainability and improving livelihoods through ecoregional approaches to integrated natural resource management, and
- facilitating rice research for impact.

The first deals with research prioritization and impact assessment based on understanding farmers' needs and livelihood strategies, and interactions among technologies, infrastructure, and institutions. The second applies ecoregional approaches at selected sites to demonstrate the use of systems models for improving rural livelihoods through efficient management of

Integrated Natural Resource Management at IRRI

Much of IRRI's research in the past has been in the area of integrated natural resource management (INRM)—albeit under different names. The ultimate goal of the Institute's research on the more efficient use of natural capital such as water and soil is to protect and improve the environment while maintaining and improving rice yields.

Long-term yield/productivity. In recent decades, IRRI and other institutions have been concerned about the sustainability of intensive irrigated rice production. Based on IRRI's own long-term research, evidence suggested that, under continuously flooded rice with three crops per year, yields declined significantly.

However, with help from a 100-strong team of researchers in China, India, Indonesia, the Philippines, Thailand, and Vietnam, exhaustive long-term experiments have shown that, in tropical and subtropical Asia, yield declines seem to be less common than previously thought. IRRI's fields proved to be one of the exceptional cases, probably because of prolonged soil wetness at the experiment station that is less commonly observed in farmers' fields.

Nevertheless, the research also pointed out that in farmers' fields there has been a general slowdown in gains in productivity, a measure of output in relation to all inputs of rice production. Most studies have found that productivity growth was healthy in the 1970s, '80s, and early '90s, but it is not possible to predict future trends. The findings point to the need for continued research on ways to improve both rice yield and productivity to ensure that rice production is profitable for farmers and that rice prices are low for consumers.

SSNM. Site-specific nutrient management or SSNM is a tactic that has been successfully tested in more than 200 on-farm experiments across Asia. The generic approach is to make fertilizer applications more efficient. By using a leaf color chart and a fertilizer calculation chart, farmers can assess whether and how much fertilizer is needed at any time during the growth of the rice plant. Average yields and profits have shown increases of 10–15%. Pilot studies at the village level began in six Asian countries in 2001. The leaf color chart, in particular, has proven to be an inexpensive guide to nitrogen fertilizer status and is catching on across Asia.

Land leveling. Proper leveling of farmers' fields has potential for a major impact on rice production in South and Southeast Asia. After flooding, level fields drain much faster than unleveled fields, reducing weed incidence and improving the timeliness of land preparation and crop establishment. Tests have shown that leveling improves yields by 15% without fertilizer and up to 28% with recommended fertilizer use. Leveling also provides a 40% reduction in weed biomass, a 5–7% increase in crop area by removal of bunds, a 10% reduction in water requirements, and an increased opportunity for direct seeding, which dramatically cuts labor needs.

Ongoing regional-level INRM research. Together with other centers within the Consultative Group on International Agricultural Research, other advanced research institutes, nongovernmental organizations, and national research institutions, IRRI is involved in two initiatives that span different ecological zones:

- *Ecoregional Initiative for the Asian Humid and Subhumid Tropics*—using a systems approach to develop research and operational methods of INRM that can be used at the regional level. The research uses a bottom-up approach, that is, focusing on the implications and cumulative effects of field and farm interventions at higher geographical levels of integration. A top-down approach is also used in tackling resource-use issues.
- *Rice-Wheat Consortium for the Indo-Gangetic Plains*—in a region (South Asia) of increasing demand for food grains, where past sources of productivity increases (e.g., modern varieties, fertilizers, irrigation investment) have been exhausted, and where soils are being degraded and groundwater depleted, to improve yields through integrated crop, soil, and residue management. A holistic approach is being taken to combine resource conservation approaches for rice with promising wheat production technologies that use reduced or zero tillage.

New INRM research. IRRI's new projects on integrated natural resource management, outlined in the Institute's latest medium-term plan, became operational in January 2001. As with the INRM research results described above, with the exception of the region-level work, the new.studies, listed below, are based at the field and farm level. They are components of one or more of the four IRRI research programs described in this section:

- Natural resource management for the rainfed and upland ecosystems
- Managing resources under intensified rice-based systems
- Enhancing water productivity in rice-based systems
- Understanding rural livelihood systems for research prioritization and impact assessment
- Ecoregional approaches to integrated resource management and livelihood improvement.

natural resources. The third aims at understanding the pathways of technology dissemination, and validation and adaptation of promising technologies through farmer-participatory research to be conducted in partnership with nongovernmental organizations, the private sector, and other extension agencies.

International collaborative research mechanisms at IRRI

IRRI is involved in a variety of international partnerships: research consortia, research networks, technology evaluation networks, joint ventures, shuttle research, bilateral collaboration, and direct consultation.

Research consortia

A consortium is defined here as a group of a limited number of national and international institutions formally organized to collaborate in research, training, and technology-generating activities designed to meet mutually agreed objectives. Three ecosystem-based consortia involving IRRI are currently active:

Rainfed Lowland Rice Research Consortium

The national research systems in Bangladesh, India, Indonesia, the Philippines, and Thailand, and IRRI are conducting targeted research at seven key sites that represent different rice production constraints in subecosystems of the rainfed lowlands: drought-prone, submergence-prone, drought- and submergence-prone, and medium-deep water.

Upland Rice Research Consortium

The national research systems in India, Indonesia, the Philippines, and Thailand, and IRRI have targeted five key sites, each of which represents a subecosystem and a major production constraint. Research is under way on the impact of drought, problem soils, weeds, land degradation, and blast disease on rice productivity, environmental security, and the well-being of upland farm families. Improving rice productivity is seen as the entry point to alleviating interrelated problems that contribute to upland degradation and damage to the lowland watersheds.

Irrigated Rice Research Consortium

This group, composed of national research institutions from China, India, Indonesia, Lao PDR, Malaysia, the Philippines, Thailand, and Vietnam, is working with IRRI to create region-wide multidisciplinary and integrated research projects to look at the interaction of crop production with the resource base, natural arthropods, pathogens, and weed communities, and the assessment of the effects of these interactions at local, regional, and global scales.

Research networks

Individual scientists from IRRI and other institutions organize and conduct research driven by a particular theme or set of research tools. Two networks are currently convened by IRRI:

- Asian Rice Biotechnology Network (ARBN), established at IRRI in 1993 to provide a vehicle for collaborative research in rice biotechnology with universities and rice breeding institutes of the Asian countries. ARBN, through its training and collaborative research activities, is providing a unique mechanism for national research institutions to gain access to relevant knowledge and biotechnology tools. The ultimate goal is to assist them in applying biotechnology to meet their own national needs in rice varietal improvement.
- Hybrid Rice Network, which aims to speed up development and use of hybrid rice technology in Asia. Current activities are in Bangladesh, India, Indonesia, the Philippines, Sri Lanka, and Vietnam, in collaboration with FAO, the Asia Pacific Seed Association, and China. The network has contributed significantly to the development and promotion of hybrid rice technology.

Technology evaluation networks

These voluntary, open, informal associations of scientists and research organizations enable members to exchange and evaluate technologies systematically and share experiences and information. Two are coordinated at IRRI:

The International Network for the Genetic Evaluation of Rice (INGER), created to evaluate promising cultivars, elite breeding lines, traditional cultivars, and genetic donors through a network of multilocational trials in different environments and subject to different stresses.

Crop and Resource Management Network (CREMNET). The objective of this network is to evaluate prototype technologies for sustainable production systems in different agroecological and socioeconomic situations.

Other mechanisms

As partners in national systems of rice-growing countries undertake more and more of the needed strategic and applied research, IRRI is shifting its agenda to conduct more anticipatory research in new partnerships. These include

- shuttle research in which IRRI and another institution exchange scientists to undertake specific phases of a project, for example, the IRRI-Japan Shuttle Project, which focuses on biotechnology, pathosystems, physiology of rice and yield, microbiology, and social science, maximizes the use of scientific research in Japan, and gives Japanese scientists exposure to tropical rice;
- bilateral collaboration with stronger national institutions in which IRRI and collaborating institutions agree on a joint work plan for specific activities of mutual benefit; bilateral work plan meetings between IRRI and such national institutions are held biennially for this purpose; and
- special projects to strengthen less developed national institutions, in which IRRI provides support in four major themes—rice research strategy and associated work plans, productivity and quality of research implementation, national research linkages to extension and farmers, and international linkages among national and international research institutions.

West Africa Rice Development Association (WARDA)

Mandate and structure

WARDA is an autonomous intergovernmental research association of 17 countries in West and Central Africa: Benin, Burkina Faso, Cameroon, Chad, Côte d'Ivoire, The Gambia, Ghana, Guinea, Guinea Bissau, Liberia, Mali, Mauritania, Niger, Nigeria, Senegal, Sierra Leone, and Togo. WARDA was founded in 1970. Since 1986, WARDA has also been a member of the CGIAR. WARDA's mission is "to contribute to food security and poverty alleviation in poor rural and urban populations, particularly in West and Central Africa, through research, partnerships, capacity strengthening, and policy support on rice-based systems, and in ways that promote sustainable agricultural development based on environmentally sound management of natural resources."

Rice is a unique and highly political commodity in Africa. Because of its ease of preparation, low preparation costs, low price, and steady supplies (through imports), rice consumption is increasing rapidly, providing food security for Africa's poor. Because of its large genetic diversity, it tolerates extreme soil and weather conditions, allowing farmers to produce food in adverse environments. Rice is also an integrating crop. Rice development often acts as an entry point for the development of the agricultural sector as a whole by enhancing household food security, reducing production risk, enhancing income, and averting natural resource degradation. WARDA will therefore seek to contribute to poverty alleviation in Sub-Saharan Africa through integrated sustainable development of the rice sector, particularly through technology development. WARDA's ultimate beneficiaries are rice producers and consumers in West and Central Africa. WARDA's more immediate beneficiaries are its research and development collaborators, such as national agricultural research and extension systems and nongovernmental organizations (NGOs).

Research areas and strategy

Africa's land resources are typically poor and fragile. In the rainfed uplands, shortened fallow periods and slash-and-burn agriculture lead to a general decline in land quality through erosion and soil-fertility mining. The increase in population pressure, limited capital, poor market access, and labor shortage aggravate the situation, resulting in low and unstable rice yields. In the rainfed lowlands, poor water control and infestation by weeds, pests, and diseases (e.g., blast, rice yellow mottle virus, African rice gall midge) are major constraints, in addition to population pressure, limited capital and investment, land-tenure issues, and lack of mechanization. In the irrigated systems, the main constraints are related to input access, limited mechanization, and knowledge on best-bet crop and natural resource management practices. WARDA's research focuses on three main rice ecologies (upland, rainfed lowland, and irrigated rice) and provides support to national research in the mangrove and deepwater ecosystems.

WARDA's research strategy first builds on past achievements, such as WARDA's New Rice for Africa (NERICAs), based on crosses between African rice (*Oryza glaberrima*) and Asian rice (*O. sativa*). Other important achievements include the use of participatory varietal selection (PVS) methods, the development of community-based seed supply systems, the introduction of high-yielding rice cultivars combined with integrated crop management practices in irrigated Sahelian systems, and improved understanding of soil degradation processes (salinization, alkalinization) under irrigation in the Sahel.

WARDA's research strategy is ecology-based and can be summarized as follows: (1) stabilization of the rainfed upland systems, (2) intensification and diversification of rainfed lowland systems, and (3) improvement of resource-use efficiency in irrigated rice systems.

Rainfed upland environment

Given the need to stabilize the rainfed upland environment and the success of the NERICAs, WARDA implements its research strategy for this environment along the following four lines:

- Increased use of NERICA seed in farmers' fields. PVS work in Guinea and Côte d'Ivoire has shown the potential of NERICAs in farmers' fields. In the coming decade, WARDA will expand the number of farmers exposed to NERICAs throughout the region through PVS.
- The development of new inter- and intraspecific rice cultivars. WARDA will enlarge the gene pool for uplands using advanced and more traditional breeding tools. In addition, participatory plant breeding will provide farmer input into WARDA's breeding program and speed up the identification of favorable plant material with resistance to drought and soil acidity.
- Validation and dissemination of integrated natural resource management (INRM) practices. INRM practices ensure preservation of the natural resource base and/or sustainable intensification. Building on previous achievements in the development of N-fixing fallow legumes and the introduction of a modest use of fertilizer, including rock-phosphate, emphasis will be placed on integrated biophysical and socioeconomic assessment of these technologies and the formulation of adequate baskets of technologies.
- Creating an enabling environment for the scaling-up of NERICAs and INRM practices. For the varietal component, a focus will be placed on community-based seed supply systems to address weak seed multiplication systems for farmers of rainfed rice. For INRM, the focus will be on strengthening the capabilities of rice development stakeholders in applying research results.

Rainfed lowland environment

Given the need to intensify and diversify the rainfed lowland environment and the success of the NERICAs in rainfed upland environments, WARDA is undertaking the following research:

- Develop new germplasm for lowland conditions. Evaluation of inter- and intraspecific crosses for lowland conditions is being expanded to obtain a range of appropriate plant material. This includes screening of available NERICAs and application of PVS. Particularly promising is the recent identification of a gene for resistance to rice yellow mottle virus, which will result in widespread availability of rice cultivars resistant to this virus in the coming decade.
- Identify INRM practices. Water management is the key factor for increased productivity and a means of controlling iron toxicity in the lowlands. WARDA assesses feasible modes of water management with respect to the biophysical environment, and farmers' means and perceptions.
- Develop options for diversification. Given the potential of vegetables to diversify rice-based lowland systems, WARDA has started to mobilize resources to address integrated rice-vegetable cropping in rainfed lowlands, such as the inland-valley bottoms. Other options that need to be explored include integration of aquaculture and livestock.

Inland valleys offer the greatest potential for expansion, intensification, and diversification in Sub-Saharan Africa. Inland valleys are the upper reaches of river systems, in which alluvial sedimentation processes are absent or of little importance. They are located upstream from river floodplains that are much wider. Inland valleys address the whole toposequence, or continuum, including valley bottoms, which may be submerged for part of the year, their hydromorphic fringes, and the contiguous upland slopes and crests extending over the area that contribute runoff and seepage to the valley bottom.

Inland valleys have a key role to play in providing options for sustainable intensification and diversification of agricultural production in Sub-Saharan Africa. Soils in lowland ecosystems are the least fragile and best able to support continuous cultivation. The soils in many valley bottoms retain residual moisture well after an initial flooded rice crop, permitting two crops per year, or aquaculture when base flow lasts long enough. The hydromorphic fringes and upland slopes and crests offer potential for other food and cash crops, and for trees and livestock. Thus, inland valleys constitute an important agricultural and hydrological asset at the local and national level, and can make a major contribution to future food security and poverty alleviation in Sub-Saharan Africa.

Inland valleys cover approximately 200 million hectares in Sub-Saharan Africa. Only a small fraction, probably less than 15%, is currently used in the subhumid and humid zones and crop yields are low, for example, rice yields are around 1 t/ha. There is tremendous potential for intensified and diversified use of inland valleys. Productivity can be improved substantially by improving water control, agronomic or breeding innovations, growing crops other than rice or crops after rice (such as vegetables or legumes), introducing aquaculture, and/or introducing low-cost water harvesting and management systems. Next to their agricultural potential, inland valleys have other important social and ecological service functions, such as water storage and/or drainage, and maintenance of biodiversity.

Inland valleys have very complex, dynamic, and diverse human, social, natural, and physical dimensions and interconnections. This complexity needs to be understood to determine options for improved and integrated natural resource management. Given the diversity of inland-valley ecosystems, a bottom-up social learning process will be critical. Only then can sustainable and lasting impact on food security in the region be achieved. A participatory learning and action research approach among inland-valley development stakeholders (farmers, change agents, extension, research) at the grass-roots level is required. This will help build bridges between indigenous knowledge and scientific expertise, and stimulate the formation of networks between farmers and other inland-valley development stakeholders. It will also be instrumental to the development of an INRM framework and curriculum for farmer learning for inland valleys in West Africa. WARDA is using integrated rice management as an entry point to INRM for inland valleys. A complete curriculum has been developed, containing a facilitators' manual and a technical manual dealing with all management interventions from land preparation to harvest and postharvest issues. The Inland Valley Consortium (10 West and Central African countries hosted by WARDA that focus on intensified but sustainable use of inland valleys) forms the foundation of the inland-valley research and development platform that is being built at regional and national levels.

Irrigated lowland environment

Given the need to improve the resource-use efficiency in irrigated lowland environments and the success of the Sahel cultivars and integrated crop management (ICM), WARDA implements its research strategy for this environment along the following three lines:

- Breeding. In the irrigated lowland environments, breeding uses the enlarged gene pool provided by inter- and intraspecific cultivars. Breeding activities focus on developing high-yielding short-duration cultivars with good grain quality and resistance to major African stresses. Emphasis is being placed on developing varieties that are competitive with weeds in direct-seeded systems.
- Further development and extrapolation of ICM practices. ICM decision-support systems developed for the Sahel are being evaluated and disseminated to other areas, in particular to irrigation schemes in northern Nigeria. ICM is being adapted for the medium-input irrigation systems in the

(sub-)humid agroecological zones. Potential spillover to the low- to medium-input rainfed lowland systems is given particular attention.
- Identification of INRM practices. To ensure the success of ICM, sound natural resource management practices need to be developed. Particular attention is paid to soil and water management at the farm and water-basin levels and preservation of the natural resource base, that is, maintenance of soil fertility and avoidance of salinity buildup. Senegal is used as a pilot case, taking advantage of the knowledge accumulated on irrigated systems.

Constraints that cut across ecosystems include the generally unstable socioeconomic environment, dysfunctional information and communication systems, poor research and development (R&D) linkages, and limited market integration. To respond to these challenges, WARDA considers development of new partnerships of great importance, including alliances with nontraditional stakeholders such as NGOs, the private sector, and farmers' organizations. Enhanced collaboration through participatory approaches will improve research priority setting and enhance the effectiveness and efficiency of R&D efforts. WARDA attaches great importance to developing labor-saving technologies and rice-based production systems that provide additional nutritional value that could mitigate the effect of HIV/AIDS on the welfare of the rural population.

To ensure the relevance of technologies and to facilitate scaling-up, end-users (such as farmers and extension agents) are involved throughout the technology development and dissemination process. Technologies need to be easy to implement by women (because many rice farmers are women). WARDA ensures that research is conducted with a holistic view, that is, the combined effect of a technology on productivity, farmer well-being, and the quality of the natural resource base will be assessed.

Implementing WARDA's research strategy

Implementing WARDA's research strategy comprises four essential elements:
- Technology generation for sustainable development
- Technology dissemination and farmer empowerment
- Policy research
- Capacity building and training of key agents of change

WARDA's rainfed and irrigated area programs result in improved technologies for the respective environments. These programs address technology generation and dissemination, from upstream strategic research targeting specific constraints to the application and integration of different technology components in farmers' fields. WARDA's Rice Policy and Development Program addresses the overall environment in which the proposed technology will be disseminated and used, from the farm level up, including higher aggregation levels such as watersheds, markets, agroecological zones, nations, and regions. This program thereby focuses on problems that cut across ecosystems, particularly with respect to policy research and scaling-up technology dissemination. The fourth principal element, capacity building and training, is not specifically linked to any particular program. Instead, it is integrated into each program under the coordination of a training unit.

The boundary between the two ecology-based programs is not rigid. First is the upland-lowland continuum in rainfed environments. Second is the anticipated intensification of rainfed lowland rice farming and the corresponding investment in water management, creating a continuum from purely rainfed to fully irrigated lowland. Third, there has been a shift since the late 1990s in the irrigated rice program from the Sahel toward the (sub-)humid zones. Consequently, strong synergies exist between the two ecology-based programs.

International collaborative research mechanisms at WARDA

The rice-producing environments of West and Central Africa are too varied, the production constraints too many and complex, and WARDA's resources too small for the association to have significant impact working alone. WARDA requires the contributions of scientists in national programs, in other CGIAR centers, and from advanced research institutions in developed countries. To ensure efficiency and to avoid duplication, these diverse resources need to be combined in a coherent and complementary regional research program.

It is with this objective that WARDA has developed the character of an "open center." This means that WARDA provides a permanent institutional framework within which to attract, focus, and facilitate the efforts of a range of collaborators. The unifying factor is that all partners contribute to solving priority regional problems. While WARDA generally serves as a catalyst to identify the priority themes and partners for collaborative research, research leadership in any given area is determined on the basis of institutional comparative advantage.

Partnership with national agricultural research and extension systems (NARES)

The goal of WARDA's partnership with national systems is to achieve the most cost-effective means of developing and transferring new rice technologies within the region as a whole. WARDA views the regional rice science infrastructure as an integrated and interdependent system. Because agroecological environments cut across political boundaries, technologies developed in any single location are usually transferable to far broader areas. The objectives of WARDA's partnership with NARES are to achieve a more complementary and efficient sharing of research tasks by allocating responsibilities on the basis of comparative advantage, and to maximize research spillover between these systems and WARDA, and among the national systems themselves.

The Reseau Ouest et Centre Africain du Riz (ROCARIZ), a Rice Research and Development Network for West and Central Africa, was created in 2000 after in-depth consultations among WARDA, the West and Central African Council for Agricultural Research and Development (WECARD/CORAF), national research systems, and the U.S. Agency for International Development (USAID). ROCARIZ is a successor network to the highly successful WARDA/NARS Rice Task Force Collaborative Mechanism for rice research and development within the subregion, which was supported financially by USAID in 1991-97. ROCARIZ maintains the task force structure with seven functional task forces on mangrove swamps, rice breeding, Sahel resource management, integrated pest management, natural resource management, rice economics, and technology transfer. ROCARIZ finances small two-year research projects through national scientists in member countries. The technology transfer task force includes the active participation of researchers, extension officers, and end-users of research results. ROCARIZ has established a forum—the Biennial Regional Rice Research Review (4Rs)—for reporting and publishing rice research and development activities within the subregion. The principal stakeholders of ROCARIZ are WARDA, CORAF member countries, national research systems, USAID, the private sector, and rice farmers. ROCARIZ values its guiding principles of trust, mutual respect, and joint ownership—essential ingredients for a successful partnership program. ROCARIZ is coordinated at WARDA headquarters and is managed by a coordinator under the guidance of a steering committee composed of national scientists, the private sector, development agents, and a WARDA representative.

WARDA's genetic resources unit comprises traditional genebank operations, but also the International Network for Genetic Evaluation of Rice in Africa (INGER-Africa). The genebank activities have taken up a higher profile with the successful use of the *Oryza glaberrima* rice species in the development of NERICAs. *Oryza glaberrima* and wild African species are priority areas for collection, conservation, and characterization. During the 1990s, INGER-Africa became a demand-responsive network providing new genetic material for NARES partners, including nominations from the NARES themselves (effectively allowing such material to undergo regional adaptation testing). INGER-Africa has also embarked upon studies of biodiversity (variety portfolio) management by farming communities.

In addition to traditional national partners, WARDA has recently been expanding its partnership mode to encompass the NGO community, the private sector, regional universities, and farmers' organizations. In particular, a community-based seed system has been promoted in Côte d'Ivoire and Guinea in collaboration with NARES, NGOs, private seed companies, and farmers' organizations.

Partnership with advanced research institutions

Many institutions in the global research community have strengths in specialized areas that are highly complementary to WARDA and national programs. WARDA proactively identifies and mobilizes these potential partners through collaborative research projects that focus on priority problems. In some instances, research is conducted within the partner institution. Alternatively, partner institutions outpost a collaborating scientist to WARDA, where she/he becomes a full member of WARDA project teams.

WARDA is currently collaborating with the following partner institutions: IRRI (Philippines), Centro Internacional de Agricultura Tropical (CIAT, Colombia), Asian Vegetable Research and Development Center (AVRDC, Taiwan, China), Centro Internacional de Mejoramiento de Maíz y Trigo (CIMMYT, Mexico), Food and Agriculture Organization of the United Nations (FAO, Italy), International Crops Research Institute for the Semi-Arid Tropics (ICRISAT, India), International Food Policy Research Institute (IFPRI, USA), International Institute of Tropical Agriculture (IITA, Nigeria), International Livestock Research Institute (ILRI, Ethiopia and Kenya), International Water Management Institute (IWMI, Sri Lanka), Aberdeen University (UK), Cemagref (France), Centre de coopération internationale en recherche agronomique pour le développement (CIRAD, France), CABI Bioscience (UK), Cornell University (USA), Danish Government Institute for Seed Pathology (Denmark), Horticultural Research International (UK), Institut de recherche pour le développement (IRD, formerly ORSTOM, France), International Centre for Insect Physiology and Ecology (ICIPE, Kenya), International Laboratory for Tropical Agricultural Biotechnology (ILTAB, USA), Japan International Cooperation Agency (JICA), Japan International Research Center for Agricultural Sciences (JIRCAS), John Innes Centre (UK), Laval University (Canada), Leeds University (UK), Natural Resources Institute (UK), Sainsbury Laboratories (UK), University of Hohenheim (Germany), University of Louvain (Belgium), University of Tokyo (Japan), Wageningen University and Research Centre (WUR, The Netherlands), and Yunnan Academy of Agricultural Sciences (YAAS, China).

Inland Valley Consortium

WARDA hosts a research consortium of 10 countries (Benin, Burkina Faso, Cameroon, Côte d'Ivoire, Ghana, Guinea, Mali, Nigeria, Sierra Leone, and Togo) and international research institutes (WECARD/CORAF, CIRAD, FAO, IITA, ILRI, IWMI, WUR). The consortium aims to enhance cooperation and complementarity among R&D institutions working with the common goal of intensifying the cultivation of inland valleys in a sustainable and environment-friendly manner. Consortium members work together to characterize driving forces, constraints, and technical needs for inland-valley development, to develop improved low-cost water management systems, to test component agronomic technologies, and to assess their socioeconomic and environmental impact. Each of these activities is conducted at key sites selected as being representative of broader domains in West and Central Africa.

NERICA Consortium for Food Security in Sub-Saharan Africa

The consortium to expand and speed up the dissemination and adoption of the "New Rice for Africa" and complementary technologies was launched at an international workshop in March 2002. Broad participation is expected, encompassing NARES; donors, including the United Nations Development Programme; Japan; World Bank; African Development Bank; Rockefeller Foundation; USAID; NGOs, including Sasakawa Global 2000, Organisation des Volontaires pour le Développement Local (OVDL), Association d'Appui et d'Aide aux Groupements Ruraux (Mali); farmers' organizations; and the private sector, including Pioneer Seed (Nigeria). The consortium will not only evaluate the available technology and help develop new technologies but will also identify constraints to effective adoption and scaling-up. It will also provide a mechanism for introducing participatory varietal selection and community-based seed systems to East and southern Africa.

Centro Internacional de Agricultura Tropical (CIAT) (International Center for Tropical Agriculture)

Mandate and structure

CIAT seeks to contribute to the alleviation of hunger and poverty in tropical developing countries by applying science to the generation of technology that will lead to lasting increases in agricultural output while preserving the natural resource base. To fulfill this mission, CIAT scientists integrate two lines of investigation:

- *Commodities*. The Center has a long and successful history of research on commodities for which its mandate is global (beans, cassava, and tropical forages) or regional (rice in Latin America).
- *Agroecosystems*. Through initiatives in the forest margins, hillsides, and savannas of tropical America, CIAT applies a wide range of expertise to research on the management of natural resources.

The Rockefeller and Ford Foundations established CIAT in 1967 with support from the Government of Colombia. The Center's 500-ha headquarters is near Palmira in the Cauca Valley of Colombia. CIAT has 70 international staff members from 30 countries, working with national agricultural research programs in Latin America, the Caribbean, Africa, and Asia.

Rice project

The rice project's goals have been extended since its creation in 1967, from improving the nutritional and economic well-being of rice growers and low-income consumers in Latin America and the Caribbean to include the development of sustainable increases in rice production and productivity. The CIAT rice team has eight internationally recruited staff and approximately 40 national staff. This multidisciplinary team works to make rice an economically viable crop that contributes to food security and acts as a vehicle to reduce poverty. The objective is to help resource-poor rice farmers, with the knowledge that the benefits extend to the entire rice-farming community.

The past three decades of collaboration among IRRI, CIAT, Centre de coopération internationale en recherche agronomique pour le développement, and national rice improvement programs have resulted in high-yielding rice varieties in farmers' fields, networks of germplasm improvement, and sharing of information. Evaluation of the new plant types developed at IRRI to increase rice yields is under way as well as the identification and use of new genes from wild germplasm to increase yield potential and broaden the genetic base of cultivated rice.

CIAT's rice research program aims to make significant contributions to environmental goals such as protecting soil and water and reducing agrochemical use by devoting its efforts to developing improved rice varieties and integrated crop management. The program also collaborates in developing rice as a component of cropping systems for the savannas and lowland rainfed tropics. Breeding to develop germplasm adapted to the acid-soil savannas and the understanding of rice/pasture/other crop associations will lead to more sustainable rice production in this ecosystem and a more rational use of pesticides.

CIAT's rice project works closely with the Center's national partners to solve problems and receives support from several national programs in the region. CIAT is a member of the Fund for Latin American Irrigated Rice (Fondo Latinoamericano de Arroz de Riego or FLAR), which is mainly supported by the public and private sector. Close collaboration with national partners and FLAR ensures continuity in rice research activities at the regional level. This process clearly shows that Latin American rice producers are aware of the value of new technologies.

The present program has three research elements: improved rice gene pools, integrated crop management, and enhancing regional rice research capacity.

Improved rice gene pools

The objective of this work is to increase rice genetic diversity and enhance gene pools for higher and more stable yields. This research has five steps:

1. Improved germplasm, which involves identifying progenitors for crossing, evaluating segregating material, and carrying out yield trials.
2. Improved populations (recurrent selection), requiring evaluation/recombination of gene pools, development of blast resistance through recurrent selection, development of indica gene pools with a higher response to anther culture, enhanced pest and disease resistance, high yield potential, and excellent quality.
3. Identification and use of genes from wild germplasm. First, useful wild germplasm and its traits are identified, followed by

interspecific crosses and the development of backcross populations. These populations undergo quantitative trait loci analysis and suitable isogenic lines are produced for release as advanced lines. Anther culture is used to speed up the process.

4. Determination of the physiological basis for yield enhancement and adaptation to abiotic stresses. These traits are characterized from wild rice and nutrient uptake is investigated under low pH and high Al conditions.
5. Developing marker-aided selection, part of a long-term strategy to give rice breeders another tool, allowing them to incorporate even more traits into enhanced gene pools.

Integrated crop management

The objective of this research is to develop methodologies to decrease unit rice production costs and environmental contamination.

1. For rice blast, the research focuses on characterizing the pathogen and on disease resistance, involving monitoring genetic diversity of the blast pathogen, testing breeding methods for improving blast resistance, dissecting blast-resistance genes in highly resistant cultivars, and using recurrent selection for blast resistance.
2. For the rice sheath blight pathogen, the research is on characterizing the pathogen and developing methods for screening for resistance. A transgenic approach to control the pathogen is also being investigated.
3. Rice lines with diversified resistance to *Tagosodes* and to rice hoja blanca virus (RHBV) are being developed in collaboration with other CIAT projects and national program breeders, who evaluate rice germplasm for resistance. Control of RHBV is attempted using transgenic resistance strategies.
4. For the rice stripe necrotic virus (RSNV), research involves studying its transmission using the vector *Polymyxa graminis*, developing control strategies, and breeding for resistance to the virus.
5. Weed control is being investigated by identifying traits for competitiveness and their heredity and seeking genotypes that emerge under weed-suppressing flooding.

Enhancing regional research capacity

The objective of this research is to assure that the needs of rice farmers are met. This is done through training and working in a participatory manner with farmers and our partners to develop technologies that overcome specific constraints.

1. Participatory development of rice varieties for the resource-poor farmers. This involves rice gardens from which farmers can select the varieties best adapted to their needs. Also, resource-poor farmers participate in selecting advanced lines from gene pools that target the limitations that these farmers experience.
2. The CIAT rice project works with FLAR and national programs to solve specific regional constraints.
3. Together with its partners, CIAT develops and makes available training materials and information.

International collaborative research mechanisms at CIAT

CIAT is strengthening private and public linkages and networking in the region's rice sector. To this end, the Center takes part, together with IRRI, in the Caribbean Rice Industry Development Network (CRIDNet) and in 1995 launched the Fund for Latin American Irrigated Rice (FLAR) with producer associations from Brazil, Colombia, and Venezuela. CIAT continues to be very active in training local scientists, particularly in germplasm-related and integrated pest management topics.

FLAR was created as a mechanism to mobilize public and private resources to maintain the momentum of irrigated rice research in the region. Currently, its members are CIAT and representatives from 10 countries: Argentina, Bolivia, Brazil, Colombia, Costa Rica, Cuba, Guatemala, Panama, Uruguay, and Venezuela. Other countries and international agencies are encouraged to join the group. FLAR's resources come from fees paid by producers, millers, and/or seed producers' associations. The fee is based on each country's annual rice production.

The research domain of FLAR is germplasm enhancement and crop management. FLAR's objectives are to focus rice-sector activities in an integrated manner, provide a permanent forum for the exchange of useful up-to-date information, share results among the partners, and strengthen the rice sector in its member countries by consolidating local and regional institutional arrangements. Participation of all actors in the rice sector is encouraged, including public-sector institutions, universities, and NGOs. FLAR plays an important role in diagnoses of national rice sectors, participates in the elaboration of national rice research plans, trains scientists to assess priorities, and promotes the integration of rice producers' associations and other nongovernmental associations.

FAO

The principal involvement of FAO in rice research and development is through the International Rice Commission, established in 1948. Its purpose is to promote national and international action in matters relating to the production, conservation, distribution, and consumption of rice. The members are FAO member countries in which rice is grown.

The Commission's functions also include undertaking cooperative projects directed at solving rice production problems. The Commission has an information role in the production of information on rice matters, and an advisory role in recommending to the members actions that appear necessary to solve these problems.

The notable growth in rice production over the past 50 years is attributable, at least in part, to the work of the Commission, either directly or indirectly, in applying technology, implementing cooperative programs, and disseminating information.

In the past decade, it has been involved in setting up several regional networks and working groups. The Commission's Secretariat currently supports

- the Inter-Regional Cooperative Research Network on Rice in the Mediterranean Climate Areas (Med-Rice),
- the Wetland Development and Management Network/Inland Valley Swamps,
- the Working Group on Hybrid Rice in Latin America, and
- the International Task Force for Hybrid Rice.

FAO has a memorandum of understanding with two international rice research centers: with IRRI to strengthen collaborative action aimed at promoting wider adoption of hybrid rice technology outside China (an agreement that was extended to include the provision of information to and joint publication of this Almanac) and with WARDA to support rapid rice technology diffusion in West Africa.

During the 1990s, member countries of the Commission and international institutions including FAO itself approved and funded nine technical cooperation projects in support of the Commission's work. The Commission's major recommendations are implemented by the Rice Development Programme in collaboration with national and international institutions and agencies. This program includes the following elements:

- hybrid rice development and use,
- rice integrated crop management,
- inland-valley swamp development and use,
- New Rice for Africa (NERICAs),
- prospering with rice, and
- support to the Special Programme on Food Security.

Rice around the world

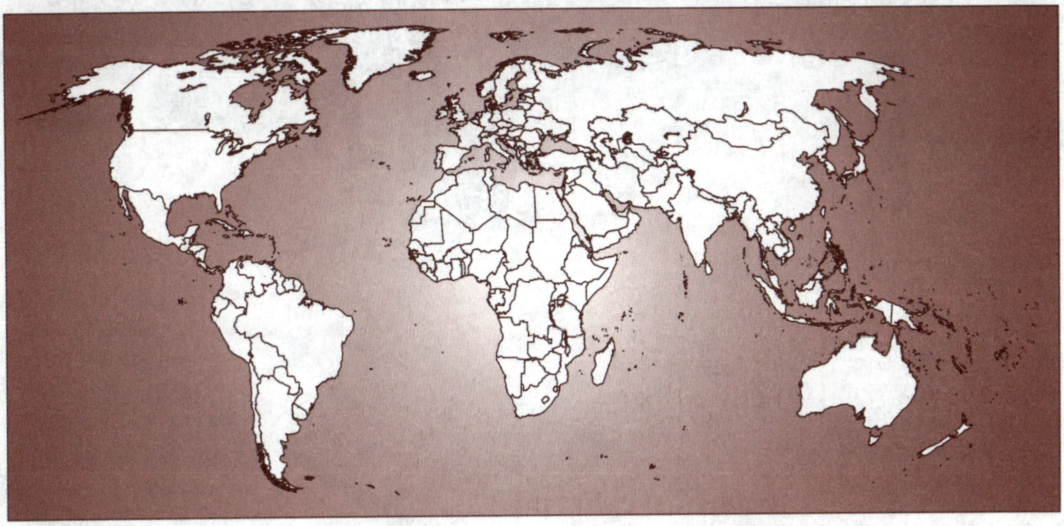

Introduction

World rice production in 2000 was about 600 million t. At least 114 countries grow rice and more than 50 have an annual production of 100,000 t or more. Asian farmers produce about 90% of the total, with two countries, China and India, growing more than half the total crop.

For most rice-producing countries where annual production exceeds 1 million t, rice is the staple food. In Bangladesh, Cambodia, Indonesia, Lao PDR, Myanmar, Thailand, and Vietnam, rice provides 50–80% of the total calories consumed. Notable exceptions are Egypt, Nigeria, and Pakistan, where rice contributes only 5–10% of per capita daily caloric intake.

The typical Asian farmer plants rice primarily to meet family needs. Nevertheless, nearly half the crop goes to market; most of that is sold locally. Only 6–7% of world rice production is traded internationally. The major rice exporters are Thailand, the United States, Vietnam, Pakistan, India, and China.

In the narratives that follow, we first present aspccts of rice production in Asia, Europe and the Mediterranean, North America, Latin America and the Carribean, and West Africa.

Next are shown the rice situation and outlook for the 10 major rice-producing countries, followed by tables and graphs of relevant data on another 54 countries that produce significant quantities of rice.

In general, the scales of the various graphs are identical from one country to another for any given data type (e.g., role of rice in the diet, level of per capita production). This consistency allows easier visual comparison across countries, but it comes at the cost of obscuring the time trends for countries whose data cluster at the bottom of the uniform scales.

There are exceptions to this scale consistency, however. They are most common in the graphs showing indices of production, area, and yield because some countries expanded production and area harvested very rapidly in percentage terms (often from a very low base) during the past 35 years. Given such extraordinary rapid growth, the use of a common scale for all countries would nearly obliterate the time trends of many countries, including most if not all of the world's major rice producers. Thus, a decision was made to use different scales for certain countries on some graphs. When the scale is not identical across all countries, a note to that effect is included for the countries with different scales. In addition to different scales for indices of production, area harvested, and yield per hectare, several countries also have different scales for per capita production (Cambodia, Guyana, Suriname).

Net trade status is defined here as net exports (i.e., exports minus imports) divided by the sum of production plus imports. For these graphs, the scale is between –100 and +100. This

Drying rice in southern Vietnam.

scale is adequate to cover most instances, but not all. For some countries (Uruguay and Australia) in some years, FAO data show that exports exceeded total production. This is possible if some of the exports in a given year came from the previous year's production.

The source of raw data for all graphs is the FAOSTAT online electronic database, and the most recent data for any given category are preliminary. All indices for production, area harvested, and yield per hectare are based on an index value of 1966 = 100.

National statistical organizations do not produce data on area and production under different rice ecosystems. The basic source of IRRI's information on rice areas by ecosystem is the surveys by R. and E. Huke[1]. The data are estimates from the returns of agricultural censuses and available Landsat imagery and other maps.

Some estimates of rice yields for different ecosystems are from personal communications with national agricultural research scientists and outposted IRRI scientists. To make the data consistent with total rice production for a country, as published by FAO, we used the yields reported by the national system for the minor ecosystems and used the residual as the yield from the major ecosystem. Production figures for each ecosystem were derived from area and yield estimates.

Some general information about the various countries was taken from the World Factbook 2000 (U.S. Central Intelligence Agency)[2] and The World Guide 2001/2002[3].

All other information is from the FAO Agrostat database and from statistical publications of individual countries.

Note that in the General Information section for each country, GNI is gross national income. It is a new term for what was formerly referred to as GNP (gross national product). PPP is purchasing power parity. PPP compensates for the fact that prices of many commodities or services differ between countries. If country A has lower (higher) prices than in the base country (usually the United States) for comparable items, the purchasing power of any given quantity of money will be greater (less) in country A than in the United States, and the GNI in PPP$ of country A will be higher (lower) than its GNI. The GNI per capita in PPP$ (as presented in the country pages for year 2000) is a better measure of income than GNI when comparing levels of material well-being across countries.

---

[1]Huke RE, Huke EH. 1997. Rice area by type of culture: South, Southeast, and East Asia, a revised and updated data base. IRRI, Manila.
[2]Web site: www.odci.gov/cia/publications/factbook.
[3]New Internationalist Publications, Oxford, England.

Rice and food security in Asia*

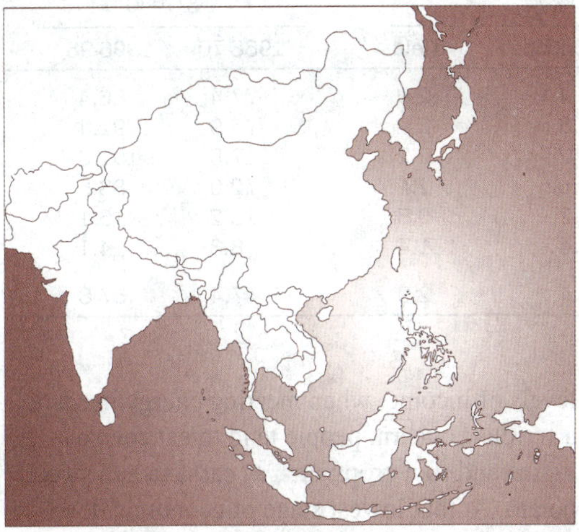

Asia accounts for 90% of the world's production and consumption of rice because of its favorable hot and humid climate. Rice continues to be grown on numerous tiny farms primarily to meet family needs. The harvest fluctuates widely because of droughts, floods, and typhoons. Maintaining self-sufficiency in production and stability in prices are important political objectives in most Asian countries. China, India, and Indonesia account for three-fourths of global rice consumption.

Increases in yields and productivity

Prior to the 1960s, rice yields (weight of grain per hectare) were low and growth of rice production in Asia was slow, originating mostly from expansion in cultivated land. Increases in productivity were occurring but only in the humid subtropics and temperate zones of East Asia, where irrigation infrastructure was already developed. The varieties grown were not fertilizer-responsive and the relatively poor market infrastructure contributed to the low-level application of chemical fertilizers.

The 1960s were characterized by a prevailing mood of despair regarding the world's ability to cope with the food-population balance. The cultivation frontier was closing in most Asian countries, while population growth rates had accelerated because of rapidly declining mortality rates.

Despite dire predictions at that time, most Asian countries have done remarkably well so far in meeting food needs of their growing population. But this is primarily because those concerns regarding the food shortages mobilized financial and scientific resources for research on food grains, which has succeeded in increasing the productivity of limited land resources. Over the last three decades, the Asian population has increased by nearly 80% but rice production has doubled (Table 1), contributing to substantial increases in individual consumption of rice and calorie intake. Several traditional rice-importing countries with severe food security problems (India, Indonesia, the Philippines, and Vietnam) achieved self-sufficiency in rice in the 1980s and Asia's rice imports declined from 60% to 20%.

More than 84% of the growth in rice production has come from an increase in productivity of rice lands, through gradual replacement of traditional varieties with dwarf and fertilizer-responsive varieties developed at IRRI and in national agricultural research institutes. The improved varieties have enabled farmers to produce two to three times more from the same parcel of land. The incorporation of insect and disease resistance into modern varieties helped stabilize yields and reduce farmers' dependence on harmful agrochemicals. The reduction in crop growth period from more than 150 days to around 110 days permitted an increase in cropping intensity and also allowed land to be used for growing nonrice crops in rice-based farming systems. Without the impressive growth in productivity, many Asian countries would have been forced to further extend cultivation into marginal lands, thus aggravating the problem of sustaining the natural resource base.

Despite an increased demand for irrigation and chemical fertilizers required for full exploitation of the potential of improved varieties, technological progress in rice cultivation led to a decline of about 20% to 30% in the cost of rice production per unit of output. Such cost-saving allowed farmers to share the

*From the paper "Sustaining Food Security in the Asian Rice Economy: Achievements and Challenges" by Mahabub Hossain, Economist and Head, Social Sciences Division, IRRI. This paper was prepared for the seminar on Sustainable Agriculture for the 21st Century, organized by the Indira Gandhi Agricultural University, Raipur, Chhattisgarh, India, 20-21 January 2001.

Table 1. Production and consumption of rice, 1968-99.

Region	Share of global rice consumption (%) 1998	Annual rate of growth (%/year) 1968-99			Per capita rice consumption (kg/year)	
		Area	Production	Yield	1968-70	1996-98
Asia	90.7	0.4	2.4	2.0	77.4	86.4
East Asia	37.8	–0.4	1.8	2.2	76.9	92.4
Southeast Asia	20.9	1.0	3.2	2.2	117.5	151.0
South Asia	30.5	0.5	2.6	2.1	72.0	83.5
West Asia	1.5	2.7	3.2	0.5	13.2	16.1
Other continents	9.3	1.0	2.4	1.4	8.3	14.1
World	100.0	0.4	2.4	2.0	47.4	57.8

Source: FAO, FAOSTAT, Dec. 2000.

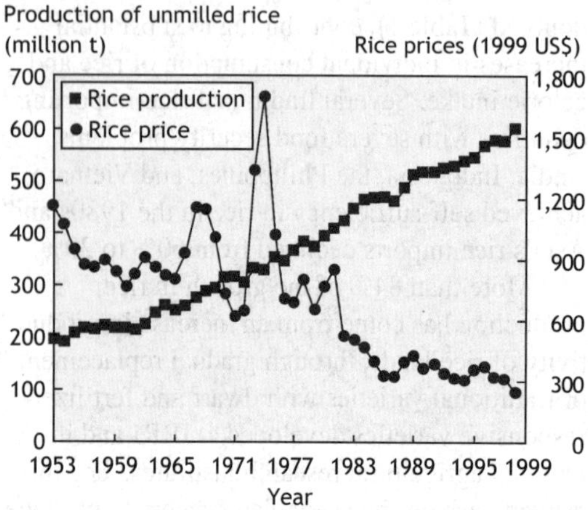

Fig. 1. Trends in world rice production and price, 1953-99.

benefits of technological progress with consumers by accepting lower prices. As rice supplies increased faster than demand, the price of rice declined relative to that of other commodities (Fig. 1). The decline in relative prices for rice benefited the rural landless and urban poor more than high-income groups and farmers who produce a surplus of rice for sale to the market because the former groups spend a much larger proportion of their incomes on rice than do the latter. As net consumers of rice, the small and marginal producers, who form the dominant group in most Asian rice-growing nations, have also gained from the downward trend in the price of rice.

Emerging trends in demand

Growth in demand for a staple grain depends on (1) the level of per capita income, (2) the rate of growth of the population, and (3) changes in price relative to those of substitute grains. At low levels of income, when meeting energy needs is a serious concern, people tend to eat coarse grains and root crops such as cassava and sweet potato. At that lowest stage of economic development, rice is considered a luxury commodity. With increasing income, demand shifts from coarse grains and root crops to rice. At high levels of income, rice becomes an inferior commodity and consumers prefer diverse foods with more protein and vitamins, such as vegetables, bread, fish, and meat.

Growing urbanization that accompanies economic growth leads to changes in food habits and the practice of eating away from home, which further reduces per capita rice consumption. The more industrialized countries of Asia, such as the Republic of Korea, Japan, and Taiwan (China), have passed through the phases of demand shifts and have experienced a decline in per capita rice consumption after reaching high levels several decades earlier (Fig. 2). Malaysia and Thailand are undergoing the same experience. But these high- and middle-income countries account for less than 10% of total rice consumption in Asia.

The annual income threshold at which consumers start substituting higher quality and more varied foods for rice is estimated at around US$1,500. This income threshold has not yet been reached in Bangladesh, China, India, Indonesia, Myanmar, the Philippines, and Vietnam. These countries account for more than 80% of total rice consumption and are going to dominate the future growth in demand. Per capita grain consumption in some of these countries is still lower than the peak levels reached by the Republic of Korea and Japan during their early phase of development (Fig. 2). In South Asia and in the Philippines and Vietnam, 30% to 50%

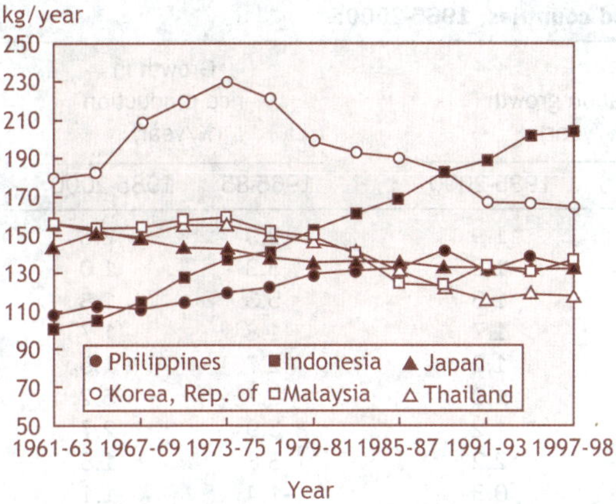

Fig. 2. Trend in per capita cereal consumption, selected Asian countries, 1961-98.

of the people live in poverty and do not have adequate income to acquire the food needed for a healthy, productive life.

However, the major force behind the future increase in demand for rice is going to be the increase in population. The Asian population is expected to increase by 35% over the next quarter century (Table 2). In the low-income countries where per capita consumption is expected to continue to grow, the population is also projected to increase by more than 30%.

The demand for marketed surplus will increase faster because of rapidly expanding urban societies that will generally increase the demand for high-cost and convenience foods. The question is whether there is enough unexploited production capacity within the existing technology and enough incentives for the producers within the existing input-output price differential to achieve the required increase in rice production and marketed surplus of rice.

Emerging trends in supply

The experience of the last three decades created a sense of complacency regarding Asia's ability to meet the growing demand for rice. Recent trends in the growth in rice prod-uction raise serious concern regarding the sustainability of past achievements. Since the mid-1980s, rice production has increased at only 1.7% per year versus 3.2% during 1975-85 and 2.9% in the decade earlier. Rice production has failed to outpace population growth in many countries (Table 3). Several factors combine to indicate that this is the beginning of a long-term trend rather than a cyclical downswing.

Constraints to further growth in rice production

Technology progress running out of steam

The Green Revolution that began in the mid-1960s has been successful mostly in the irrigated ecosystem where yield increased from 3.0 to 5.8 t/ha over the last three decades (Fig. 3). Almost all this land has already been covered with mod-

Table 2. Projections of population growth in major rice-producing and -consuming countries in Asia, 1995 to 2025.

Country	Population 2000 (million)	Annual growth rate (% per year)		Projected population in 2025 (million)	Percent increase 2000-25
		1995-2000	2020-25		
China	1,277.6	0.9	0.5	1,480.4	16
India	1,013.7	1.6	0.9	1,330.4	31
Indonesia	212.1	1.4	0.8	273.4	29
Bangladesh	129.2	1.7	1.0	178.8	38
Vietnam	79.8	1.6	1.1	108.0	35
Thailand	61.4	0.9	0.5	72.7	18
Myanmar	45.6	1.2	0.8	58.1	27
Japan	126.7	0.2	–0.5	121.2	–4
Philippines	76.0	2.1	1.1	108.3	42
Korea, Rep. of	46.9	0.8	0.3	52.5	12
Pakistan	156.5	2.8	1.5	263.0	68
Asia (excl. China)	2,045.0	1.7	1.0	3,242.7	35

Source: United Nations, 1999. World Population Prospects: The 1998 Revision.

Table 3. Population growth versus rice production in selected countries, 1965-2000.

Country	Rice harvested area in 2000 (000 ha)	Population growth (%/year)		Growth in rice production (%/year)	
		1965-85	1985-2000	1965-85	1985-2000
India	44,600	2.2	1.9	2.9	2.6
China	30,508	1.9	1.2	3.3	1.0
Indonesia	11,523	2.3	1.6	5.3	1.8
Bangladesh	10,470	2.8	1.7	1.8	1.7
Thailand	10,000	2.6	1.2	2.7	1.5
Vietnam	7,650	2.3	1.9	2.7	5.2
Myanmar	6,000	2.2	1.2	3.9	2.7
Philippines	3,900	2.8	2.2	3.8	1.8
Japan	1,800	1.0	0.3	–1.4	–1.1
Korea, Rep. of	1,059	1.8	0.9	2.6	–1.0
Malaysia	674	2.5	2.4	1.7	1.0
World	153,458	1.9	1.5	2.9	1.3

Source: FAO, FAOSTAT database, Dec. 2000.

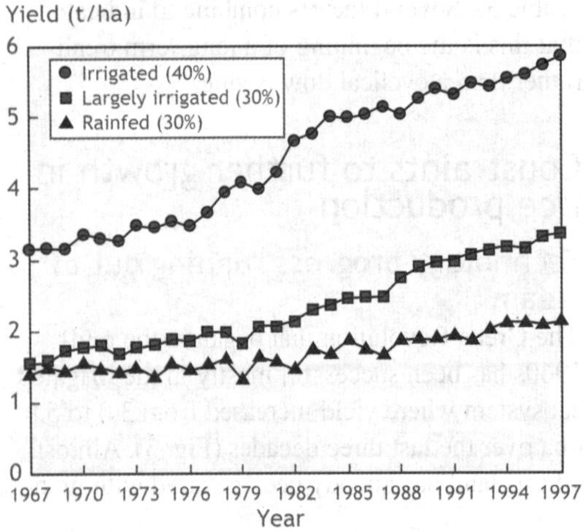

Ecosystem	Average yield (t/ha)			Growth rate (%/yr)	
	1967-69	1984-86	1995-97	1967-85	1985-97
Irrigated	3.2	5.0	5.7	2.7 (0.2)	1.3 (0.1)
Largely irrigated	1.6	2.4	3.3	2.2 (0.2)	2.7 (0.2)
Rainfed	1.5	1.8	2.1	0.9 (0.3)	1.8 (0.3)

Note: The numbers within parentheses are the standard error of the estimated growth rate. Source: IRRI World Rice Statistics and FAOSTAT database, 2000.

Fig. 3. Trend in rice yield under different growth conditions, 1967-97. Table gives comparative rates in selected periods.

ern varieties and the best farmers' yields are already approaching the potential that scientists are able to attain with today's knowledge in that environment. With intensive monoculture of rice in the irrigated systems using high doses of industrial chemicals, the natural resources are under stress, and scientists find it difficult to sustain the high yields. In Japan and the Republic of Korea, rice yield has remained stagnant at around 6.0 to 6.5 t/ha after reaching that level in the late 1960s and mid-1970s, respectively. In the humid tropics, the maximum achievable yield on farms is less than 6.0 t/ha because of increased pest pressure, frequent cloudy days with below-optimal sunshine, and susceptibility of the crop to floods, droughts, and strong winds. In regions with good irrigation infrastructure, this potential yield "ceiling" is about to be reached.

Some technologies are being developed that may help raise land productivity and input-use efficiency in the irrigated ecosystem, and thereby contribute to a further increase in rice supplies. In 1989, IRRI began to design a new rice plant, one that would make it possible to grow an irrigated rice crop with up to 30% higher yield. It is designed to increase nutrient efficiency with fewer, larger panicles per plant, to reduce unproductive tillers, and to increase photosynthesis efficiency through erect and thick leaves. Field evaluations of the breeding lines have been going on for some time with new problems, such as poor grain filling and high incidence of stem borer, being detected. The new plant architecture also needs to be matched with agronomic practices: planting method, nitrogen application, and weed control. It may take a few more years before this technology reaches farmers.

Another available technology is hybrid rice

for the tropics. The hybrids have a yield advantage of 15% to 20% over the currently inbred high-yielding varieties. Rice hybrids were developed in China. The increase in rice yields in China during 1975-90 was due largely to the diffusion of hybrid varieties to 50% of the rice area. The Chinese hybrids are not suitable for the tropical climates in Southeast and South Asia. IRRI scientists have developed suitable hybrid lines for the tropics and these are now being used by scientists in India and Vietnam to develop varieties for release to farmers. However, with the high cost of seeds and low prices (because of inferior grain quality), hybrid rice varieties do not have an advantage over the high-yielding varieties with regard to profitability. This will constrain the adoption of hybrid rice by farmers unless breeders are successful in developing better quality hybrids with higher heterosis. Another constraint may be that, when using hybrids, farmers need to change seeds every season, which is an unconventional practice.

Yields in the rainfed ecosystem have increased only marginally, from 1.4 to 2.1 t/ha over the last few decades. There is a vast potential for increasing production from the rainfed system by reducing the large yield gap. But rice scientists have yet to succeed in developing appropriate high-yielding varieties that can withstand prolonged drought, temporary submergence, and other climatic stresses common in the fragile rainfed environments. The probability of success in making scientific breakthroughs in this area is low.

Growing scarcity of land and water

Natural resource constraints to increasing rice production are becoming severe for most of the low-income countries in Asia. The per capita availability of arable land has been declining rapidly with growing populations. China now supports 17 persons per hectare of arable land, Bangladesh 13, Vietnam 11, and India, Indonesia, and the Philippines 8 to 10. Only Thailand, Myanmar, and Cambodia have favorable endowments of land, with 2 to 4 persons per hectare.

The area under rice cultivation is expected to decline with economic prosperity and urbanization as the demand for land for nonagricultural uses increases. There will also be economic pressure to release rice land in favor of vegetables, fruits, and fodder, whose markets become stronger with economic progress.

The scope for further irrigation of rainfed lands is limited because of the increasing cost of irrigation and environmental concerns regarding its adverse effects on waterlogging, salinity, fish production, and the quality of groundwater.

Water is also becoming a scarce commodity. In absolute terms, annual water withdrawals are by far the greatest in Asia, where agriculture accounts for 86% of total annual withdrawal compared with 38% in Europe and 49% in North and Central America. The per capita availability of water resources declined by 40% to 60% in most Asian countries during 1955-90 and is expected to decline further.

Sustaining interest in rice farming

Even if all the physical and environmental constraints to production are overcome, many countries of South and Southeast Asia will still face the problem of sustaining farmers' interest in rice cultivation. The expansion of industry and the services sector in urban areas, increasing nonfarm activities in rural areas, and rapidly rising labor productivity are contributing to a long-term upward trend in wage rates. In Japan, Taiwan (China), and the Republic of Korea, rural-urban migration of the agricultural labor force has caused a continuous decline in the farming population, making it difficult to sustain rural communities in some areas.

The increasing cost of rice production in middle- and high-income countries is preventing them from exporting surplus rice that has resulted from declining domestic consumption. Instead, it has lowered production. In Japan, the rice harvest reached 18.8 million t in 1967, after which it gradually fell to 10 million t in 1999. In Taiwan (China), the harvest reached 3.6 million t in 1976, whereas current production is below 2.0 million t. Meanwhile, growing scarcity of land, labor, and water continues to drive up the cost of rice production in spite of more efficient use of inputs (through improved crop management practices) and saving of labor (through mechanization). The unit cost of production is 10 times higher in Japan and seven times higher in the Republic of Korea than in Thailand and Vietnam, the major rice exporters in the world market (Table 4). The Japanese government has encouraged farmers to divert rice land to other crops, but not all rice land is suitable for other crops.

Table 4. Comparison of domestic rice prices and the cost of rice production in selected countries, 1987-89.

Country	Paddy yield (t/ha)	Cost of production (US$/t)	Domestic farm-gate price of paddy (US$/t)
Japan	6.5	1,987	1,730
Korea, Rep. of	6.6	939	957
USA	6.3	220	167
Vietnam	4.6	100	130
Thailand	1.8	120	141
Philippines	2.6	124	160
Indonesia	5.8	118	132
Bangladesh	4.6	138	180

Source: IRRI for Bangladesh and Vietnam. For other countries, FAO, Economic and Social Development Paper 101, Rome, 1991.

Under political pressure from farm lobbies, the government had to protect the domestic rice market by increasing prices and providing farm subsidies, to keep a balance between incomes of rice farmers and incomes of urban labor households. Rice production had been adjusted to domestic demand through manipulation of trade, pricing, and subsidy policies.

Ongoing negotiations within the World Trade Organization related to liberalizing rice trade may further dampen incentives for rice production, particularly in these middle- and high-income Asian countries. Large land-surplus countries (e.g., Australia, United States) can reap economies of scale using modern technology (such as mechanization and precision farming methods) because of the large size of rice farms. If domestic markets in middle- and high-income Asian countries are opened for competition, the price of rice will decline substantially, providing incentives to consumers to acquire imported food staples and forcing farmers to abandon rice cultivation in favor of more lucrative economic activities.

An important way of gaining competitive strength in the face of liberalization of rice trade is by consolidating tiny holdings into large farms. "Smart farming" in large-scale holdings, as currently practiced in developed countries, may contribute to vertical integration of the rice industry, more efficient use of machinery, and a reduction in the large number of part-time farmers whose income must be maintained at least at the level of urban labor households. The main constraint to the consolidation of holdings into efficient and competitive large-scale farming in Asia is the high price of land.

Sustaining food security

Food security means access by all people at all times to adequate food required to live an active, healthy life. Its essential elements are both availability of food and the ability of the people to acquire it. National self-sufficiency in food may not ensure food security because the poor may not have enough income to buy adequate food. India, for example, is self-sufficient in rice because it does not need to import it. Yet, one-third of the Indian population cannot afford to buy basic food. Maintaining a stable supply of basic staples (through production or imports) at affordable prices for the poor, while maintaining incentives for production growth for farmers, is a key element in sustaining food security.

Obviously, a country does not need to be self-sufficient in domestic food production to achieve or sustain food security. Singapore and Hong Kong (China) produce very little food but have better records of food security than the major rice-growing countries in the region. Malaysia meets almost one-quarter of its rice needs through imports. It is a prudent policy because Malaysia maintains domestic production with subsidies; resources tied up in rice cultivation can be used more productively in alternative economic activities. What is important for food security is achieving food self-reliance. This requires (1) a favorable export growth at the national level, permitting food-deficit countries to import food from food-surplus countries that can produce it at a lower cost; and (2) at the household level, developing productive employment that enables the population to buy adequate food. Most countries in East and

Southeast Asia are fortunate in this respect. With growing economic prosperity and alleviation of poverty, they are able to fulfill these conditions. In fact, as the cost of rice production increases with growing wage rates, land prices, and scarcity of water, it makes sense to readjust resources away from labor-intensive rice cultivation, if improving economic efficiency is the primary consideration.

But what will happen if every country in Asia abandons the production of staple grains and reallocates resources to more profitable economic activities and opts for sustaining food security through international trade? In Japan and the Republic of Korea, consumers now pay for domestic rice a price many times higher than that in the world market. But, who will produce the exportable surplus for them? Will rice supply to the world market increase substantially in response to higher prices? What would be the political response in rice-exporting countries to international transactions in staple food when trade generates scarcity in the domestic market? What would be the impact of rising food prices on inflation and other macroeconomic variables? The answers to these questions have important implications for the strategy for sustaining food security through trade for the affluent Asian nations.

The international rice market is very small. Only 6–7% of the crop is traded in the world market versus 20% for wheat and 11% for coarse grains. Unpredictable floods, droughts, and typhoons cause shortages and surpluses from year to year. These produce wide fluctuations in marketable surplus and import needs. However, the world rice market is becoming less volatile.

The other factor to consider is the influence of the giant economies of Asia—China and India—on the world rice market. The volume of international trade is equivalent to only 13% of rice consumption in China, or 8% of the combined consumption of India and China. If these countries decide to meet only 5% of their rice needs through imports, the additional demand would swamp the world market. When rice production in Japan fell by 25% because of abnormal weather from October 1993 to April 1994, the price of quality rice in the world market surged in response.

Given an adequate increase in rice prices, there is some potential for expansion of the rice-growing area in the humid tropics of Africa, where only 15% of 20 million hectares of potentially suitable rice land is currently cultivated, and in Latin America, which has an additional 20 million hectares of land suitable for rice. The unit cost of production and the marketing margin are many times higher in Africa and Latin America than in Asia, and the demand for rice has been growing faster in other continents than in Asia, so the exportable surplus available for Asia from these continents could be quite small.

Only Myanmar and Cambodia could produce an additional exportable surplus to meet potential shortages in other Asian countries and this would require substantial investment for land reclamation, expansion of irrigation, technologies for improvement in rice quality, and development of marketing infrastructure. Also, any gains may be offset by likely declining exports from Thailand and Vietnam, as farm wages and the opportunity cost of family labor increase in Thailand, and as internal demand grows in Vietnam.

Provided there is free trade in rice, high-income food-deficit countries can always acquire rice from the market, even when there is scarcity, because they can afford the resulting higher prices. It is the poor consumers in low-income countries who will suffer when there is a scarcity and it will have far-reaching effects on their domestic economy. Since rice is a major component of the food basket, the increase in prices will contribute significantly to inflation and put upward pressure on industrial wages, as the organized labor force bargains for sustaining growth in real incomes. Industrial profits will shrink and the competitive strength of the economy in the production of labor-intensive goods will erode.

These are the reasons why rice is regarded as a strategic commodity in Asia, and why maintaining stability in rice prices is a key political objective for the government in many low-income countries. When prices soar, the government may intervene in the market to protect the interest of the nation. Imposing a ban on exports of staple foods when there is a scarcity in the domestic market is not a rare phenomenon. Thus, it is in the interest of every nation to sustain a safe capacity of domestic production of staple food, whatever the cost of production.

Rice in Europe and the Mediterranean*

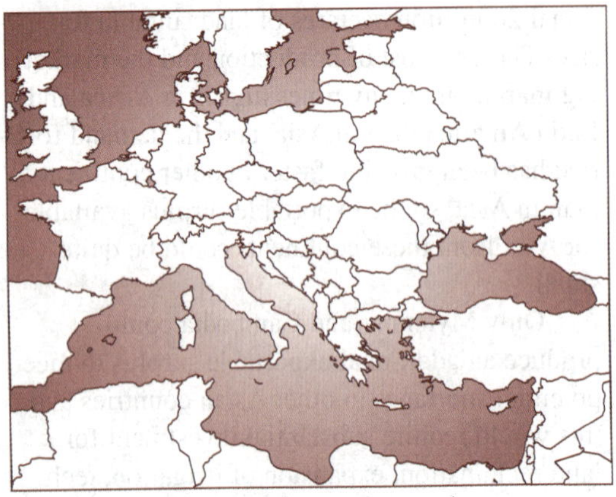

Rice in Europe is grown under a Mediterranean climate characterized by warm, dry, clear days and a long growing season favorable to high photosynthetic rates and high rice yields. Compared to tropical and subtropical rice-growing areas, the climate is cool, but warm summer nights during panicle development, when pollen formation takes place, help to avoid cold-induced floret sterility. Low relative humidity throughout the growing season reduces the development, severity, and importance of rice diseases. However, cool weather and strong winds during stand establishment may cause partial stand loss and seedling drift.

Rice is mostly grown on fine-textured, poorly drained soils with impervious hardpans or claypans. These soils are principally in three textural classes: clays, silty clays, and silty clay loams ranging from 8% to 55% clay. A few of the soils are loam in the surface horizon but are underlain with hardpans. The pH is from 4 to 8 and organic matter from 0.5% to 10% (this last value on only a limited surface area). These soils are well suited to rice production because their low water permeability enhances water-use efficiency. In some regions (the Camargue in France and Ebro Delta in Spain), soils are saline or very saline. Most of the irrigation water for European rice comes from rivers (the Po in Italy, Ebro in Spain, Rhone in France, Tejo in Portugal, etc.) and lakes. It is estimated that less than 5% of rice irrigation water is pumped from wells—in areas where surface water is not available or as a supplement to surface supplies. The high cost of pumping well water prevents its widespread use in rice production. Surface water and most groundwater are of very good quality for rice irrigation.

In all European countries, rice is commonly cultivated with a permanent flood with short periods during which soil is dried to favor rice rooting (in the early stages) or weed control treatments. The conventional irrigation system is also known as a "flow-through" system because water is usually supplied in a series from the top-most to the bottom-most basin and is regulated by floodgates by means of removable boards.

The main rainfalls occur during the first stages of the crop (April-June) and during the harvesting period. Average temperatures range from 10–12 °C during rice germination to 20–25 °C during crop flowering.

Seedbeds are commonly prepared by plowing in autumn or springtime at a depth of 20 cm and incorporating residues of the previous crop into the soil. To favor weed germination to control weeds better before rice planting, the soil is sometimes prepared by adopting minimum tillage practices. Precision land grading, obtained with laser-directed equipment, is an agronomic practice that has greatly contributed to better water management, and consequently to better crop stand establishment and weed control.

Fertilization is mostly aimed at restoring the main plant nutrients removed by crops. Because of flood conditions, nitrogen is principally absorbed in ammoniac form. This nutrient is commonly supplied at 80–120 kg/ha, 50% in preplanting and 50% in postplanting, using urea or other ammoniac fertilizers. Phosphorus and potassium are supplied entirely in the preplanting stage at 60–80 and 100–150 kg/ha, respectively.

About 80% of the rice area is cultivated with japonica varieties and the remainder with indica or indica-type varieties (mainly Thaibonnet). Rice is planted from mid-April to the end of May and harvested from mid-September to the end of October. Since the beginning of the 1960s, rice

*Contributed by Mr. Aldo Ferrero, Dipartimento di Agronomia, Selvicoltura e Gestione del Territorio, Via Leonardo da Vinci, 44, 10095 Grusliasco (TO), Italy.

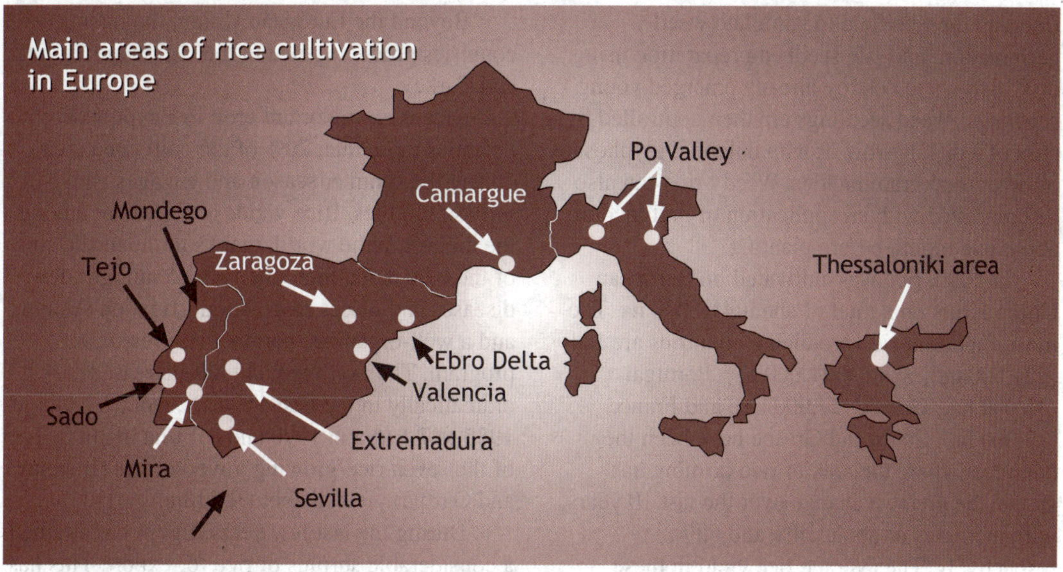

has been direct-seeded mechanically. Most is broadcast-planted in flooded fields and only 40,000 ha (almost all in Italy) are row-planted in dry soil. Rice planted in dry soil is commonly managed as a dry crop until the crop reaches the 3–4-leaf stage; after this period, the rice is flooded as in the conventional system with continuous flooding. In these conditions, rice has no competitive growth advantage over weeds, which can compete with the crop from the beginning of stand establishment.

During rice cultivation, water is commonly kept at a depth of about 4–8 cm and drained away 2–3 times to improve crop rooting (about 20 days after planting), to reduce algal growth, and to allow application of herbicides, which require dry soil. Rice fields are commonly drained toward the end of August to allow harvesting.

Rice yield is frequently affected by infestation of animals, fungus diseases, and weeds. The main animal pests are crustaceans such as *Triops cancriformis* and insects such as *Hydrellia griseola, Limonia modesta, Chironomus cavazzai, Donacia dentata,* and *Rhopalosiphum padi*, which are commonly controlled with soil drainage or organophosphorus products. The main fungus diseases are *Rhizoctonia* spp., *Pythium* spp., *Pyricularia oryzae,* and *Drechslera oryzae*. These diseases are controlled by treating crop seeds with iprodione, carbendazim, and mancozeb, spraying rice plants with tricyclazole (against *P. oryzae*), or using resistant varieties.

Weeds are reported to be the pests causing the greatest damage to rice in Europe. The major species are *Echinochloa crus-galli, E. phyllopogon, E. crus-pavonis, E. colona, Oryza sativa* var. *sylvatica* (red rice), *Heteranthera rotundifolia, H. reniformis, H. limosa, Alisma plantago-aquatica, A. lanceolatum, Bolboshoenus maritimus,* and *Schoenoplectus mucronatus.*

Alisma plantago-aquatica and *S. mucronatus* have shown numerous cases of resistance to ALS inhibitor herbicides. The infestation trend of these species is considered stable or slightly expanding in spite of weed control treatments (commonly 2–3 applications). Without weed control, crop losses at a yield of 7 to 8 t/ha were estimated to be as high as 92%.

The main herbicides applied for weed control are against *Echinochloa* spp.: molinate, propanil, thiocarbazil, dimepiperate, quinclorac, cyhalofop-butyl, and azimsulfuron; against *Heteranthera* spp.: oxadiazon; against *Alismataceae* and *Cyperaceae* species: bensulfuron-methyl, cinosulfuron, ethoxysulfuron, azimsulfuron, metosulam, MCPA, and bentazone. Red rice is controlled with a combination of preventive, cultural, and chemical practices. The most common preventive means are planting certified and weedy-rice-free seed or turning to rotational crops such as maize or soybean. The cultural practices are mainly based on the adoption of the stale seedbed practice, applied by preparing the seedbed early in the season (March) and then flooding the rice field to stimulate weed germination, and late seedbed preparation using disk harrows to destroy already emerged young seedlings. Weed seedlings are then controlled by a pass of a disk harrow or with dalapon, glyphosate, or other total graminicides. Weedy rice can also be controlled with an application of pretilachlor, about one month before planting.

In 2000, rice was cultivated on European Union farms on a total of about 410,000 ha. The most important rice-producing countries are Italy (221,000 ha), Spain (111,000 ha), Portugal (31,000 ha), Greece (27,000 ha), and France (19,000 ha). Spain and Greece have been the countries where the area of rice farming has shown the greatest change over the last 10 years, with increases of about 24% and 68%, respectively. The average rice yield in these countries is 6.6 t/ha, although farm yields of 7–7.5 t/ha are frequently recorded.

Beyond the European Union, the main countries producing rice in the region are Egypt and Turkey.

Rice is an important crop in Egypt, where it occupies more than 20% of the cultivated area during the summer season and engages about 1 million families. Rice yields in Egypt are among the highest in the world, thanks to the fertile soil of the Nile delta, high intensity of sunlight, few diseases and insect pests, good irrigation system, and a well-organized national rice research program. The average yield has increased dramatically in the past 15 years, from 5.7 t/ha in 1985 to 7.1 t/ha in 1990 and 9.1 t/ha in 2000. Two of the seven rice-growing governorates (Beheira and Garbia) yielded about 9.5 t/ha.

During the last few years, Egypt has attained a considerable surplus of rice for export. This has been achieved by releasing and spreading new early and high-yielding varieties such as Giza

175, Giza 176, Giza 181, Giza 177, Giza 178, Sakha 101, and Sakha 102; the transfer of appropriate technology to the farming community to improve crop management; and the improvement of productivity in the saline areas in the North Delta.

The development of the rice economy has been influenced by several policy changes resulting from the implementation of the economic reform program. This program is principally based on liberalization of crop production through cancellation of area allotments and phasing out of input subsidies. Despite large, optimistic land reclamation projects, the potential for increasing the area planted to crops is limited. The main limitation is the availability of irrigation water, as rice is one of the most water-consuming crops.

In Turkey, rice is grown mainly in seven geographical regions: Marmara-Thrace, Black Sea, Central Anatolia, Southeastern Anatolia, Mediterranean, East Anatolia, and the Aegean. The largest production areas are in the provinces of Edirne and Samsun.

Rice is cultivated under continuous flooding conditions. Planting is carried out using pregerminated seed (about 200 kg/ha) broadcast by hand on flooded soil. Fertilization is usually done with nitrogen and phosphorus fertilizers at 100–150 and 80 kg/ha, respectively. Potassium is normally not supplied because of the high availability of this nutrient in Turkish soils.

Harvesting is from mid-September to the end of October and is commonly carried out when rice grains reach 20–24% water content. Rice is harvested by hand and then treshed with tresher equipment or with a combine harvester.

The rice land areas and yields in Egypt and Turkey and the other non-European Union countries producing rice are shown in Table 1.

More information on European and Mediterranean rice-growing countries can be found in the individual country sections.

Med-Rice is the FAO Inter-Regional Cooperative Research Network on Rice in the Mediterranean Climate Areas. It was created in 1990 as one of the interregional and regional networks on rice and field projects, supported by the International Rice Commission (IRC) and the Rice Development Programme (RDP). The objective of Med-Rice is to promote scientific exchanges among rice scientists working in the Mediterranean area and in the other regions with a Mediterranean climate.

Table 1. Rice lands and yields in non-European Union countries, 2000.

Country	Area (ha)	Yield (t/ha)
Egypt	660,000	9.1
Turkey	60,000	5.6
Morocco	5,600	4.5
Romania	1,500	2.5
Russia	175,800	2.5
Bulgaria	3,000	2.3
Hungary	3,088	2.3

Source: FAOSTAT.

The network began as a response to the need to collaborate and coordinate research on rice in view of its increasing cultivation and consumption in Europe. Some of the important subjects are quality and competitiveness of European rice; resistance to blast, stem borers, and diseases; control of red rice, a weed that competes with cultivated rice; cataloguing of rice genetic resources in the region; and a databank of knowledge on all aspects of rice cultivation for the purpose of improved management and rice yields. These are all being investigated through cooperative research programs among member institutions of the network.

Twelve countries participate in Med-Rice: Egypt, France, Greece, Hungary, Italy, Morocco, Portugal, Romania, Russia, Spain, Turkey, and the United Kingdom. Activities of Med-Rice include scientific meetings, cooperative research programs, and publications ranging from reports and proceedings to a newsletter (Medoryzae). Scientific activities fall under five working groups, on agronomy, biotechnology, economy, selection, and technology. The network has its own Web site (http://medrice.agraria.unito.it).

Rice in North America

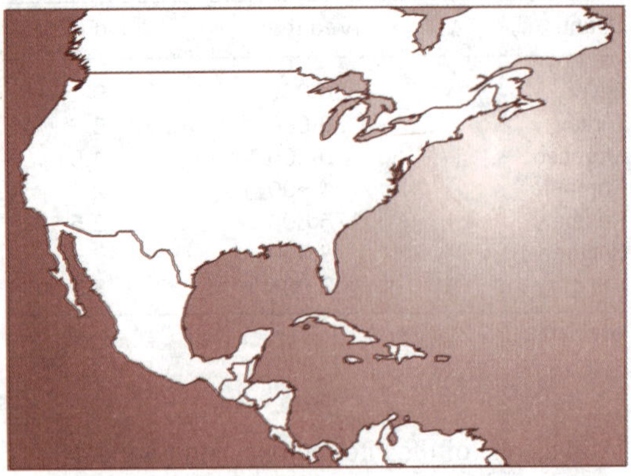

Rice in North America is grown in the United States and Mexico. "Wild" rice, not a true rice but sometimes grown in the same manner as real rice, occurs in southern Canada and the U.S., and is mentioned briefly here. The following description focuses on the major producer, the U.S.

The United States*

The U.S. population was approximately 276 million in 1999. The country is highly industrialized. The populace is employed principally in manufacturing and service industries, with only 2% directly involved in forms of agriculture, which itself makes up 2% of the gross domestic product. The 1997 Census of Agriculture reported a total of 9,291 farms that produced rice that year.

Carolina Gold, the first rice variety

Carolina Gold was the first rice variety to be grown commercially in what is now the United States. It is thought to have arrived in 1681 when a ship from Madagascar took shelter in Charles Towne (now Charleston, S.C.) during a storm. The ship's captain left a bag of rice seeds he had collected in Madagascar. The seeds thrived and became known as Carolina Gold. A sister selection was called Carolina White. The Gold Coast of South Carolina and Georgia was named because of the golden fields of rice that ships saw when approaching the coast. Exports of the Carolina rice made Charles Towne the wealthiest city in England's American colonies. The vast Carolina rice plantations disappeared after the U.S. Civil War (1861-65) and so did the Carolina sisters. But, in the 1990s, scientists of the Centro Internacional de Agricultura Tropical (CIAT) found Carolina Gold and White growing along the upper Amazon in South America.

U.S. rice production in 1999-2000 was 9.3 million t, accounting for about 1.5% of total world production. The rice was grown on 1.42 million ha with an average yield of 6.6 t/ha.

Rice is grown in seven states, but three states—Arkansas, California, and Louisiana—together account for about 80% of U.S. rice area and production.

Arkansas is the main U.S. rice-growing state, producing about 46% of U.S. production. Arkansas harvested 4.3 million t on 660,000 ha in 1999. California was second, harvesting 1.7 million t on 206,000 ha; followed by Louisiana, 1.4 million t on 251,000 ha; Mississippi, 0.8 million t on 132,000 ha; Texas, 0.7 million t on 105,000 ha; and Missouri, 0.45 million t on 75,000 ha. Florida and some other states also grow small amounts of rice.

Exports

The U.S. exported about 2.8 million t of rice, mostly milled, in 2000. That is about 12% of all world exports, making the United States the fourth largest exporter of rice (after Thailand, Vietnam, and China).

The main markets for U.S. rice are Canada, Haiti, Japan, Mexico, Saudi Arabia, and Turkey.

Consumption

Although low compared with most Asian countries, U.S. rice consumption more than doubled over the past 20 years, to an all-time high of 12.3 kilograms per person in 2000.

Direct food use accounts for 63% of U.S. consumption; processed foods, including pet food and baby food, 22%; and beer, 15%.

Part of the increase in direct food use is because of a general interest in rice for improving diet and health, plus marked increases in Asian and Hispanic populations, who prefer rice. Imported rice constitutes 11% of direct food

*Contributed by Dr. Thomas Hargrove.

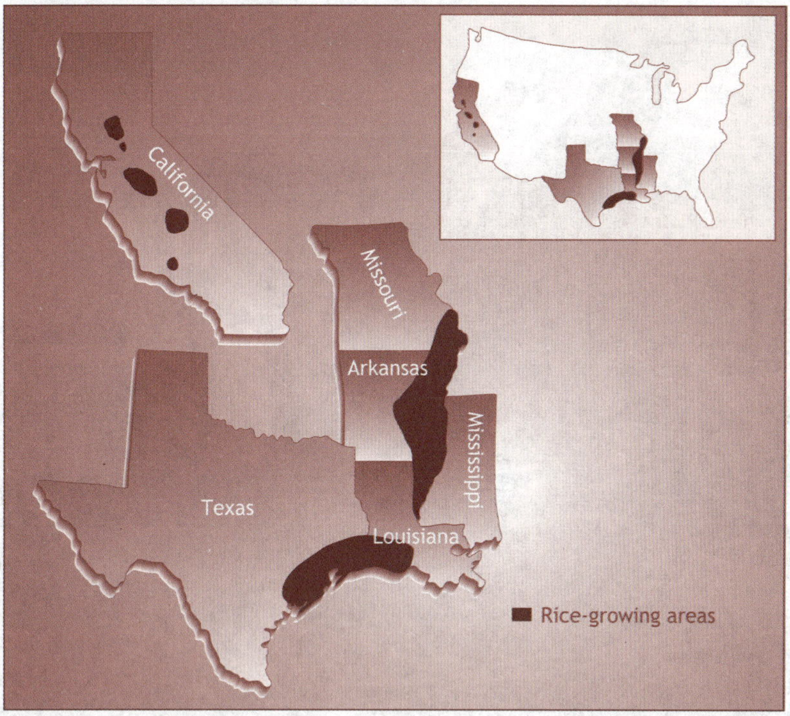

Map courtesy of Dr. T. Hargrove.

consumption; most is aromatic Thai jasmine and Indian and Pakistani basmatis, consumed by ethnic Asians. Arborios are also imported from Italy.

Brewing

The brewing company Anheuser-Busch is the largest purchaser of U.S. rice, buying about 8% of the annual crop. The brewing giant owns its own rice mills in Arkansas and California. Budweiser, its most popular beer brand, uses rice as an adjunct. Rice and corn flour are used in other Anheuser-Busch beers. Coors is also a rice-based beer.

Wild rice

American "wild rice," favored by gourmets, is not rice at all; it is *Zizania aquatica*, a semiaquatic grass native to the Great Lakes region of the U.S. and Canada. Native North Americans have gathered and eaten wild rice for thousands of years. The early native inhabitants of the Great Lakes region increased the natural production of wild rice by rolling seeds into a ball of clay and dropping the seeded balls into the water. It is still harvested wild, although domestication in Minnesota began in the 1950s—perhaps the first cereal to be domesticated by humans since the time of the Pharaohs.

Wild rice came to California in 1972 when a rice farmer in northern California's Sacramento Valley planted some seed brought from Minnesota in an ice chest. Commercial production began in 1977. Wild rice is now being grown commercially in Minnesota, California, Utah, and Oregon, and in Canada. In the U.S., wild rice is now grown in much the same way as "real" rice, in flooded fields, with yields of up to 1.6 t/ha in Minnesota and twice that amount in California reported in the early 1990s. In Canada, commercial production is mainly from leased lakes that are seeded; the leaseholder is given exclusive harvesting rights and much of the harvesting is done using airboats.

Rice environments

The rice production data above show that rice in the U.S. is grown in three principal areas: the Grand Prairie and Mississippi River Delta of Arkansas, Louisiana, Mississippi, and Missouri from 32° to 36° N; the Gulf Coast of Florida, Louisiana, and Texas from 27° to 31° N; and the Sacramento Valley of California from 38° to 40° N. The climate varies from semiarid California,

Large-scale mechanized rice farmers in the United States use equipment such as this minimum till air drill that plants 65 rows at a time with a holding capacity of 10,000 pounds of seed. Source: Jay Cockrell, *Texas Rice Newsletter*.

with less than 50 mm of rainfall during the growing season, to the humid subtropical Gulf Coast of Louisiana, Texas, and Florida, where rainfall may total 700–1,000 mm. In all these environments, rice is grown as a single crop per year, but can be ratooned in the warmest, southernmost regions of the Gulf Coast states.

Approximately 40% of Texas and southwestern Louisiana rice area is ratooned annually. All rice in the U.S. is irrigated and direct-seeded. In California, pregerminated seed is seeded into standing water by aircraft. Southwestern Louisiana is also wet-seeded. Dry seeding with a mechanized grain drill is the most common method of planting in the southern U.S.

Rice is grown on natural flatlands. Nearly 100% of these flatlands in California and approximately 40% in the southern U.S. have been further leveled by laser-directed machinery. In rice monocrop systems, the land may be leveled to a slope of 0.02 to 0.05 m/100 m. In rice–row crop systems, grades of 0.1 to 0.2 m/100 m are required for drainage or irrigation of the rotation crop. Precision leveling has greatly facilitated water management and is considered second only to the introduction of semidwarf varieties as contributing to increased rice yields.

In California, where the rain-free environment and high latitude provide maximum solar radiation and low disease pressure, farm yields average 8–9 t/ha. In the humid southern rice environments, disease (primarily blast and sheath blight), warm nights, cloudy days, and frequent thunderstorms at heading limit average yields to about 6.5 t/ha.

Problems and opportunities

Although U.S. yields are high, several problems constrain production.

- In the humid subtropical climates of the Gulf Coast and Mississippi River Delta areas, diseases (particularly blast and sheath blight) limit yields.
- Because all areas are direct-seeded, weeds and poor stand establishment are significant problems.
- In the southern U.S., many share-crop arrangements are short term and tenant farmers are reluctant to spend capital for long-term improvements to productivity.
- An indirect production constraint is embodied in concern about agriculture's role in environmental degradation. All U.S. agriculture now operates within a stringent and costly regulatory environment. One critical concern is maintaining high-quality surface water and groundwater with respect to potable water, the health of aquatic organisms, and recreational uses.
- In some areas, especially California, degradation of air quality from burning rice straw is highly regulated and rice straw disposal is a production problem.

Rice research and extension are an integral part of the U.S. Department of Agriculture and University Land Grant system, supplemented by private research and development, primarily by commercial seed and agricultural chemical producers. All of the principal rice-growing states have well-staffed and equipped public-sector rice research stations. Scientific exchange among these institutions is linked by the U.S. Rice Technical Working Group. Linkages to IRRI and other international programs on rice are scientist-to-scientist, mostly on an ad hoc basis.

The challenges for U.S. rice production are to maintain high yields and quality as well as the sustainability of the rice-based cropping system, in the context of maintaining and improving soil, air, and water quality in an increasingly regulated environment. Improved technology and equipment for land and irrigation management and for harvesting and handling high-quality rice; varietal improvement through the integration of genetic engineering and conventional breeding programs; integrated pest and crop management; and the development of sophisticated pest control technology will be key elements in future production opportunities.

Summary data on U.S. rice production are given on page 231.

Mexico

Maize is the main staple of the 97 million inhabitants of Mexico, but the country produces about 450,000 t of rice annually. Rice is grown in at least 17 states, the three major states being Sinaloa, Campeche, and Veracruz, each of which contributes a little over 20% of the nation's total rice production. Other states producing significant amounts of rice (2% to 6%) are Tabasco, Colima, Tamaulipas, Morelos, Nayarit, Michoacán, and Jalisco.

Production has varied over the past two decades with little discernible trend. Annual production in 1980, 445,000 t, was almost the same as that in 2000. However, the harvested area has decreased from 127,000 ha in 1980 to 98,000 ha in 2000, indicating a gradual improvement in yield.

Consumption of rice is low, about 8 kg per capita per year, but is not sustained by domestic production. Imports of rice grew from less than 100,000 t in 1980 to more than 400,000 t in 1999, almost equal to national production. Most is imported from the U.S., for which Mexico is the largest export market.

More details on Mexican rice production are given on pages 194-195.

Rice in Latin America and the Caribbean

Rice is a staple food crop in Latin America and the Caribbean (LAC). The region's per capita annual consumption increased from about 9 kg of milled rice in 1924-28 to about 30 kg in 1993-95. Rice consumption is concentrated in the tropical countries of the region, which have a total population of 320 million. About 40% live below the FAO poverty line. Tropical Latin Americans consume an average of 37 kg of milled rice yearly—equal to about 1.3 cups of cooked rice daily. After sugar, rice is their single most important source of daily calories, supplying 11.5% of daily caloric intake. In Brazil, Colombia, Panama, Guyana, and the Dominican Republic, rice provides 25% more calories than any other crop.

Rice is also a leading source of protein for the poorest 20% of the tropical population, supplying more per capita than beans, beef, or milk. Rice is income elastic in the region: consumers tend to increase consumption as their incomes rise.

From 1967 to 1995, increasingly efficient production, triggered by the adoption of modern semidwarf varieties that were more input-sensitive, caused the real price of rice to decline by 50%. Massive resulting social benefits went to the urban poor. Rice is particularly important from the standpoints of growth and equity. Rice is preferred by the poor because it is cheap, nutritious, appealing, easy to prepare, and easy to store and transport.

Poverty in Latin America and the Caribbean is extensive: 31% of the total population is poor and 21% is desperately poor. Most of the poor live in urban areas. Poor urban dwellers spend about 15% of their income on white rice—their cheapest source of energy, carbohydrates, and protein. Their well-being is therefore affected by the amount, quality, security of supply, and price of the rice they eat.

Pushed by large debt burdens, fiscal and trade imbalances, and high inflation rates during the 1970s and 1980s, most of the region's countries have developed self-sufficiency policies for rice production to maintain low and stable prices for urban consumers.

Source of agricultural development

Traditionally, rice was a leading pioneer crop for area expansion and colonization until the 1980s when the trend in agriculture reverted to more intensive practices as a result of more open trade practices and the need to increase efficiency and competitiveness. During 1990-2000, rice production in LAC expanded annually at 3.0% fueled by a 3.8% annual growth in yield while cultivated area contracted annually at 0.8%. Most of the decline in area occurred in upland rice. Irrigated rice area continued a steady increase. Higher yields were the result of the shift to irrigation as well as the continuous release of improved varieties.

Such growth in production has provided many opportunities for reactivating local rural economies. Moreover, local feed and food agroindustries used nearly 4 million t of rice by-products per year by in the mid-1990s.

Urbanization and economic liberalization are forcing the integration of regional rice markets and agribusiness. Demand is increasingly high for healthy, diversified, rice-based convenience foods. Most countries do not rely on rice imports to meet domestic needs.

In 1993-95, 3.8% of the world's rice production and 3.4% of the cultivated area were in Latin America and the Caribbean. From 1966 to 1995, rice production increased from 9.8 million t to 20.7 million t. In 1993-95, 6.7 million ha in Latin America were under rice. Of these, 3.1 million ha were upland, 1.1 million ha rainfed lowland rice, and 2.5 million ha irrigated rice.

About one million farmers in the region depend on rice as their main source of energy, employment, and income. Of these, about 0.8 million are resource-poor smallholders, planting less than 3 ha. They cultivate rice manually, producing only 6% of the total rice output in Latin America and the Caribbean. The other 0.2 million rice growers produce 94% of the rice, having larger (15–50 ha on average) mechanized farms.

Geographical features of rice cultivation in Latin America

More than two-thirds of Latin America's arable lands are within lowland ecosystems. Rice is well adapted to the wet soils common in the lowlands. Opportunities exist for lowland rice expansion in the vast wetlands of Brazil (with a potential of 24 million ha), the Andean countries (Venezuela, Ecuador, Peru, and Bolivia), the River Plate Basin (Argentina, Uruguay, and Paraguay), Guyana, and Central America.

Another ecosystem with potential is the high-rainfall savannas, with aerobic upland acid soils, found in Brazil, Venezuela, Colombia, Bolivia, Guyana, and Suriname. Rice is more tolerant of acidity than are other grain crops. New technologies (more input-responsive rice varieties) and production systems (upland rice-pasture cultivation) can encourage the establishment of improved pastures and relieve pressure for food production from more fragile environments.

Rice grown under irrigation provides 59% of the total production on just 37% of the total rice area, with average yields of 5.0 t/ha. For rainfed lowland rice, the corresponding figures are 22%, 16%, and 3.9 t/ha, respectively; and for upland rice, 19%, 46%, and 1.3 t/ha, respectively.

Brazil accounts for 65% of all the rice area in Latin America and the Caribbean. The rice area in Brazil is 62% upland, grown on acid soils, whereas, outside Brazil, most rice is irrigated on richer soils. Brazil produces 52% of all irrigated rice, 38% of all rainfed rice, and 92% of all upland rice in the region.

In contrast with other rice-growing regions in the world, rice cultivation on larger farms in tropical Latin America and the Caribbean is predominantly mechanized. Direct seeding and purchased inputs are used across ecosystems. These features fit fairly well with the characteristics of the region—abundant flat lands and scarce labor—and enhance its comparative advantage in efficient rice production compared with more labor-intensive cultivation systems elsewhere in the world.

A direct-seeded culture

In contrast to Asia, where rice is commonly transplanted, rice in Latin America is mostly direct-seeded, a practice that developed because of increasing labor costs. The Asian transplanting system is used on only 6% of the region's total rice area.

About two-thirds of irrigated rice lands are minor schemes developed by farmers who divert water from streams, rivers, and wells. The new short-duration varieties, developed regionally, and capital inputs and water have helped farmers deal successfully with weed populations and with the more demanding water, fertilizer, and pest management needs for direct-seeded rice.

Research challenges

Environmental concerns require reduced agrochemical use and alternatives to forest clearing for crop production. To achieve these objectives, rice scientists must continue working on (1) developing higher-yielding wetland rice varieties, (2) developing more productive varieties for upland rice, and (3) reducing production costs and environmental hazards through genetically resistant varieties and better crop management practices to achieve higher efficiency in the use of inputs. The last point encourages farmers to maintain or increase the cultivated area, despite low rice prices, and to reduce agrochemical use.

Research and development in Latin America

Nearly 100 publicly funded agricultural research and development institutions work on rice in

Experimental rice plots in Uruguay.

Latin America and the Caribbean. Each follows one of four models: (1) as part of a research department in the Ministry of Agriculture, (2) as part of a decentralized agricultural research institute, (3) as an association with a rice development project, or (4) as a decentralized rice research institute.

Rice research is also carried out by farmers' organizations, private companies, universities, and regional institutions, their opportunities having increased since the internationalization of most countries in the region during the 1990s. Helping researchers to benefit from collaboration and to improve their capacity are networks such as INGER (International Network for Genetic Evaluation of Rice), FLAR (Fondo Latinoamericano de Arroz de Riego), and CRIDNet (Caribbean Rice Industry Development Network). Fom 1975 to 1995, 250 rice varieties were released in the region. About 70% of these varieties were introduced to the countries through INGER.

The high return to international and national rice research in Latin America and the Caribbean is equivalent to an annual interest rate of 69%. This is extremely attractive, compared with the interest rate of about 10% earned on commercial investments.

Rice in West Africa

Revolutionary change in the preferences of West African consumers has created a wide and growing imbalance between regional rice supplies and demand. The major trends in consumption, production, and imports of rice are illustrated in Table 1. Since 1973, regional demand has grown at 6.0% annually, driven by a combination of population growth (2.9% growth rate) and substitution away from the region's traditional coarse grains. The consumption of traditional cereals, mainly sorghum and millet, has fallen by 12 kg per capita, and their share in cereals used as food from 62% in the early 1970s to 50% in the early 1990s.

In contrast, the share of rice in cereals consumed has grown from 15% to 25% over the same period, and from 12% to 18% in calorie terms from the 1960s to the end of the 1990s. Much of this dramatic shift occurred in the late 1970s and '80s. After decreasing to around 2.5%, per capita rice consumption has begun to increase again at more than 3% annually since the late 1990s. Accounting for population growth, total rice consumption has increased at nearly 6% per year during the last five years, meaning that it will have increased 2.5-fold by 2010.

The most important factor contributing to the shift in consumer preferences away from traditional staples and toward rice is rapid urbanization and associated changes in family occupational structure. As women enter the workforce, the opportunity cost of their time increases and convenience foods such as rice, which can be prepared more quickly, rise in importance. Similarly, as men work at greater distances from their homes in the urban setting, a greater proportion of meals is consumed from the market, where the ease of rice preparation has given it a distinct advantage.

These trends have meant that rice is no longer a luxury food, but has become a major source of calories for the urban poor. Urban consumption surveys in Burkina Faso, for example, have found that the poorest third of urban households obtains 33% of its cereal-based calories from rice. For that same group, rice purchases represent 45% of its cash expenditures on cereals, a share that is substantially higher than for other income classes. Similar findings have been obtained in several other West African nations, demonstrating that rice availability and rice prices have become a major determinant of the welfare of the poorest segments of West African consumers who are the least food-secure.

Production and imports

In comparison with the rapid growth in demand, regional rice production rose at 4.6% annually from 1973 to 2000. Although this rate was high compared to the performance of other major crops, it meant that regional rice production only barely exceeded population growth, and was meeting only two-thirds of the increments in demand. The source of the increases in rice production carries the important danger signal that such growth is not likely to be sustainable. Regional rice yields, which average only 40% of the world mean, have risen at only 1.5% per year since 1983. The major source of growth has been the expansion of cultivated area, which has grown at a remarkable 3.7% annually over the period.

The widening gap between regional supply and demand has been met by imports. The rapid increase in demand and much slower growth in production from 1973 to 1983 contributed to a dramatic jump in imports,which rose at more than 20% annually from 0.6 million t in the early 1970s to 2.2 million t a decade later. Since 1983, growth in imports has decelerated as domestic production has improved, leading to a much more modest 2.3% annual increase in imports, which averaged 2.8 million t in the early 1990s. Imports reached more than 3 million t in 1999, costing

Table 1. Production, consumption, and imports of rice in West Africa, by country, 1999, and average annual growth rates, 1995-99.

Country	Production[a]		Consumption[b]		Imports[b]	
	1999 (t)	Growth 1995-99 (%)	1999 (t)	Growth 1995-99 (%)	1999 (t)	Growth 1995-99 (%)
Benin	37,198	21.2	110,996	−2.9	129,200	−7.2
Burkina Faso	94,209	2.9	208,103	18.6	142,116	22.9
Cameroon	67,470	17.6	100,945	−7.0	60,003	−16.6
Chad	138,282	15.0	81,029	6.5	2,285	−21.0
Côte d'Ivoire	938,481	-2.7	1,165,806	7.4	429,000	1.5
Gambia, The	31,700	13.7	113,676	9.0	94,265	8.4
Ghana	209,750	1.0	166,107	−11.2	69,131	−9.8
Guinea	750,000	4.4	498,354	−3.7	125,000	−19.0
Guinea Bissau	80,300	−11.9	102,379	−4.3	62,230	1.4
Liberia	196,300	36.7	148,696	24.8	40,700	7.7
Mali	719,600	11.7	537,442	14.4	55,000	4.7
Mauritania	51,878	−0.4	161,243	12.4	122,300	15.0
Niger	60,921	4.5	60,608	−4.1	58,463	10.0
Nigeria	3,277,000	2.9	2,585,224	6.7	687,925	23.1
Senegal	364,000	23.8	653,651	5.8	625,160	9.1
Sierra Leone	247,235	−8.7	423,801	2.0	243,200	0.0
Togo	81,061	24.9	217,911	60.6	238,000	111.1
West Africa	**7,345,385**	**3.8**	**7,335,971**	**5.8**	**3,183,978**	**6.5**

[a]Unmilled paddy rice. [b]Milled rice.
Source: FAO online database.

more than $800 million in scarce foreign exchange. Imports of this magnitude represent a major brake on broader development efforts.

Rice economy liberalization and privatization

The acceleration in per capita rice consumption since the mid-1990s is due to the liberalization of rice imports combined with a downward trend of the price on the world market. Rice trade liberalization has also opened the door to the importation of low-quality rice that can be purchased by the poorest groups.

On the supply side, liberalization has led to a greater difference in crop yields. While production efficiency is improving in irrigated areas where local farmers' organizations have taken over previously public institutions, the disruption of input supply and the unfinished reform of the rural financial systems have resulted in a stagnation of yield in other areas. Most of the increase in irrigated rice production will rely on improved resource-use efficiency and rehabilitation of existing irrigation schemes. In the short term, the largest share of production will be in rainfed rice-based systems, which require labor-saving technologies. For the mid term, intensification in rainfed lowlands through the adoption of appropriate water management technologies offers a large and sustainable potential to increase rice production.

The privatization of the government-managed rice commodity chain has also allowed the emergence and fast growth of small-scale processing units in place of parastatal industrial mills. This has resulted in decreasing rice processing costs, thus improving the overall competitiveness of the local rice vis-à-vis imported rice. But this change has also led to a degradation of average local rice quality, which is becoming quite variable. Consumers now favor lower-quality imported rice, which does not have the same organoleptic attributes as local rice, but which is cleaner and more homogeneous in appearance, thus requiring less time for preparation. The future of the West African rice economy depends highly on the improvement of supply-to-demand linkages.

Rice as a women's crop

In many areas of West Africa, rice is produced primarily by women farmers, thus producing an important share of their income. Women's income tends to benefit children and other vulnerable groups more than does the income of men. Despite this fact, past efforts to develop and transfer new rice technologies have most often bypassed women farmers. Thus, although rice

Rice in Côte d'Ivoire.

research can be particularly effective in improving the welfare of rural groups at risk, it needs to be explicitly structured and focused to deal with complex gender issues.

Rice, environmental degradation, and sustainable intensification

In highly populated areas of West Africa, rising cropping intensity in fragile upland ecosystems has already begun to degrade the resource base, with environmental damage and loss in production potential. Rice has a key role to play in providing options for sustainable intensification. Rice is uniquely well adapted to flooded lowland ecosystems where the soils are least fragile and best able to support continuous cultivation. The development of profitable lowland rice technologies is therefore a central element in strategies to induce farmers to reduce pressure on rapidly degraded uplands by shifting cultivation to lowland ecosystems. Relative to other cereals, rice also responds more to improved management and to higher inputs of nutrients, water control, and labor, and is thus favored as production systems intensify.

Rice-growing environments

Rice in West Africa is grown in several ecosystems and in a wide range of production systems.

The humid and subhumid "continuum" environment

This continuum is composed of several contiguous ecosystems in which rice can be grown within the warm subhumid and warm humid tropics of Africa.

Rainfed upland ecosystem

Rice in this ecosystem is sown on approximately 2.2 million ha, representing 48% of the total rice area and 29% of regional production. Although there is large scope for area expansion regionally, in locations where access to good arable soils is limited, expansion of upland rice cultivation can occur only by shifting out of other crops, by reducing the fallow period, or by exploiting soils less well suited to rice cultivation. Farm-level yields are low, averaging about 1 t/ha, and vary as a function of local soil and rainfall conditions. Moderate technical potential for increasing yield exists through improved management. Progressive farmers applying moderate input levels can now achieve yields of 2.5 t/ha, and yields of 4 t/ha and greater are commonly achieved on research stations.

Rainfed lowland/hydromorphic ecosystem

The rainfed lowland ecosystem comprises floodplains and inland valleys. Overall, rice is grown on about 1.4 million ha of rainfed lowlands representing 30% of the total rice area and 36.5% of regional production. Average yield across the rainfed lowlands is 2.0 t/ha. In the floodplains, rice can be grown on residual and water-table moisture in the broad, flat areas adjacent to rivers prone to seasonal flooding. Enormous technical potential exists to expand rice production in inland-valley lowlands, as currently only about 10–15% of these areas are cultivated. It is estimated that Sub-Saharan Africa has approximately 130 million ha of inland-valley lowlands and their hydromorphic fringes within which rice can be cultivated. Of this total, West Africa may have about 20–30 million ha. The diversity of inland valleys is large. Although regional yields average around 1.4 t/ha, they vary according to local soil, landform, and hydrological conditions. Potential yields in unfavorable lowlands are around 2.5 t/ha, increasing to more than 5 t/ha in favorable lowlands.

Lowland irrigated ecosystem

This ecosystem covers about 142,000 ha in West Africa, representing 3% of the total area and 5.5% of regional production. The large technical potential for area expansion is constrained primarily by high investment costs. Current yields average around 3.0 t/ha. Progressive farmers employing moderate to high input management can obtain yields of 5 t/ha, and yields of 7 t/ha can be achieved on research stations.

The Sahelian irrigated environment

Irrigated rice in the Sahel forms the second most important rice-producing environment in West Africa, covering approximately 345,000 ha (7.5% of total rice area) and producing 20.5% of regional production. Irrigation potential is much greater, estimated at more than 3 million ha along the Senegal, Niger, Black Volta, Chari, and Logone rivers. This enormous technical potential has attracted large public investments: the first irrigation schemes were constructed in the Sahel in the 1920s. Until the recent introduction of privatization policies, the state has played a lead role in developing and managing irrigated rice schemes, with over 75% of the areas currently under parastatal control. The remainder is managed by a growing and increasingly dynamic private sector.

Because irrigated rice is an introduced production system in the Sahel, there are no traditional varieties or cultural practices upon which to build. Management skills in water control and rice cultivation vary widely. Mean paddy yields are currently about 4.5 t/ha, but vary widely from as low as 1 t/ha, rising to 4–6 t/ha if water is available throughout the season, to as high as 8 t/ha with modern input-responsive varieties and optimal management. Although rice-rice double cropping is currently practiced on only about 20% of the total area, by using varieties with appropriate duration and adaptation, an annual yield potential of 15 t/ha can be achieved.

The mangrove swamp rice environment

Rice is also grown on approximately 147,000 ha of mangrove swamps, representing 3% of the total area and producing roughly 4% of the region's output. Located on tidal estuaries close to the ocean, most mangrove swamps experience a salt-free growing period during the rainy season when freshwater floods wash the land and displace tidal flows. As a result, the rice-growing period is directly related to distance from the ocean, varying between less than four months in the nearest estuaries to more than six months in those more distant. Soils are generally more fertile than in the other environments since they benefit from regular deposits of silt left during annual flooding. However, the soils are also characterized by high salinity and sulfate acidity. Lower rainfall during the last two decades has reduced seasonal flushing substantially, further accentuating both problems. Although West Africa has approximately 1 million ha of

Women farmers inspect their crop in The Gambia.

potentially cultivable mangrove swamps, the high labor costs associated with clearing and potentially negative environmental effects pose major constraints to further area expansion.

The deepwater/floating rice environments

Deepwater and floating rice represent a large but increasingly marginalized production system for which area and production figures are generally poor and vary widely. Estimates of sown area in the mid-1970s varied from 187,000 to 630,000 ha. It is generally believed that, with the control of river flooding because of the construction of dams and with lower rainfall, areas under these ecosystems have probably declined during the last 20 years. The estimated area under deepwater rice in West Africa at the end of the 1990s was 373,000 ha, or 8% of the rice-growing area, producing 5% of production at an average yield of 1.0 t/ha.

The major zones of production are located along the Niger valley around Mopti in Mali and Birnin Kebbi in Nigeria, and in northern Guinea. Deepwater rice ecosystems can be defined as those where flooding achieves a depth of from 60 to 100 cm, and floating rice systems as those where flooding exceeds 100 cm. These production systems are among the least developed in West Africa, with very little use of fertilizer or mechanization and dominated by the use of *Oryza glaberrima* and tall traditional *O. sativa* cultivars. As a result, yields are among the lowest in the region, and highly variable across sites and years. Research to improve these systems has made little progress.

The top 10 rice-producing countries

1. China
2. India
3. Indonesia
4. Bangladesh
5. Vietnam
6. Thailand
7. Myanmar
8. Japan
9. Philippines
10. Brazil

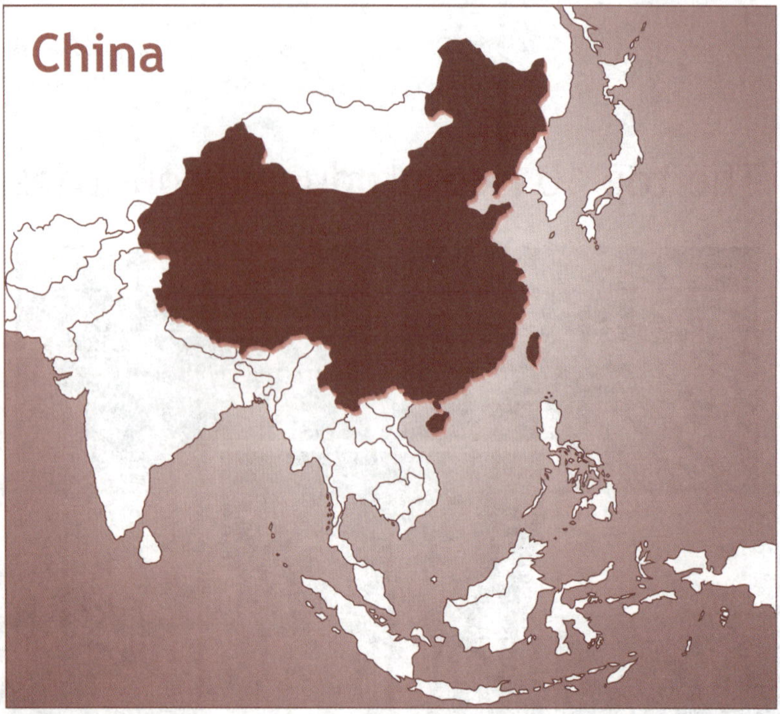

China

General information

- GNI per capita PPP$, 2000: 3,920
- Internal renewable water resources: 2,800 km^3
- Main food consumed: rice, wheat, meat, maize, roots, and tubers
- Rice consumption, 1999: 90.7 kg milled rice per person per year

Production season

	Planting	Harvesting
Early	Feb May	Jun-Oct
Intermediate	Feb-May	Jun-Oct
Late	Jun-Jul	Oct-Nov
Main season, North	Apr-Jun	Sep

China is situated between 18° and 54° N latitude and 73° and 135° E longitude. Marked by topographical variety and complexity, China's landmass is made up of mountains (33%), plateaus (26%), basins (19%), plains (12%), and hills (10%).

China lies in four AEZs, all of which are subtropical and include some temperate areas. They are AEZ 5, warm arid and semiarid subtropics with summer rainfall; AEZ 6, warm subhumid subtropics with summer rainfall; AEZ 7, warm/cool humid subtropics with summer rainfall; and AEZ 8, cool subtropics with summer rainfall. Rice is produced primarily in AEZs 6 and 7.

China's climatic features include a pronounced monsoon climate with a hot summer and a cool winter, marked seasonal variations in precipitation, and a distinctive continental climate with large annual temperature fluctuations. The climate types are so varied and complex that high rainfall, cold waves, and typhoons are all important climatic phenomena. China can be divided from the coastal areas to the northwest interior into four regions according to moisture regime: (1) the humid region south of the Qinling Mountains and Huaihe River, comprising 32% of total land area, (2) the subhumid region including most of northeastern and central China, 15%, (3) the semiarid region, 22%, and (4) the arid region, 31%.

China is the most populous country in the world. The 1999 population on the mainland was nearly 1.3 billion, with about 66% living in rural areas. Three-fourths of the nation's population is concentrated in the northern, northeastern,

eastern, and south-central areas, which make up only 44% of the nation's land area. The remaining one-fourth of the population is dispersed in the southwestern and northwestern parts. Because of an active family planning program, including some restrictions on family size, annual population growth slowed from 2.6% per annum in the late 1960s to just 0.9% per annum in the late 1990s.

Recent developments in the rice sector

Rice is the staple food of China and accounts for about 35% of total grain production, which was 345 million t in 2000 (converting paddy to its milled rice equivalent). However, wheat is more important in some areas, especially in the North. Although rice is still a large part of people's diets, its importance has declined considerably in the past 15 years. Since 1985, the share of total calories obtained from rice has fallen from 38% to 30%. During this same period, the share of total calories coming from wheat has declined only slightly, from about 22% to 19%. In terms of protein, rice and wheat each accounted for slightly more than one-fourth of total protein intake in the late 1980s, but during the 1990s this share declined sharply, and it now stands at slightly less than one-fifth for each cereal.

China is the world's largest rice producer, accounting for 32–35% of total world production (India has a larger rice area harvested, but lower per hectare yields). Except for Japan and the Republic of Korea, rice yields in China are the highest in Asia, due in part to favorable growing conditions. Rice area harvested has declined from its peak of 37 million hectares in the mid-1970s to just over 30 million ha today. The decline in area has been due to both economic reforms that reduced government requirements to grow rice and economic development that increased the opportunity cost of land. In recent years, much of the fall in rice area has occurred in coastal provinces such as Guangdong and Zhejiang. Hunan is the largest rice producing province, and most rice production is in the Yangtze River Valley (or further south) where ample supplies of water are available. However, rice production in northern China has increased substantially in recent years, with its share of national production nearly doubling from 7.6% in 1989 to 13.3% in 1999. Much of this increase has come in Heilongjiang and the other two northeastern provinces of Jilin and Liaoning, but production has also expanded noticeably in Henan and Shandong.

China regularly imports and exports rice each year. Imports exceeded exports in 1995 and 1996, but China has been a net exporter since then. In 1998 and 1999, it was the world's fourth largest rice exporter (in gross terms, not net), and its exports helped to stabilize world market rice prices in the face of a strong El Niño that severely disrupted production in Indonesia and the Philippines.

Rice environments

More than 90% of the rice area in China is irrigated, with only relatively small areas being cultivated under rainfed conditions. However, rice-growing conditions vary because of topography and weather. In southeastern China, high temperature and adequate rainfall make an ideal environment for rice during a long growth period, and many areas grow two crops of rice per year. In the Yangtze River Valley, much of the land is planted to a rice-wheat rotation. In northeastern China, low temperature, a short growth period, little rainfall, and lack of water limit the rice area. The varieties grown in this area are typically japonica and are considered to be of higher quality than rice grown in other areas. There are some scattered rice areas in arid and semiarid regions of northwestern China.

Production constraints

Total area harvested to all crops continues to increase in China. Area harvested to rice has declined during the past 25 years, however, because of crop diversification. (Rice formerly accounted for 26% of all crop area harvested in the mid-1970s, but more recently the share is just 20%.) At the same time, population continues to grow by about 13 million people per year. Until per capita rice consumption begins to decline because of rising wealth accompanied by dietary diversification (as has happened in Japan, the Republic of Korea, Malaysia, and Thailand), rice yields will need to increase to meet consumption demand without resort to imports.

Water shortages in the north are another important production constraint. Although northern China has only 24% of the nation's water resources, it contains more than 65% of China's cultivated land. While water shortages

New rice plant type in Zunnan Province.

constrain production in some areas, flooding is also a problem. Land salinization and soil erosion also pose challenges for continued development.

Although the population is still growing, increased labor demand in urban areas is drawing many people out of agricultural production and hurting yields. As a result, labor-saving technologies such as direct seeding (or seedling throwing) are becoming more common. Future trade liberalization under the World Trade Organization may also affect grain production, especially for wheat and maize, because domestic prices for these grains are substantially above world prices. For rice, however, domestic prices are approximately equal to world prices, and no large influx of imports is anticipated.

Production opportunities

Chinese scientists recently became the first in the world to prepare a draft sequence of the genome for the indica race of rice. Indica rice is by far the most widely planted in Asia, and this achievement has the potential to create many benefits for Asian producers and consumers of rice.

In seeking to alleviate the main constraints of land and water, Chinese scientists have also made substantial progress in improving yields and water productivity. China has developed the most successful varieties of hybrid rice in the world, and more than one-third of total rice area is currently planted to hybrids. More recently, scientists have developed irrigation techniques for rice that reduce water consumption by allowing intermittent drying of the paddy field, without sacrificing grain yields. The successful adaptation of aerobic rice (rice that is grown as an upland crop but still exhibits a substantial response to nitrogen fertilizer) to new areas would also allow rice to be grown in water-short environments. In the future, construction of canals from southern to northern China may also help to alleviate water shortages in the Yellow River Basin.

In other fields, there is also potential for improved fertilization strategies that increase nitrogen-use efficiency by improving splitting of nitrogen and reducing levels of nitrogen application. Opportunities also exist for reducing pesticide applications to improve farmer health and the quality of drinking-water supplies.

Since 1980, China and IRRI have cooperated on several research projects of mutual concern such as the exchange of rice germplasm to strengthen breeding programs; hybrid rice research to exploit heterosis in rice; shuttle breeding to speed the development of rice varieties with high yield potential, good quality, multiple resistance to insects and diseases, and wide adaptability; and natural resource management studies to improve fertilizer- and water-use efficiencies.

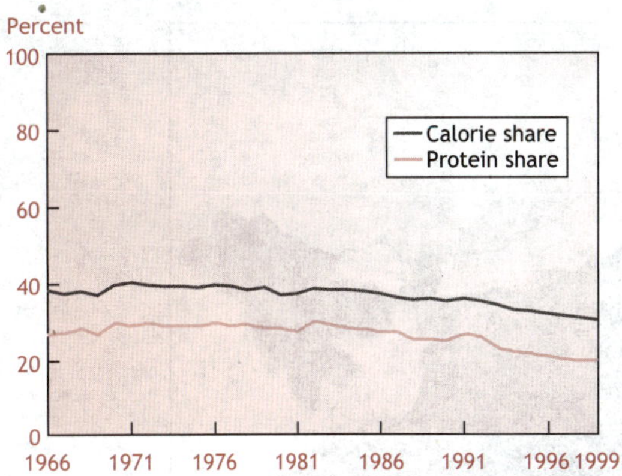

Share of calories and protein from rice, 1966-99

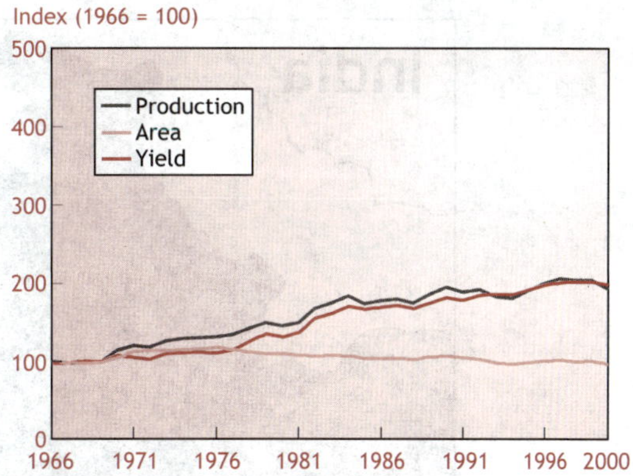

Indices of rice production, area, and yield, 1966-2000

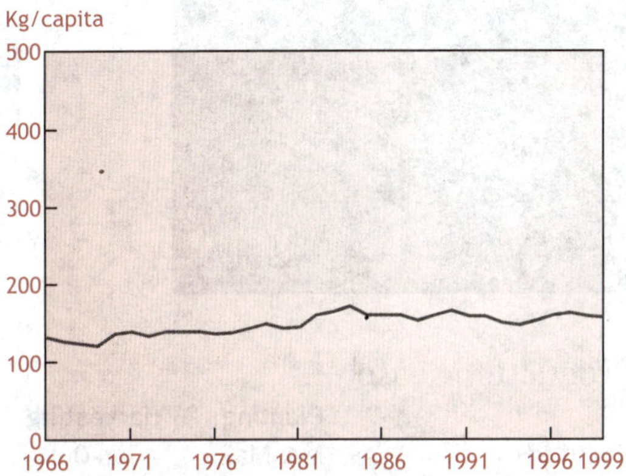

Per capita production (paddy terms), 1966-99

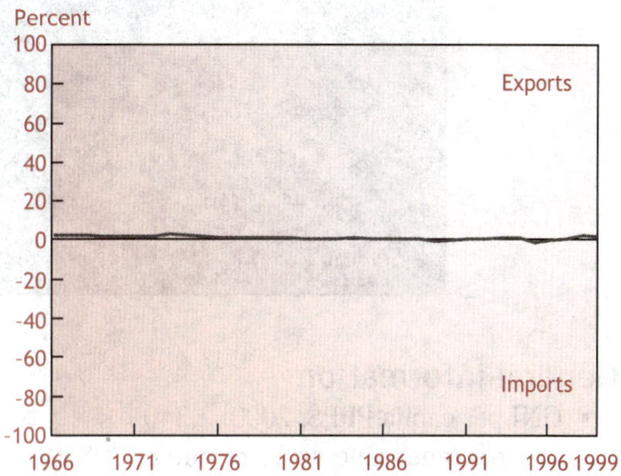

Net trade status, 1966-99

Basic statistics, China

	1985	1990	1995	1998	1999	2000
Rice						
Area harvested (ha)	32,633,684	33,518,971	31,107,479	31,571,500	31,637,100	30,503,100
Yield (t/ha)	5.2	5.7	6.0	6.4	6.3	6.2
Production (t)	171,318,871	191,614,680	187,297,968	200,571,557	200,403,308	190,168,300
Rice imports (t)	na	na	na	246,892	172,106	na
Paddy imports (t)	0	465	173	229	258	na
Rice exports (t)	1,045,848	405,381	235,934	3,791,615	2,819,010	na
Paddy exports (t)	0	203	1,308	1,889	4,953	na
Others						
Population, total ($\times 10^3$)	1,075,936	1,161,382	1,227,170	1,262,817	1,274,107	na
Population, agriculture ($\times 10^3$)	783,923	834,688	850,299	854,536	855,167	na
Agricultural area ($\times 10^3$ ha)	495,897	531,398	534,701	535,566	na	na
Irrigated agricultural area ($\times 10^3$ ha)	44,584	47,967	49,859	52,582	na	na
Total fertilizer consumption (t)	16,851,600	27,027,408	35,181,200	35,077,600	na	na
Tractors used in agric. (no.)	861,364	824,113	685,202	704,070	na	na

Source: FAOSTAT online database.

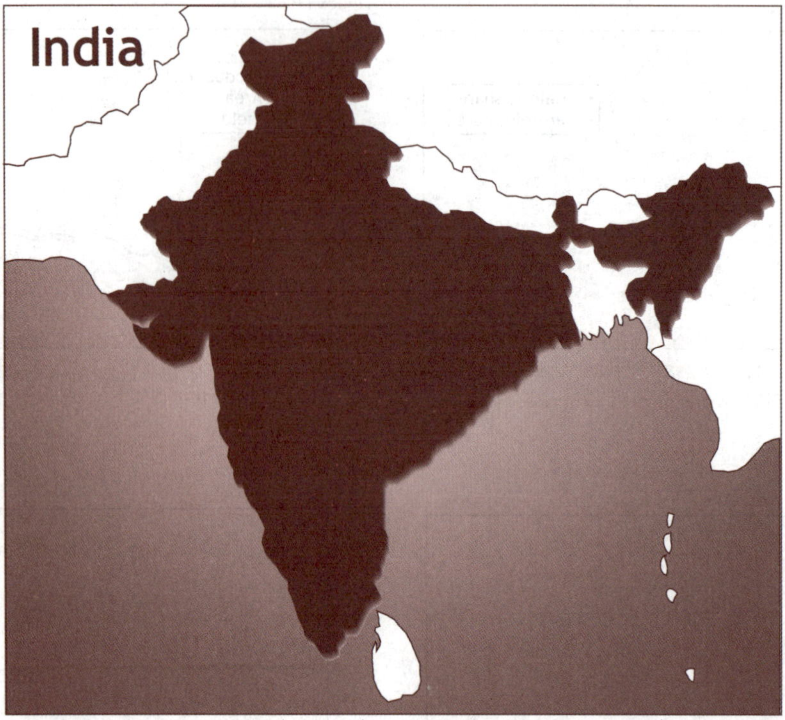

General information

- GNI per capita PPP$, 2000: 2,340
- Internal renewable water resources: 1,850 km^3
- Incoming water flow: 235 km^3
- Main food consumed: rice, wheat, sugar and honey, millet and sorghum, oil and fat
- Rice consumption, 1999: 74.2 kg milled rice per person per year

Production season

	Planting	Harvesting
Kharif early	Mar-May	Jun-Oct
Kharif medium	Jun-Oct	Nov-Feb
Rabi	Nov-Feb	Mar-Jun

India extends between 8°4′ and 37°6′ N latitude and 68°7′ and 97°25′ E longitude. It measures about 3,214 km from north to south between the extreme latitudes and about 2,933 km from east to west between the extreme longitudes. It has a land frontier of about 15,200 km. The mainland comprises four regions: the great mountain zone, plains of the Ganges and the Indus rivers, the desert region, and the southern peninsula. On its northern frontiers, India is bounded by the Great Himalayas, which are three almost parallel ranges interspersed with large plateaus and valleys, such as the Kashmir and Kullu valleys, that are very fertile.

India lies in five agroecological zones: AEZ 1, characterized by warm and semiarid tropics; AEZ 2, warm subhumid tropics; AEZ 5, warm arid and semiarid subtropics with summer rainfall; AEZ 6, warm subhumid tropics with summer rainfall; and AEZ 8, cool subtropics with summer rainfall. Most rice is grown in AEZs 1, 2, and 6.

The climate of India can be described as tropical monsoon type. There are four seasons: winter (December-February), summer (March-May), rainy southwestern monsoon (June-September), and postmonsoon, also known as northeastern monsoon in the southern peninsula (October-November). The beginning of winter and summer periods differs in different regions.

Four broad climatic regions are identified based on rainfall. The whole of Assam and the west coast of India lying at the foot of the Western Ghats and extending from the north of Mumbai (earlier Bombay) to Thiruvanthapuram (earlier Trivandrum) are areas of high rainfall.

The Rajasthan desert extending westward to Gilgit is a region of low precipitation. In between are two areas of moderately high and low rainfall. The area of high rainfall is a broad belt in the part of the peninsula merging northward with the Indian plains and southward with the coastal plains. The low rainfall area is a belt extending from the Punjab plains across the Vindhya mountains into the western part of the Deccan region, widening considerably in the Mysore plateau.

India is the world's second most populous nation, with a population in 2000 of 1,014 million growing at 1.7% per year. The rural population in 2000 was 726 million.

Recent developments in the rice sector

Agriculture is the backbone of India's economy, providing direct employment to about 67% of the working people in the country. It forms the basis of many premier industries of India, including the textile, jute, and sugar industries. Agriculture contributes about 29% to GDP; one-fourth of India's exports are agricultural products.

Rice is the staple food of 65% of the total population in India. It constitutes about 52% of the total food grain production and 55% of total cereal production. Food grains consist of cereals such as rice, wheat, sorghum, pearl millet, and maize as well as pulses. Food crops grow on nearly 70% of the gross sown area. Important commercial crops are cotton, jute, sugarcane, and tobacco.

Rough rice production reached 134 million t in 2000 from 112 million t in 1990, growing at 1.9% annually. The growth rate has slowed down significantly from 3.4% per year during the 1980s, mainly from the sluggish performance in the progressive states such as Punjab, Andhra Pradesh, Tamil Nadu, and Haryana, as many districts in these states are approaching the economically optimum yield with the available technologies.

The rough rice yield has increased from 2.61 t/ha in 1990 to 3.01 t/ha in 2000, an annual growth of 1.4%. In Punjab and Tamil Nadu, where almost the entire rice land is irrigated, yield has reached 5.26 t/ha and 5.38 t/ha (1998), respectively. Yield fluctuates widely in Bihar and Orissa, states that suffer from drought and floods often in the same year, making rice cultivation a highly risky economic activity.

Adoption of modern technology

Since 1965, India has released about 640 improved rice varieties: 54% of them for irrigated areas, 27% for the rainfed lowland, and 19% for upland areas. The coverage of modern high-yielding rice varieties reached 78% of the rice harvested area by 1999. The rate of adoption varies from 67% in Assam in the northeast to more than 90% in Andhra Pradesh, Tamil Nadu, and Kerala in the south. Since a large area in the irrigated states of Punjab and Haryana in the northwest is allocated for the production of high-quality basmati rice, traditional varieties account for a significant portion of rice land in these states.

Swarna (MTU 7029), a derivative of Mahsuri, is the most popular improved rice variety that is grown in a large number of states. In 1999, it was grown on about 12% of India's rice land. The other popular varieties are Vijeta (MTU 1001), Samba Mahsuri (BPT 5204), Mahsuri, Lalat, IR64, and IR36.

Data on fertilizer sales show a large regional variation in the use of nutrients. NPK use varies from less than 50 kg/ha in Assam, Orissa, and Madhya Pradesh (mostly rainfed areas) to more than 140 kg/ha in Punjab, Andhra Pradesh, and Haryana (irrigated land). A report on the cost of cultivation of principal crops in 2000 notes a heavy use of pesticides in rice cultivation in Punjab and Andhra Pradesh but very little in other states. Mechanization of agricultural operations is prevalent in Punjab and Haryana and is gaining ground in Andhra Pradesh, western Uttar Pradesh, and West Bengal, but is almost absent in other states.

External trade

In the 1960s, India imported 0.7 to 1.0 million t of rice annually to meet the deficit in domestic demand. India became self-sufficient in rice in 1977 with imports of small amounts in years of crop failures. The latest large imports were 0.5 million t in 1984, 0.7 million t in 1988, and 0.47 million t in 1989. Since then, imports of rice have been limited to below 100,000 t.

India exports a small amount of high-quality basmati (aromatic) rice on a regular basis. Exports of rice jumped from 0.9 million t in 1994 to 4.9 million t in 1995 in response to the large increase in demand in the world market. However, India could not sustain exports at that level because of the low quality of indica rice, of which a substantial surplus is produced in Punjab and Andhra Pradesh. Although there is a large unmet demand for staple food grains in the poverty-stricken states of eastern India because of a lack of purchasing capacity of low-income households, the disposal of the surplus rice procured by the government has become a major concern for India.

Rice environments

Rice environments in India are extremely diverse. India has the largest area under rice in the world. Of the 45 million ha of harvested rice area, about 28% are rainfed lowland, 46% irrigated, 12% rainfed upland, and 14% flood-prone. In some traditional wheat-growing states, such as Punjab, Haryana, and Uttar Pradesh, rice production has increased substantially since the late 1960s with the introduction of modern high-yielding rice varieties that induced farmers to undertake commercial cultivation of rice. In Punjab, for example, rice production increased from 0.9 to 13.1 million t, and in Uttar Pradesh from 4.4 to 19.4 million t from 1968 to 1999. This rapid expansion was possible because of the favorable irrigation infrastructure.

Production constraints

Since the major portion (55%) of the area under rice in India is rainfed, production is strongly tied to the distribution of rainfall. In some states, erratic rainfall leads to drought during the vegetative period, but later the crop may be damaged by submergence caused by high rainfall. In the eastern states, damage from flash floods is quite high.

Other constraints relate to the land and soil. Soil acidity is a problem in southern and eastern India, whereas, in northern India, soil salinity and alkalinity are the problem. Low soil fertility and P and Zn deficiency are widespread.

Nearly all of the rainfed area suffers from a lack of infrastructure. Moreover, most farmers cannot afford the inputs necessary for full exploitation of the yield potential of modern varieties. Crop residues are used as livestock feed and for thatching of roofs of houses; animal dung is used for fuel, and is not available to compensate for the loss of nutrients in the cultivation of modern varieties.

Stem borers, brown planthopper, gundhi bug, leaffolders, green leafhopper, and gall midge are major insect pests causing large yield losses. Bacterial blight, blast, sheath blight, and brown spot are important diseases. With increases in wage rates, weeds are becoming a major factor constraining productivity and profitability in rice farming.

Production opportunities

Much of India's agricultural growth, particularly in major cereals, can be traced to an agricultural strategy adopted in the late 1960s. The strategy included

- provision of a package of inputs consisting of short-duration, high-yielding modern varieties, fertilizers, and improved agricultural practices in areas of assured water supply;
- supply of credit from public institutions to finance working capital needs of farmers; and
- declaration of a minimum price before planting at which surplus grains are to be procured by the government.

To extend the production package to less favored areas in order to achieve more balanced regional growth, agroclimatic zonal planning is applied. India has been divided into 21 agro-climatic regions based on homogeneity in rainfall, temperature, soil, topography, and water resources. Rice research priorities have shifted from the irrigated ecosystem in the northwest and southern region to the predominantly rainfed ecosystem in eastern and northeastern India. Strategic research to increase the productivity of rice is being done in collaboration with the International Fund for Agricultural Development and the International Rice Research Institute in six states in eastern India that account for two-thirds of the total rice area. The Rice-Wheat Consortium for the Indo-Gangetic Plains is studying the problem of sustainability of high yields in rice and wheat by examining system-level issues.

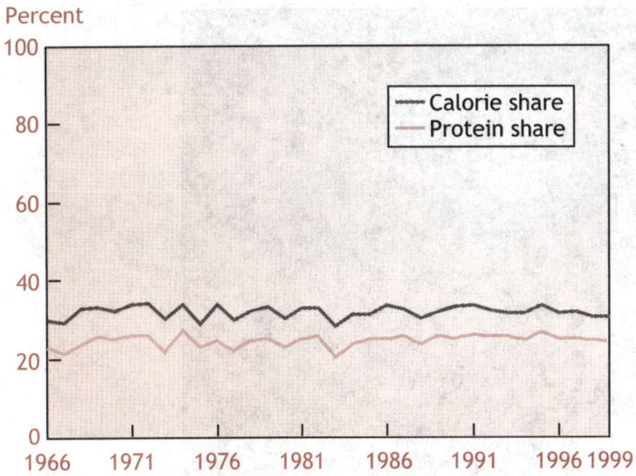

Share of calories and protein from rice, 1966-99

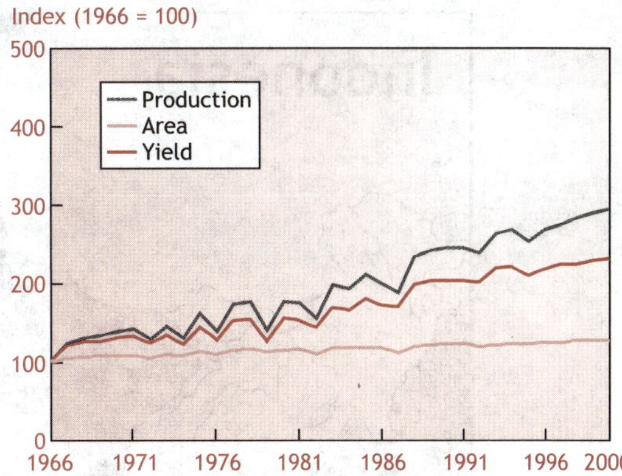

Indices of rice production, area, and yield, 1966-2000

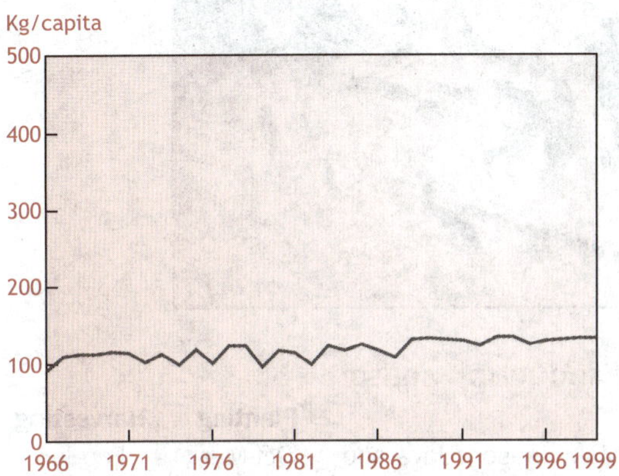

Per capita production (paddy terms), 1966-99

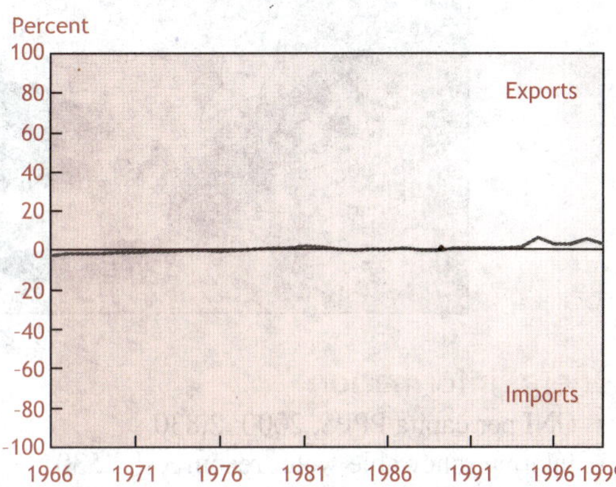

Net trade status, 1966-99

Basic statistics, India

	1985	1990	1995	1998	1999	2000
Rice						
Area harvested (ha)	41,137,200	42,686,608	42,800,000	44,598,000	44,607,000	44,600,000
Yield (t/ha)	2.3	2.6	2.7	2.9	3.0	3.0
Production (t)	95,817,696	111,517,408	115,440,000	128,928,000	132,300,000	134,150,000
Rice imports (t)	61,100	66,038	52	6,635	50,094	na
Paddy imports (t)	0	0	80	3	3	na
Rice exports (t)	315,070	505,027	4,913,156	4,962,941	2,571,000	na
Paddy exports (t)	1,020	7	2,444	1,878	0	na
Others						
Population, total (×10³)	767,842	850,785	933,665	982,223	998,056	1,014,000
Population, agriculture (×10³)	10,475,832	506,548	534,245	548,794	553,227	na
Agricultural area (×10³ ha)	180,949	181,040	180,780	180,600	na	na
Irrigated agricultural area (×10³ ha)	41,779	45,144	53,000	59,000	na	na
Total fertilizer consumption (t)	8,504,300	12,584,000	13,876,100	16,797,500	na	na
Tractors used in agric. (no.)	607,773	988,070	1,354,864	1,550,000	na	na

Source: FAOSTAT online database.

Indonesia

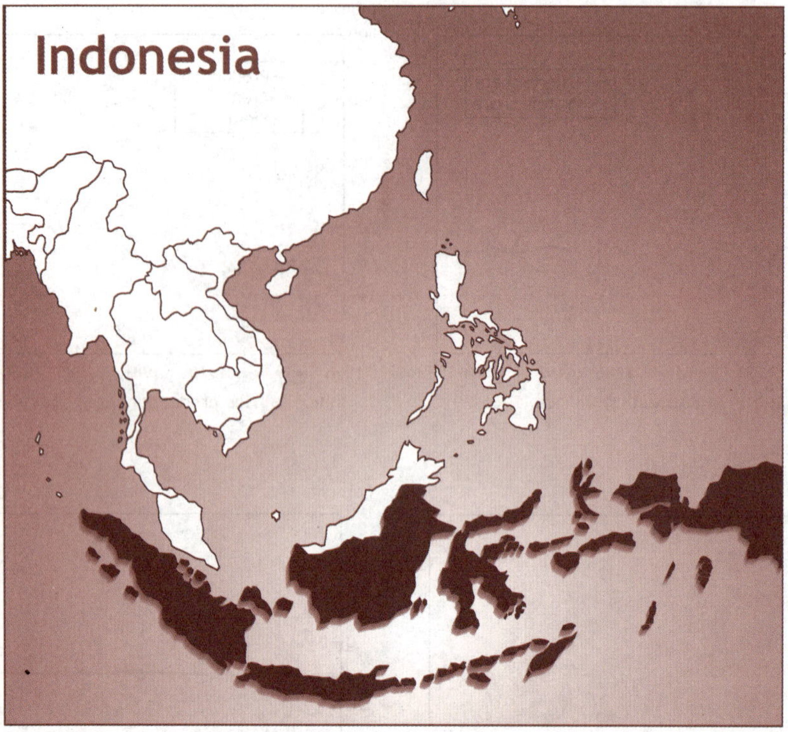

General information

- GNI per capita PPP$, 2000: 2,830
- Internal renewable water resources: 2,530 km^3
- Main food consumed: rice, oil and fat, nuts, roots and tubers, maize
- Rice consumption, 1999: 154 kg milled rice per person per year

Production season

	Planting	Harvesting
Main season, Java and South Sumatra	Oct-Mar	Feb-Jun
Main season, Sulawesi	May-Jun	Aug-Oct
Main season, Sumatra	Jul-Sep	Nov-Dec

The Indonesian archipelago extends from 6° N to 11° S latitude and from 95° to 141° E longitude or about 2,000 km from north to south and 5,000 km from east to west. There are more than 13,000 islands, including five of the world's largest: Sumatra, Kalimantan (Indonesian part of Borneo), Irian Jaya (western New Guinea), Sulawesi (Celebes), and Java.

Indonesia lies within AEZ 3, characterized as the warm humid tropics. Most of Indonesia has a moist tropical climate, with abundant rain and high temperatures. Annual rainfall ranges from 1,000 to more than 5,000 mm per year, with more than 90% of the country receiving average rainfall of more than 1,500 mm. December through March are the months with the highest rainfall.

Indonesia is the world's fourth most populous country, with about 215 million people as of 2001. Because of rapid economic growth and an active family planning program, population growth has declined from 2.4% per annum during the late 1960s and early 1970s to 1.5% recently, and the United Nations forecasts this rate to decline to less than 1% by 2015. The mean population density is $111/km^2$, but, on Java, where nearly 60% of the people reside, the population density is approximately $980/km^2$. The share of the population in urban areas has grown to 40%, with the remainder in rural areas. About half of the economically active population is in agriculture, although many of these people derive a substantial share of their income from non-agricultural activities. The incidence of malnutrition among children age 2–5 (as measured by weight for age) is estimated at 34%.

Recent developments in the rice sector

The agricultural sector (including forestry and fisheries) contributed approximately one-sixth of GDP in 2000, with rice production responsible for about one-fourth of this contribution. Rice constitutes nearly 40% of total crop area harvested and is the country's staple food, accounting for slightly more than half of caloric intake and nearly half of protein intake. Rice accounts for 20% of total expenditures for the poorest quarter of the urban population. Even in rural areas, many poor people are net consumers of rice, since 45% of rural households on Java do not own any land and another 20% own less than 0.25 ha. In the 1970s and early 1980s, Indonesia was the world's largest rice importer, often importing one-fourth of total supplies on the world market. From 1967 to 1986, however, total rice production tripled, probably the most extraordinary growth rate of staple food production in human history, and self-sufficiency was achieved by the mid-1980s. This rapid production growth was achieved through the adoption of modern high-yielding varieties and fertilizers by farmers. Adoption of these varieties was greatly facilitated by government programs such as fertilizer subsidies and rice price stabilization around the long-term trend of world prices. Large investments in rural infrastructure such as irrigation, roads, and schools also played a critical role.

While per hectare yields more than doubled from the mid-1960s to the mid-1980s, the national average yield today is no greater than it was ten years ago. This stagnation has been the primary factor behind Indonesia's return as the largest importer on the world market in the second half of the 1990s. Despite continued urbanization, rice harvested area increased steadily during the past decade, from about 10.5 million hectares in 1990 to nearly 12 million ha in 1999.

Rice policy has changed substantially in the aftermath of the Southeast Asian financial crisis that led to the resignation of President Suharto in 1998. A substantial increase in rice prices in late 1998 led to significant increases in poverty, and this precipitated shifts in policy. While the national logistics agency Bulog formerly had monopoly control over all rice import and export decisions, adoption of an International Monetary Fund stabilization program led to the entry of private-sector traders, who are currently allowed to make import decisions subject only to a tariff. Bulog's successful defense of farm-gate floor prices for nearly 30 years also ended because of political and institutional constraints. Recently, all subsidies on fertilizer were eliminated.

Rice environments

As of the mid-1990s, 54% of the rice area in Indonesia was irrigated, with 35% rainfed lowland, and 11% upland. Nearly all of the irrigated area can be planted to two or more crops of rice per year, and much of the rainfed area has favorable growing conditions. Most irrigated lowland rice areas are located in floodplains. However, they can also be found on mountainsides wherever there is water. Rainfed lands are on both floodplains and undulating landscapes. Uplands are mostly on undulating landscapes. Current estimates are that about 80% of rice land is planted to modern varieties. On Java, most rice is transplanted. Labor use in rice production is relatively high, typically exceeding 200 person-days per hectare per crop.

Production constraints

In irrigated and favorable rainfed lowlands, especially on Java, the relatively heavy application of nitrogenous fertilizers makes nutrient imbalance a serious problem. Indonesia is also particularly vulnerable to the vagaries of the El Niño Southern Oscillation (ENSO) phenomenon. In years when surface water temperatures rise substantially in the western Pacific Ocean, signaling an El Niño event, rice production suffers a serious shortfall, with most of the effects coming from a reduction in rice area planted (as opposed to lower yields). The reduction in area occurs even in systems that are usually irrigated, as lower rainfall leads to a reduced availability of irrigation water.

In upland rice areas, erosion is a serious problem because on steep slopes the fields are neither bunded nor terraced. This can cause serious sedimentation problems in lowland irrigation systems. Alley cropping as well as terracing are being introduced in some areas, but these cultural practices have not yet been widely adopted by farmers. Upland soils are also more weathered and leached, leading to problems of Al toxicity and P nutrient deficiencies that combine

Weighing the rice harvest in West Java.

to reduce yields. Soil acidity is serious in tidal swamps because of acid-sulfate soils. That is accompanied by Fe toxicity as well as some deficiencies of P and micronutrients.

Production opportunities

Indonesia has developed a cadre of researchers capable of undertaking rice research and collaborating with colleagues in other countries. The Research Institute for Rice (RIR), located in Sukamandi, West Java, is the main institute conducting biophysical rice research. Some trials and assessments on rice are also conducted by the regional institute of the Assessment Institute for Agricultural Technology (AIAT) at the provincial level. The Center for Agro-Socio-Economic Research (CASER), located in Bogor, has a long tradition of conducting socioeconomic research on rice and the agricultural sector more broadly.

Future research efforts will need to focus on several areas:

- achieving higher and sustainable rice yields through integrated crop and resource management
- breeding of varieties with higher yield (Indonesia is a member of the hybrid rice network) and actively testing and developing the new plant type
- further breeding of varieties resistant to pests and diseases using marker-aided selection and other techniques
- development of varieties tolerant of drought and soil toxicities
- development of varieties and other strategies to stabilize yields
- development of improved nutrient management strategies
- crop diversification

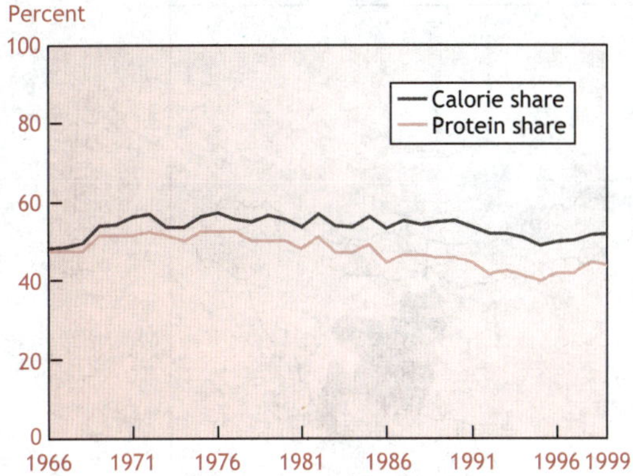

Share of calories and protein from rice, 1966-99

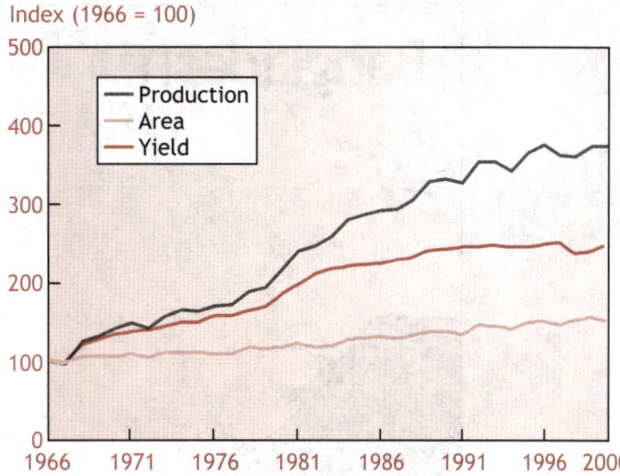

Indices of rice production, area, and yield, 1966-2000

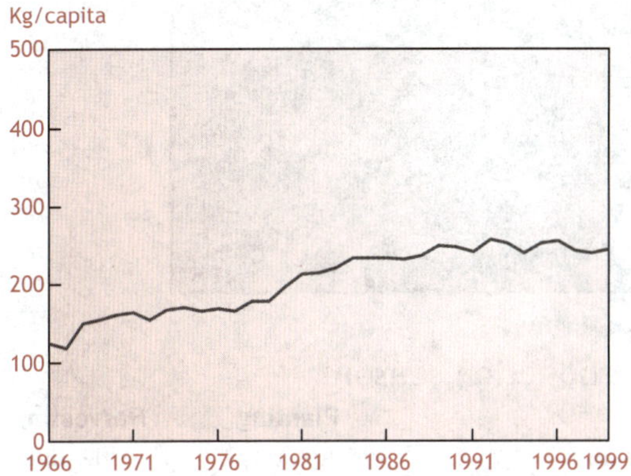

Per capita production (paddy terms), 1966-99

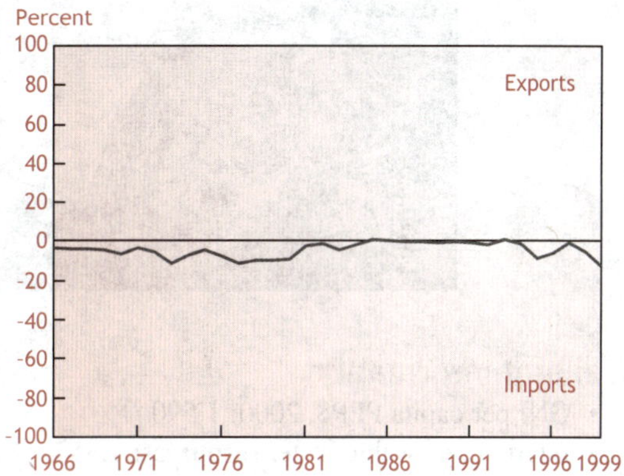

Net trade status, 1966-99

Basic statistics, Indonesia

	1985	1990	1995	1998	1999	2000
Rice						
Area harvested (ha)	9,902,293	10,502,357	11,438,760	11,716,499	11,963,204	11,523,068
Yield (t/ha)	3.9	4.3	4.3	4.2	4.3	4.4
Production (t)	39,032,944	45,178,752	49,744,140	49,199,844	50,866,388	51,000,000
Rice imports (t)	33,853	49,577	3,157,700	1,894,958	4,748,060	na
Paddy imports (t)	0	0	1,051	461	9,539	na
Rice exports (t)	258,712	1,911	5	1,981	2,701	na
Paddy exports (t)	381,096	100	0	57	1	na
Others						
Population, total ($\times10^3$)	167,332	182,812	197,464	206,338	209,255	na
Population, agriculture ($\times10^3$)	87,419	93,085	93,591	93,679	93,651	na
Agricultural area ($\times10^3$ ha)	39,350	45,083	41,980	42,164	na	na
Irrigated agricultural area ($\times10^3$ ha)	4,300	4,410	4,687	4,815	na	na
Total fertilizer consumption (t)	1,971,800	2,387,000	2,529,200	2,772,900	na	na
Tractors used in agric. (no.)	12,033	27,955	59,991	70,000	na	na

Source: FAOSTAT online database.

General information

- GNI per capita PPP$, 2000: 1,590
- Internal renewable water resources: 1,357 km^3
- Incoming water flow: 1,000 km^3
- Main food consumed: rice, wheat, oil and fat, sugar and honey, pulses
- Rice consumption, 1999: 168.2 kg milled rice per person per year

Production season

	Planting	Harvesting
Aus	Apr-May	Jul-Aug
Aman	Apr-May	Nov-Dec
Boro	Dec-Feb	Apr-May

Bangladesh lies in the northeastern part of South Asia between 20° and 26° N latitude and between 88° and 92° E longitude. The country is bounded by India on the west, north, and northeast; by Myanmar on the southeast; and by the Bay of Bengal on the south.

Except for the hilly regions in the southeast and some in the northeast, and patches of highlands in the central and northwest regions, Bangladesh for the most part consists of low, flat, fertile land. About 230 rivers and their tributaries, with a total length of about 24,140 km, flow across the country down to the Bay of Bengal. The alluvial soil is continuously enriched by heavy silt deposited by the rivers through frequent flooding during the rainy season.

Bangladesh is in AEZ 3, characterized as warm humid tropics, with a length of growing period >230 d for most parts of the country. The country enjoys a subtropical monsoon climate. Summer, monsoon, and winter are the most prominent of six distinct seasons. Winter, which is pleasant, extends from November to February, with minimum temperature ranging from 7 to 13 °C; in summer, maximum temperature ranges from 24 to 41 °C.

The monsoon starts in June and lasts until October. This period accounts for 80% of the total annual rainfall, which varies from 1,200 to 2,500 mm. Maximum rainfall is recorded in the coastal areas and in the northern Sylhet and Mymensing districts, adjacent to Assam and Meghalaya, India. Minimum rainfall is observed in the districts of Jessore, Kushtia, and Rajshahi in the western parts of the country.

Recent developments in the rice sector

Nearly 80% of the land area of the country has been brought under crop cultivation. Only 15% of the land area is under forests. In 1999-2000, nearly 50% of the net cropped land was double cropped, and 13% triple cropped. The cropping intensity was 175% in the mid-1990s. Cropping intensity is low, however, in the salinity-affected coastal areas and in the flood-prone depressed basins.

Agriculture, the main occupation of the people, employs 63% of the active labor force. It contributed 30% to GDP in 2000, 57% of which came from crop production. Rice accounts for about 77% of total cropped area and two-thirds of the value added in crop production. The emphasis of government policy and research has been on achieving food grain production self-sufficiency with positive support for the distribution of modern agricultural inputs such as chemical fertilizers and irrigation water. This policy support and major achievements of the public-sector research and extension agencies have enabled the country to achieve a record 5% per year growth in cereal production from 1996 to 2000. The current development strategy focuses on agricultural and crop diversification through reallocation of resources away from the production of rice, to be achieved through a continuous increase in the productivity of land and labor, and timely supply of high-quality seeds and fertilizer.

Rice environments

The major rice ecosystems are upland (direct-seeded premonsoon aus), irrigated (mainly dry-season boro), rainfed lowland (mainly monsoon-season transplanted aman, 0–50 cm), medium-deep stagnant water (50–100 cm), deepwater (>100 cm), tidal saline, and tidal nonsaline.

The rice area has remained almost constant since the independence of Bangladesh in 1971, but there has been a major shift in rice cultivation. Over the last three decades, the area under high-yielding boro rice has increased from 0.8 to 3.4 million ha, at the expense of the very low-yielding and risky deepwater aman and upland aus rice crops. Over this period, the area under aus rice has declined from 3.4 to 1.3 million ha and that of deepwater aman rice from 2.1 to 0.7 million ha. However, aman rice still covers 5.7 million ha. Recently, some aus rice land has been diverted for the cultivation of high-value vegetable and fruit crops.

Modern varieties (MVs) make up about 95% of boro (irrigated) rice. Transplanted aman is about 60% MVs, aus about 40%. Deepwater rice is exclusively local varieties. The reallocation of

Bangladeshi farmer heads for his field.

land from traditional varieties to MVs is the main source of growth in rice production and yield. The present average rice yield of about 3.4 t/ha increased at 2.2%/yr from 1990 to 2000. Rice production reached 36 million t in 2000, an increase of 2.5%/yr over the last decade, and 5%/yr over the last five years in spite of a devastating flood in 1998.

Production constraints

Sustainability is always a problem wherever intensified cropping systems are practiced and crop residues are removed for fuel and feed. Cow dung, a traditional source of fertilizer, is being diverted to meet an acute shortage of fuel in rural areas. The use of chemical fertilizers has increased rapidly along with the spread of MVs, from 11 kg NPK/ha in 1970 to 110 kg NPK/ha in 2000. With the removal of fertilizer subsidies in the late 1980s, fertilizer use became unbalanced, with too much use of N and too little P, in response to an unfavorable trend in the relative price of P and K, which are mostly imported.

Drought is a frequent problem, but supplemental irrigation during the late monsoon could alleviate it. Subsurface groundwater is available almost everywhere in the country. Irrigation by small-scale tube wells and low-lift pumps began in the late 1970s when government control over the procurement and distribution of modern agricultural inputs was abolished. The spread of tube wells has increased more rapidly since the late 1980s when the importation of agricultural machinery was liberalized. In 1999, rice constituted nearly 80% of the total irrigated area; 70% is irrigated with shallow tube wells and power pumps owned and operated by farmers. Overexploitation of groundwater is becoming an environmental concern with adverse effects on the supply of drinking water; there are suspected links to arsenic-contaminated water.

Flooding occurs annually, but causes serious damage only about once every 10 years. Normal flooding is simply a part of the ecosystem and helps to maintain soil quality. The flood-prone areas are ideally suited for boro rice, as water is available during the dry season and the cost of irrigation is low.

Soils in coastal areas are affected by salinity. Most soils are low in organic matter (many less than 0.5%) and consequently low in N. Zinc and S deficiencies are common; replacement amounts of P and K are insufficient.

Marketing infrastructure is adequate for rice but inadequate for other agricultural commodities, especially perishables. The prices of both rice and nonrice crops fluctuate seasonally because of the lack of access to international markets and occasional good or poor harvests that affect the demand-supply balance within the economy. The price of rice is now too low to provide incentives to farmers to sustain growth in production. When food grain production approaches self-sufficiency, farm-gate prices of rice go down quickly. A policy to move stored grain routinely into market channels and replace it with fresh stocks is needed to stabilize rice prices.

The main challenge to food self-sufficiency in Bangladesh is sustainability of production in view of the many man-made, biotic, and abiotic constraints. Population is a bigger problem than food production inasmuch as food production is basically keeping pace with population growth. Population density is 900 persons/km^2, one of the highest in the world. Bangladesh has made notable progress in population control since the late 1980s. The 2001 population census reports a growth rate of 1.6%/yr from 1991 to 2001 compared with 2.4%/yr for 1981-91.

Production opportunities

Rice research and development are effective, but could be streamlined with more effective linkages between research and extension.

Among the measures that would help stabilize rice supply and encourage agricultural growth are the spread of shorter-duration MVs to intensify cropping; further development of drainage and irrigation facilities; development of varieties tolerant of salinity, drought, and submergence to raise productivity in coastal and flood-prone areas; and reducing the yield gap in irrigated areas with the spread of knowledge-intensive crop and natural resource management practices.

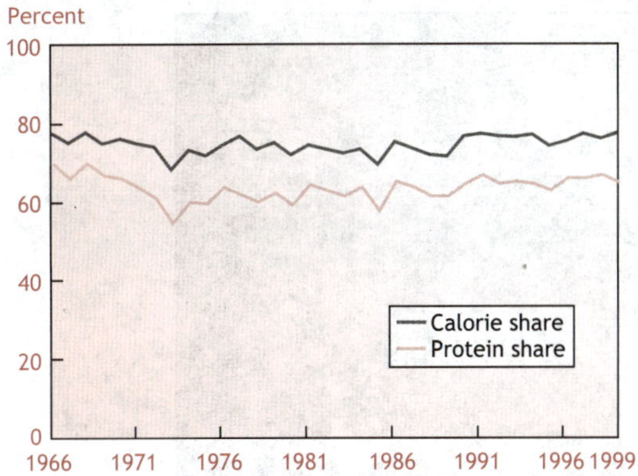

Share of calories and protein from rice, 1966-99

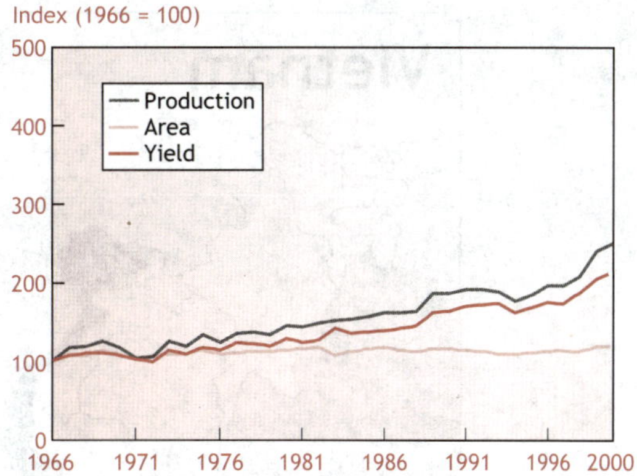

Indices of rice production, area, and yield, 1966-2000

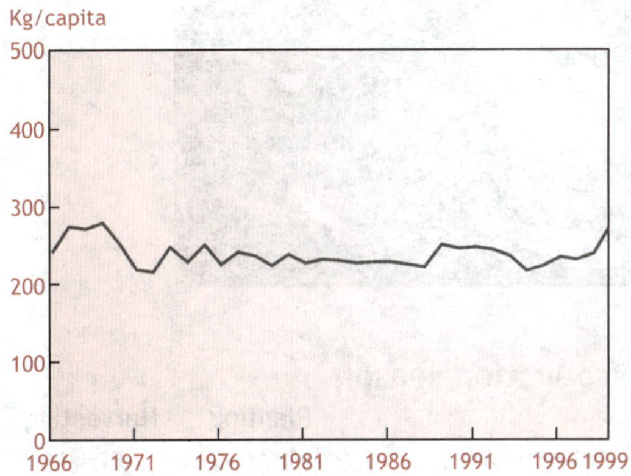

Per capita production (paddy terms), 1966-99

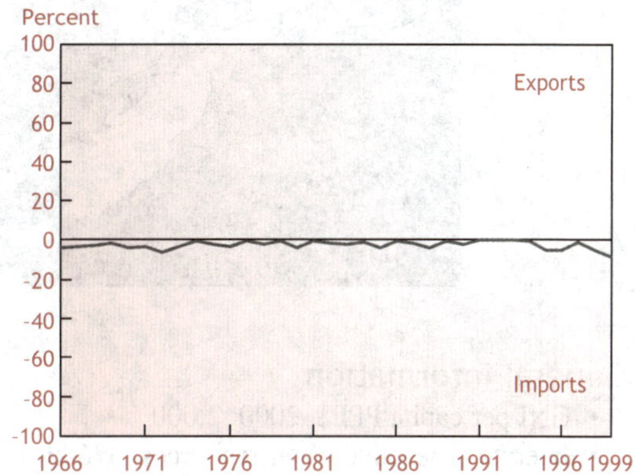

Net trade status, 1966-99

Basic statistics, Bangladesh

	1985	1990	1995	1998	1999	2000
Rice						
Area harvested (ha)	10,398,170	10,435,340	9,951,700	10,115,630	10,708,000	10,700,000
Yield (t/ha)	2.2	2.6	2.7	2.9	3.2	3.3
Production (t)	22,556,288	26,777,904	26,398,000	29,708,000	34,426,800	35,820,800
Rice imports (t)	677,323	380,062	995,946	1,127,208	2,215,322	na
Paddy imports (t)	0	0	579,601	168,472	0	na
Rice exports (t)	0	0	58	105	170	na
Paddy exports (t)	na					
Others						
Population, total (×10³)	99,373	109,465	118,616	124,774	126,947	na
Population, agriculture (×10³)	68,496	71,460	71,868	71,985	72,001	na
Agricultural area (×10³ ha)	9,735	10,037	8,748	8,932	na	na
Irrigated agricultural area (×10³ ha)	2,073	2,936	3,429	3,844	na	na
Total fertilizer consumption (t)	540,682	933,022	1,191,097	1,171,000	na	na
Tractors used in agric. (no.)	4,900	5,200	5,300	5,400	na	na

Source: FAOSTAT online database.

Vietnam

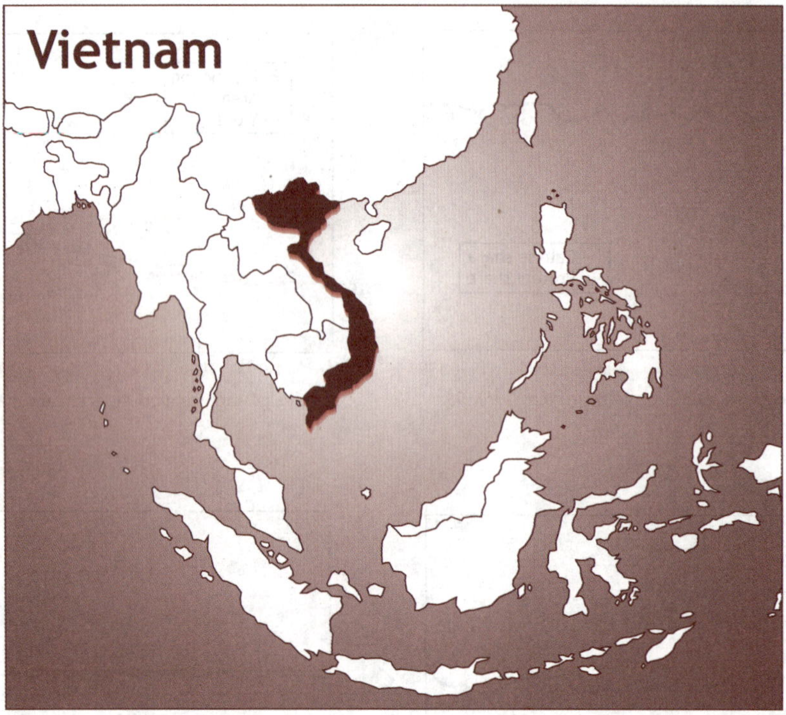

General information

- GNI per capita PPP$, 2000: 2,000
- Internal renewable water resources: 376 km^3
- Main food consumed: rice, roots and tubers, meat, sugar and honey, fruits
- Rice consumption, 1999: 170.3 kg milled rice per person per year

Production season

	Planting	Harvesting
Main season	May-Aug	Sep-Dec
Winter-spring	Dec-Feb	Apr-Jun
Summer-autumn	Apr-Jun	Aug-Sep

Vietnam is located along the eastern margin of the Indochina peninsula in Southeast Asia, extending from 8° to 23° N latitude. It is bounded by Cambodia, Lao PDR, China, and the South China Sea.

It is in AEZ 3, characterized as warm humid tropics. Over 30% of the country is forested and about 17% is cultivated for seasonal crops, with another 5% under permanent crops. Climate varies from humid tropical in the southern lowlands to temperate in the northern highlands. There are two monsoon seasons: the northeastern winter monsoon and the southwestern summer monsoon. Destructive typhoons sometimes develop over the South China Sea during hot weather. Mean annual sea level temperatures decline from 27 °C in the south to 21 °C in the extreme north.

Mean annual rainfall ranges from 1,300 to 2,300 mm. Rainfall is usually evenly distributed in June to October or November. In the Mekong Delta, the summer monsoon brings 5–6 months of rainfall above 100 mm/mo. October is the wettest month of the year.

The population of Vietnam was about 79 million in 1999 with an average density of 245/km^2. The population grew at 1.7%/yr during the 1990s. Eighty percent of the population is rural and concentrated in the two rice-growing deltas: the Red River Delta in the north and the Mekong River Delta in the south. Total labor force was 42 million, with two-thirds engaged in agriculture. The agricultural labor force grew by 17% during the 1990s compared with a 23% growth in total labor force.

Recent developments in the rice sector

The gross domestic product was estimated at US$27.2 billion for 1998. Agriculture continues to play a dominant role, contributing 21% to the GDP and 30% of total export earnings.

Rice is the single most important crop. It is cultivated on 4.2 million ha out of 5.7 million ha of arable land. The planted area for rice increased from 5.6 million ha in 1980 to 7.7 million ha in 2000. Cropping intensity has reached 183%, the highest in the world. The rapid increase in rice area and the intensity of rice cropping have been made possible by heavy investment in flood control, drainage, irrigation that turned the flood-prone ecosystem in the Mekong River Delta into an irrigated ecosystem, and the development of very short duration rice varieties.

In 1981, Vietnam departed from the collective agricultural production system by introducing the group-oriented contract system of production. That was changed to individual contracts, beginning in 1986. The average farm size is very small. The number of farm households was estimated at 9.5 million in 1994, with 8.4 million having a size of less than 1 ha. By official estimates, the average small-farm household's share of income from the crops it harvests has risen from 20% before the 1986 reforms to around 60% in the mid-1990s.

Vietnam achieved an impressive growth in rice production after the policy reforms in 1986. Total output increased from 15.1 million t in 1987 to 32.6 million t in 2000, a growth of 6.1%/year. Much of the growth came from the expansion of the rice harvested area, as farmers shifted land from a long-duration single-cropped deepwater rice to double- and triple-cropped short-duration, high-yielding modern varieties, particularly in the south. There has also been an impressive growth in rice yield, from 2.70 t/ha in 1987 to 4.25 t/ha in 2000, a growth of 3.3%/year.

As a result of the spectacular growth in rice production, Vietnam has been a major rice exporter since 1989. Initially, it captured the international market for low-quality rice, but over time the milling quality has improved. Exports of milled rice have continuously increased from 1.4 million t in 1989 to 4.6 million t in 1999. Vietnam is now the second largest exporter of rice in the world market, after Thailand.

Rice environments

About 52% of Vietnam's rice is produced in the Mekong River Delta and another 18% in the Red River Delta. The other major rice-growing regions are the northeast and the north-central coast. The northern provinces of Vietnam have a total rice area of 2.5 million ha or about a third of the total rice planted area. Almost 85% of the total area is irrigated lowland, 12% is shallow rainfed, and 4% is intermediate rainfed. The dominant cropping pattern is spring-summer rice.

The Red River Delta, which is extremely densely settled and has very small landholdings, has long been practicing double rice cropping with highly labor-intensive rice cultivation methods. The winter and spring season rice crops cover almost the same area (530,000 ha), with a yield of 5.2 and 5.7 t/ha, respectively.

The Mekong River Delta has three major cropping seasons: spring or early season, autumn or midseason, and winter, the long-duration wet-season crop. The largest rice area is cropped during the autumn season (1.95 million ha), followed by spring (1.45 million ha), and only a small area is cropped in winter (0.6 million ha). The rice yield is highest in the spring season (5.3 t/ha), and lowest in the winter season (3.3 t/ha). Farmers in this region adopt a direct-seeding method of crop establishment to save labor costs. Fifty-two percent of the rice in the Mekong River Delta is grown in irrigated lowlands, with the remaining 48% grown under rainfed conditions.

Soils in the Mekong River Delta are highly variable, but alluvial, acid-sulfate, and saline soils predominate. Acid-sulfate soils cover some 1.6 million ha, or 40% of the delta, mainly in the Plain of Reeds, Long Xuyen Quadrangle, and Ca Mau Peninsula. The soil is rich in humus and total N, but low in P. In addition, Al and Fe toxicities limit yield.

Alluvial soils, prevalent in 30% of the Mekong River Delta, are concentrated along the banks of the Tien (Mekong) and Hau (Bassac) rivers. This is the best soil in the delta, with humus content of 2%, total N of 0.1–0.25%, and medium P and K. Two to three crops can be grown on these soils each year.

Coastal saline soils occupy about 20% of the total area. The soils are rich in humus, N, and clay (55–60%), but have a high salt content.

Rice harvest transport on the Mekong River system.

Production constraints

The major constraints are flooding at the end of the rainy season and drought in the dry season.

Small farm size, a problem that is expected to increase even further because of population pressure, is a major constraint. The low and declining price of rice with the increase in rice harvest provides inadequate income for the sustenance of farming families. The low profitability in rice farming is a major disincentive to sustaining growth in productivity. Farmers have been trying to diversify into vegetables, fruit trees, and fish cultivation but without much success because of the lack of markets.

The current level of physical infrastructure is inadequate to support potential increases in agricultural production. Two-thirds of the farms have no access to drying areas; most of the crop is sun-dried. Storage space is about 1 million m^3 or 67% of the total needed. Transportation for moving the crop to market is inadequate. Energy is also in short supply. Although electricity is available in most provinces, most rural households do not have access to it.

Production opportunities

The increased production of rice and productivity of rice-based farming systems remain the country's primary goals. Studies in the Mekong River Delta have focused on various rice-based farming systems models: rice-fish integrated with fruit trees, rice-shrimp in saline areas, rice-fish in deepwater areas, and rice–cash crops in the remaining small amount of floating-rice area.

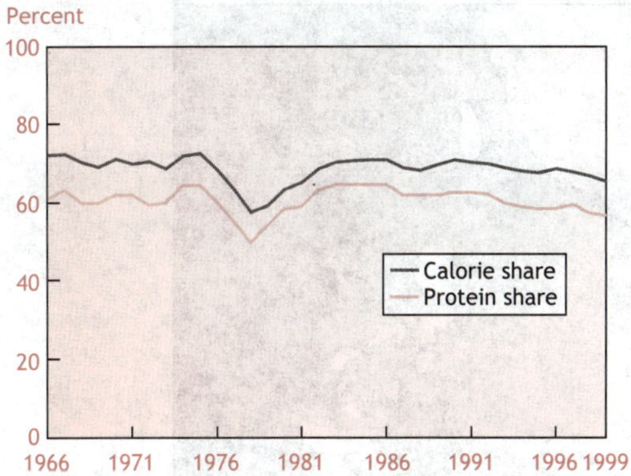

Share of calories and protein from rice, 1966-99

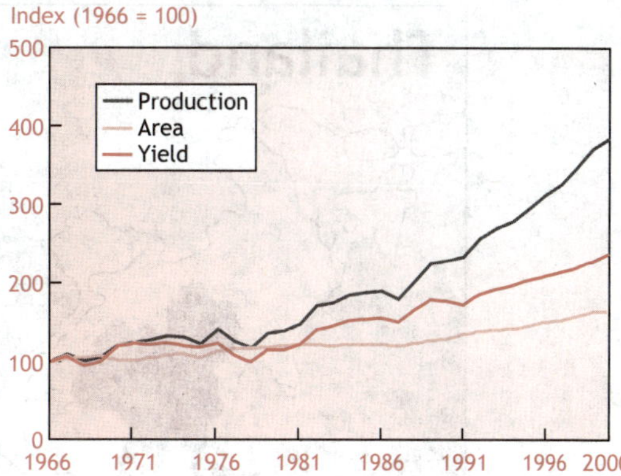

Indices of rice production, area, and yield, 1966-2000

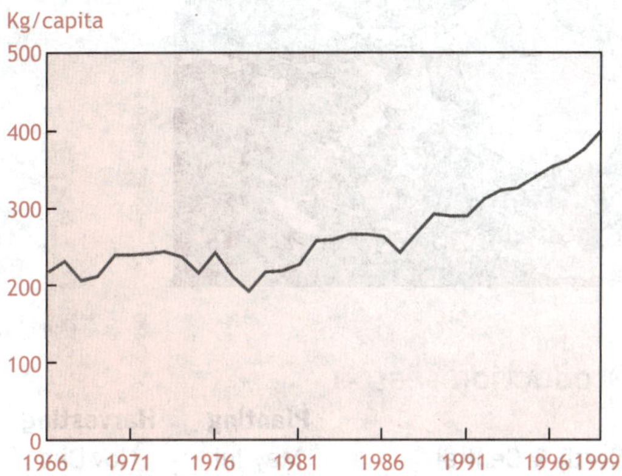

Per capita production (paddy terms), 1966-99

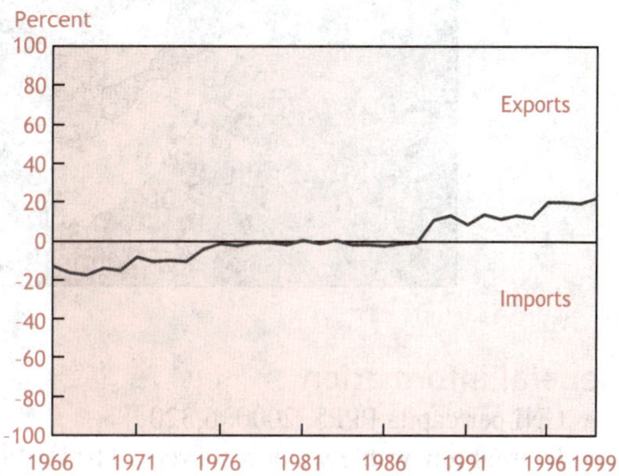

Net trade status, 1966-99

Basic statistics, Vietnam

	1985	1990	1995	1998	1999	2000
Rice						
Area harvested (ha)	5,703,900	6,027,700	6,765,600	7,362,700	7,648,100	7,654,900
Yield (t/ha)	2.8	3.2	3.7	4.0	4.1	4.3
Production (t)	15,874,800	19,225,104	24,963,700	29,145,500	31,393,800	32,554,000
Rice imports (t)	336,100	1,900	11,000	1,300	5,200	na
Paddy imports (t)	na	na	na	na	na	na
Rice exports (t)	59,400	1,624,000	1,988,000	3,700,000	4,600,000	na
Paddy exports (t)	na	na	na	na	na	
Others						
Population, total ($\times10^3$)	59,898	66,689	73,866	77,562	78,705	na
Population, agriculture ($\times10^3$)	43,275	47,546	51,232	52,869	53,330	na
Agricultural area ($\times10^3$ ha)	6,750	6,726	7,085	7,892	na	na
Irrigated agricultural area ($\times10^3$ ha)	2,500	2,900	3,000	3,000	na	na
Total fertilizer consumption (t)	385,600	544,484	1,214,000	1,947,400	na	na
Tractors used in agric. (no.)	31,620	25,086	97,817	122,958	na	na

Source: FAOSTAT online database.

Thailand

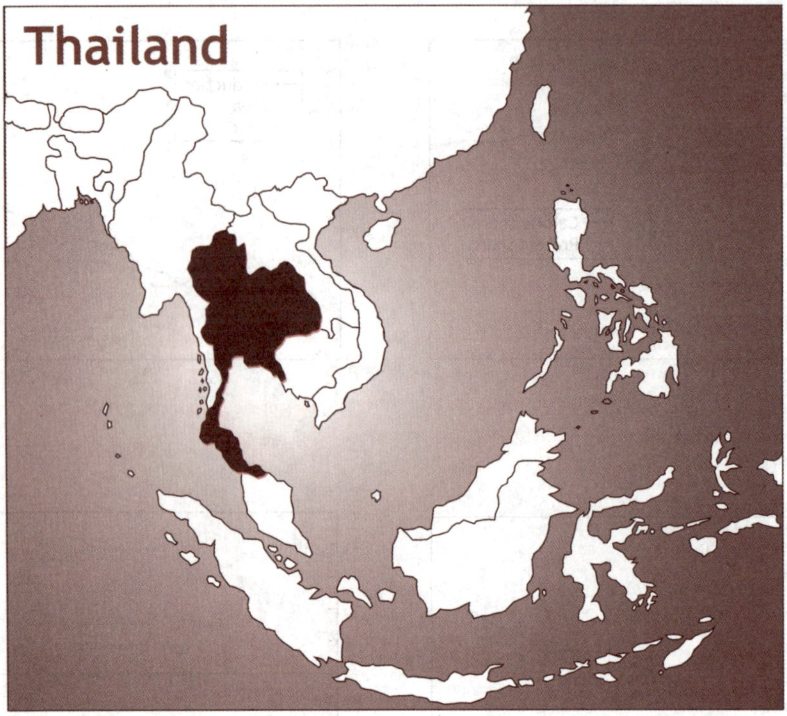

General information

- GNI per capita PPP$, 2000: 6,320
- Internal renewable water resources: 110 km^3
- Incoming water flow: 69 km^3
- Main food consumed: rice, sugar and honey, oil and fat, meat, nuts
- Rice consumption, 1999: 100.8 kg milled rice per person per year

Production season

	Planting	Harvesting
North & Central, major season	May-Jul	Nov-Dec
North & Central, minor season	Dec-Jan	May-Jun
South, major season	Sep-Nov	Mar-May
South, minor season	Apr-May	Aug-Sep

Thailand lies between 5° and 21° N latitude and between 97° and 106° E longitude. It is a peninsular country in Southeast Asia sharing boundaries with Myanmar in the west, Lao PDR and Cambodia in the northeast, and Malaysia in the south. The South China Sea touches the east coast, while the Indian Ocean and Andaman Sea border the west coast. Thailand has a land area of 51 million ha, of which one-third is cultivated for annual crops and about 7% is under permanent crops.

Thailand is in AEZ 2, characterized as warm subhumid tropics. Four seasons are recognized: southwest monsoon from May through September, a transition period from the southwest to the northeast monsoon during October, the northeast monsoon from November through February, and a premonsoon hot season during March and April.

Temperatures in the Central Plain during the rainy season (May to November) average 27 °C, with only 8–10 °C between the daily minimum and maximum. There is a brief cool period (December and January) with temperatures as low as 2–3 °C in the northern highlands.

Rapid economic growth in nonagricultural sectors over the two decades prior to the financial crisis in 1997 greatly reduced the relative importance of agriculture as a contributor to the gross domestic product and export earnings. But agriculture is still the dominant economic activity in rural Thailand and is now recognized as a critical sector to cushion the adverse effect of the financial crisis. The population was 60.9 million in 1999 and grew at 0.9%/yr during the 1990s. Despite the rapid transformation to an industrialized economy, 78% of the population is still rural. The average rural person has a low level of schooling because most of the highly educated people migrate to cities in search of better living conditions.

The GDP was estimated at US$111 billion in 1998, of which only 11% originates from agricultural activities. However, agriculture employs nearly 56% of the total labor force, which shows the large disparity in the productivity of labor between agriculture and nonagricultural sectors. The agricultural labor force grew by only 3.4% during the 1990s vis-à-vis a 17% increase in the total labor force.

Recent developments in the rice sector

Rice is the most important crop of the country. Even though declining in relative importance, it still occupies about 55% of the total arable land. Rice farmers in the northeast, the main rice-growing region and the home of the famous Jasmine rice, are generally subsistence farmers, selling only their excess production. The main surplus production is from the central region and the north, where the average farm size is three times larger than in the northeast, and the production environment is favorable. Rice is the staple food of the entire population regardless of income. The average annual per capita consumption is 100.8 kg of milled rice, which has been declining since the mid-1980s.

The rice planted area grew rapidly from 6.9 million ha in 1968 to 9.8 million ha in 1988, and since then has fluctuated between 9.0 and 10.0 million ha depending on the relative price of rice in the world market. Rice production increased from 12.4 to 21.2 million t during the first two decades of the Green Revolution. Rice yield increased relatively slowly compared with that in other Asian countries, from 1.79 t/ha in 1968 to 2.15 t/ha in 1988, because of the predominance of the rainfed rice ecosystem and farmers' preference to grow high-quality, low-yielding traditional varieties that fetch a premium price in the domestic and the world market. Since 1988, rice production and yield have increased very little. The present rice yield of 2.33 t/ha is one of the lowest in the world. However, farm mechanization has spread widely since the mid-1980s, which has led to a rapid increase in labor productivity and helped sustain the profitability of rice farming and its competitiveness in the world market.

Thailand is the major rice exporter in the world market, currently exporting about 6.5 million t of milled rice per year. The exports continue to grow despite the stagnation in domestic production because of the declining trend in the domestic demand for rice. Thailand has a reputation for high-quality, long-grain white rice, which usually commands a substantial price advantage over lower grades.

Rice environments

Administratively as well as geographically, Thailand is divided into four regions: central, north, northeast, and south. Each region has different rice-growing environments.

Northeast region. Almost one-third of the land area of Thailand is located in the northeast region. Nearly one-half of the rice land is located in this region and the average size of the rice farm is smaller than in other regions. Soil erosion and drought during the dry season are acute. The water-holding capacity of the soil is extremely poor. Less than 20% of the total irrigated land of the country is located in this region, with less than 10% planted to rice in the dry season.

Central region. The central region is an intensively cultivated alluvial area. During the rainy season, rice covers the major part of the region. The central region accounts for about one-fifth of the total cultivated rice land of the country in the wet season. The average farm size is large, and a larger proportion of the rice land has access to irrigation facilities that allow many farmers to grow two rice crops during the year. Almost 75% of the dry-season rice grown under irrigated conditions is located in this region. Farm operations are almost entirely mechanized, and farmers adopt the direct-seeding method of crop establishment to save labor.

Northern region. The northern region contains almost one-third of the land area of Thailand. Upland rice is grown in the lower altitudes of high hills and in upland areas. Lowland rice is grown mainly in lower valleys and on some terraced fields where water is available. The northern region has about 20% of the total rice land in the country.

Southern region. The southern region, touching the west and east coasts of the peninsula, constitutes about 14% of the total area of the country. The region has only 6% of the total rice land. The soil is acidic. With limited rice fields under cultivation, there is always a shortage of rice for local consumption.

Production constraints

In addition to the physical and climatic problems associated with the different rice ecosystems, the main problems that farmers face are stagnating yields and labor shortages.

The average rice yield in the wet season is stagnant at about 2 t/ha. But the high-quality, aromatic Thai rice varieties, which command high prices on the world market, are traditional varieties. So, the low yields do not necessarily reflect low input supply, poor water control, or any of the other constraints normally associated with low yields. Still, maintaining productivity of the system, even at a 2 t/ha plateau, is of some concern.

The greater constraint is a shortage of labor during peak periods, especially in the central region where industrial employment is higher than in the other rice-producing regions. Mechanization of agricultural operations is almost complete and combine harvesters are widely used. The price of rice fell sharply after the devaluation of the currency in 1997. Low rice prices provide disincentives to increasing production.

Production opportunities

The Rice Research Institute of the Department of Agriculture is directly involved in rice research, primarily in increasing productivity through variety improvement and crop, soil, and water management. Related divisions have mandates for research and extension activities for crop protection and machinery development.

The major research efforts are on variety improvement, better crop establishment methods for rainfed and irrigated areas, optimum fertilization rates for organic and inorganic fertilizers, efficient chemical and biological pest control, improved rice farming machinery, and basic research on seed improvement and processing.

Crop substitution and diversification are targeted for areas not well suited to growing rice. In addition, more emphasis is now given to crop rotation, organic farming, and other measures aimed at sustainability of rice production.

Rice on its way to market in rural Thailand.

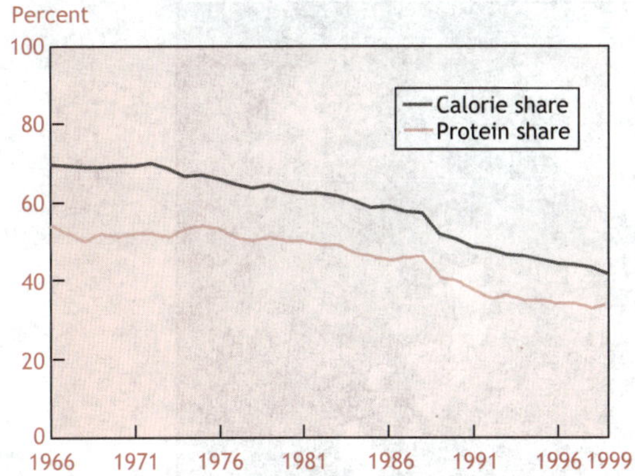

Share of calories and protein from rice, 1966-99

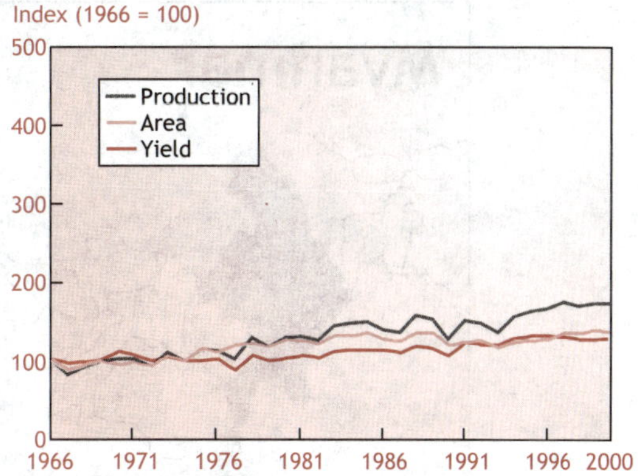

Indices of rice production, area, and yield, 1966-2000

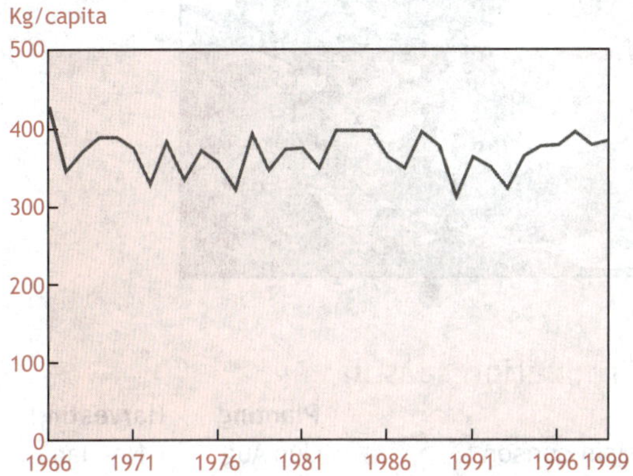

Per capita production (paddy terms), 1966-99

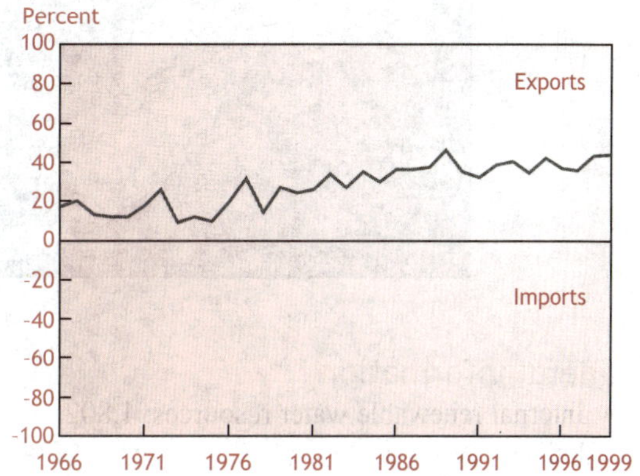

Net trade status, 1966-99

Basic statistics, Thailand

	1985	1990	1995	1998	1999	2000
Rice						
Area harvested (ha)	9,833,074	8,791,885	9,112,951	9,900,000	10,080,000	10,048,000
Yield (t/ha)	2.1	2.0	2.4	2.3	2.3	2.3
Production (t)	20,263,872	17,193,216	22,015,500	22,784,436	23,313,000	23,402,900
Rice imports (t)	0	0	68	836	1,406	na
Paddy imports (t)	0	0	0	1	250	na
Rice exports (t)	4,061,715	4,017,079	6,197,990	6,537,492	6,838,900	na
Paddy exports (t)	1,500	0	0	12	0	na
Others						
Population, total ($\times 10^3$)	51,146	55,595	58,610	60,300	60,856	na
Population, agriculture ($\times 10^3$)	30,940	31,622	30,933	30,428	30,252	na
Agricultural area ($\times 10^3$ ha)	20,577	21,383	21,210	21,175	na	na
Irrigated agricultural area ($\times 10^3$ ha)	3,822	4,238	4,590	4,749	na	na
Total fertilizer consumption (t)	433,562	1,043,791	1,507,000	1,660,863	na	na
Tractors used in agric. (no.)	31,415	57,739	148,841	220,000	na	na

Source: FAOSTAT online database.

Myanmar

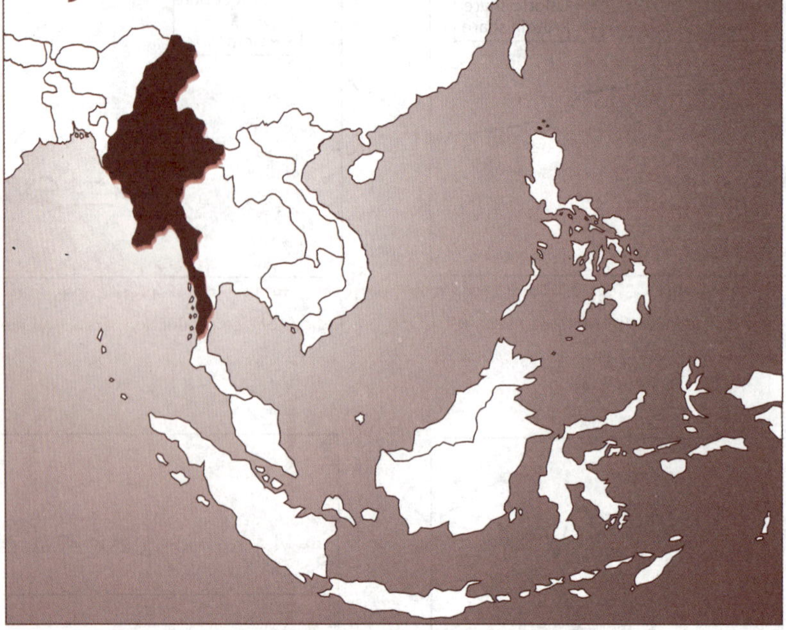

General information

- Internal renewable water resources: 1,802 km^3
- Main food consumed: rice, oil and fat, pulses, sugar and honey, nuts
- Rice consumption, 1999: 210.6 kg milled rice per person per year

Production season

	Planting	Harvesting
Main season	Jun-Aug	Nov-Jan
Dry season	Nov-Dec	Apr-May

Myanmar is the largest country of the Southeast Asian mainland. It lies between 10° and 29° N latitude and 92° and 101° E longitude. Myanmar has a common boundary with Bangladesh, India, China, Lao PDR, and Thailand. There are six distinct regions: the western, northern, and eastern mountain ranges; the delta area of the Ayeyarwady and Sittoung rivers; the coastal strips; and the central plain or dry zone. The land area is about 66 million ha, of which only 9.6 million ha are under cultivation for annual crops. There is a large potential for reclamation of substantial arable land.

Myanmar is largely in AEZ 2, characterized as warm subhumid tropics. The southern coast, where most of the rice is grown, however, is under AEZ 3, humid tropics. Generally, the country receives rain from the southwest monsoon, which normally starts in May and ends in October. Being on the windward side, coastal strips and deltaic regions receive heavy rains, ranging from 2,560 to 6,150 mm annually. The leeward central plain or dry zone receives lower and erratic rainfall, ranging from 700 to 1,200 mm annually.

About 72% of the 45 million population in 1999 lived in rural areas and most of the people were engaged in agriculture. The population grew at 1.2%/year during the 1990s. About 70% of the rural labor force (26.3 million) is employed in agriculture. The agricultural labor force grew by 18% during the 1990s compared with 23% for the total labor force.

Recent developments in the rice sector

Myanmar has a predominantly agricultural economy based on rice production. The agricultural sector, including livestock and fisheries, accounts for 53% of the country's GDP. Rice is the single most important crop, grown on 6.0 million ha or about two-thirds of the country's total cultivated area. Rice production employs 40% of the total labor force and consumes 70% of total commercial fertilizers. Rice maintains its position as the main staple food crop, accounting for 97% of total food grain production. The importance of rice as a major foreign exchange earner, however, has declined over time. Before World War II, Myanmar was the largest rice-exporting country in the world. However, because of stagnation of production since the early 1960s, Thailand took the place of Myanmar in the export market, as exports declined from 1.7 million t in 1962 to 0.3 million t in 1975. Annual exports had fluctuated from 0.5 to 0.9 million t during 1976-86, a period of respectable growth in rice production. Exports in recent years have fallen below 100,000 t. Myanmar has the capacity to substantially increase rice exports, but it is constrained by poor grain quality, inadequate processing and marketing infrastructure, and an underdeveloped trading system.

To meet the increasing demand for rice and to sustain exports, Myanmar embarked on a program to increase rice production through area expansion, yield increase, and crop intensification in the early 1990s. The area under rice increased from 4.8 to 6.0 million ha during 1990-94, which raised annual rice production from 14.0 to 17.9 million t within a 4-yr period. But the growth in production could not be sustained. Rice yield increased from 3.17 t/ha in 1994 to 3.33 t/ha in 2000.

Rice environments

The rice ecosystems of Myanmar include irrigated lowland, rainfed lowland (including late-sown and Mayin area), deepwater, and upland. Late-sown rainfed lowland is the area sown during the monsoon period; Mayin area can be transplanted only after the monsoon when floodwater recedes. The rainfed lowland area is by far the largest, accounting for 52% of total rice land. Irrigated lowland accounts for about 18%. Deepwater rice is planted on about 24% of rice land; upland rice makes up only about 6%. The area under dry-season irrigated rice has been growing in recent years with the spread of small-scale irrigation through low-lift power pumps.

Rainfed lowland and deepwater rice are confined to the delta region and coastal strip of Rakhine State. Irrigated lowlands are mainly in Mandalay, Sagaing, and Bago divisions. The upland area is mostly in Mandalay, Sagaing, and Shan states. Some upland area in Shan State occupies sloping land and is exposed to low temperatures starting in November.

Production constraints

Problems in rainfed lowland production include inadequate labor for transplanting, inadequate supply and high cost of fertilizer, and flooding in low areas. Farmers who practice wet seeding face problems of insect pests, poor crop establishment, and weeds in upper fields. Modern varieties are cultivated widely with very little application of chemical fertilizers, contributing to a decline in soil fertility.

In the upland areas at Aungban and Kyaukme (hilly regions), problems are weeds, insect pests (stem borer and white grub), low soil fertility, soil erosion, poor crop establishment, drought, and reduced fallow period.

Problems encountered in deepwater rice are unfavorable water conditions, weeds (particularly wild rice), soil physical problems, and pests (stem borer and white grub). The high cost of labor, declining rice yield, excess water, and low soil fertility are the main problems in irrigated rice ecosystems.

In the hilly region, the problems are mainly related to infrastructure—lack of good roads for transporting rice, and, in the main rice-growing areas, inadequate storage and postharvest facilities, which lower grain quality, particularly for export. The high cost of imported fertilizers and low rice prices have reduced fertilizer consumption and are considered the major constraint to the growth of rice production in Myanmar.

Production opportunities

Significant yield increases were recorded during 1977-82, from 1.95 to 3.15 t/ha. Several factors explain this impressive performance. Foremost is the introduction of high-yielding modern varieties (MVs) and technologies developed at the Central Agricultural Research Institute. Fifty MVs were

Women farmers harvesting rice in Myanmar.

released in Myanmar from 1966 to 1990. Of these, 11 were produced by Myanmar scientists using selected genetic materials from the International Network for Genetic Evaluation of Rice (INGER), and from local rice germplasm collections. The rest were direct selections from INGER nurseries tested in Myanmar. Increased plant densities and increased fertilizer usage, particularly for MVs, also had a significant effect on production. However, yield has fluctuated around 3.0 to 3.3 t/ha since 1983 mainly because of the inadequate supply of chemical fertilizers.

To overcome the labor shortage, farmers practice the wet-seeding method of crop establishment, do late transplanting using overaged seedlings, exchange labor, and grow cash crops. They substitute gypsum and farmyard manure for expensive fertilizers, fertilize selectively, and wet-seed the crop or change cultivars.

Choosing suitable cultivars and constructing fishponds are the solutions to the problem of excess water in lower areas. To control soil erosion, farmers construct erosion-control structures above fields and drainage and diversion canals around fields.

The production gains achieved by improved rice technology in the rainfed lowlands must be maintained. The government also considers it important to develop new technology for the coastal and saline rice environment and for the upland ecosystem, which experience stress from drought, erosion, weeds, insect pests, nematodes, and low temperature. The technology needs are being met in part through the Myanmar-IRRI collaborative research program.

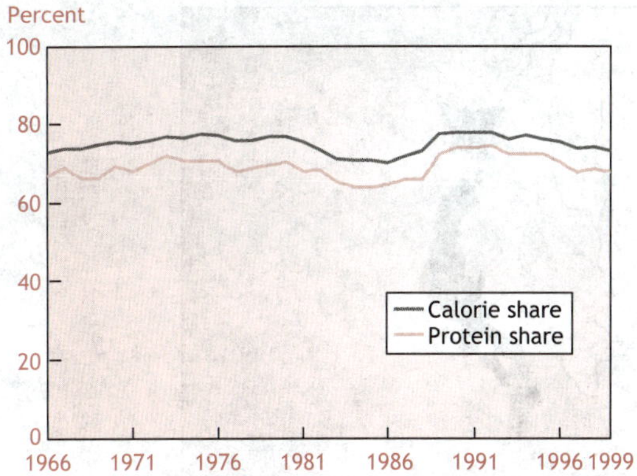

Share of calories and protein from rice, 1966-99

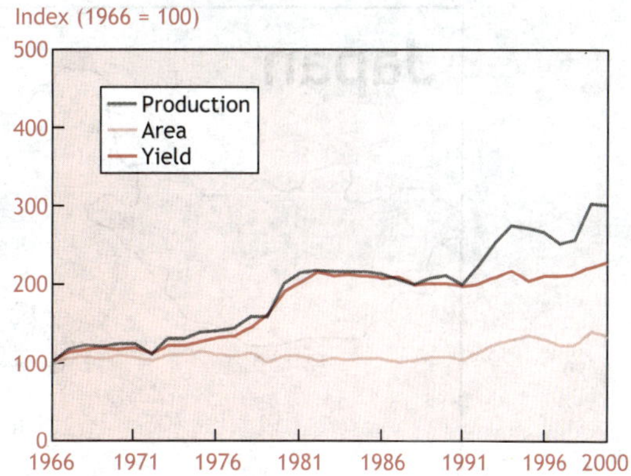

Indices of rice production, area, and yield, 1966-2000

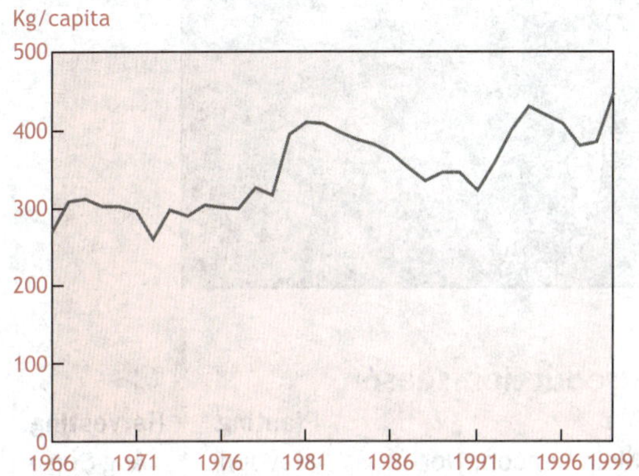

Per capita production (paddy terms), 1966-99

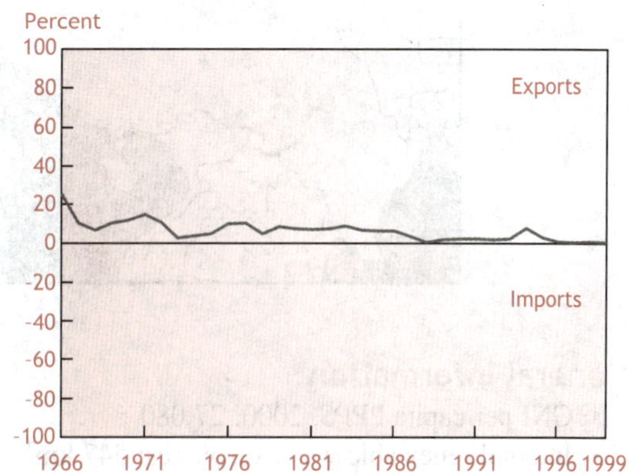

Net trade status, 1966-99

Basic statistics, Myanmar

	1985	1990	1995	1998	1999	2000
Rice						
Area harvested (ha)	4,660,800	4,760,000	6,032,700	5,458,500	6,210,787	6,000,000
Yield (t/ha)	3.1	2.9	3.0	3.1	3.2	3.3
Production (t)	14,317,048	13,971,800	17,956,900	17,076,728	20,124,708	20,000,000
Rice imports (t)	0	0	0	19	19	na
Paddy imports (t)	na	na	na	na	na	na
Rice exports (t)	581,500	213,600	353,800	86,966	36,000	na
Paddy exports (t)	na	na	na	na	na	na
Others						
Population, total ($\times10^3$)	37,544	40,520	42,877	44,497	45,059	na
Population, agricuture ($\times10^3$)	27,990	29,688	30,770	31,520	31,776	na
Agricultural area ($\times10^3$ ha)	10,429	10,428	10,450	10,505	na	na
Irrigated agricultural area ($\times10^3$ ha)	1,085	1,005	1,555	1,592	na	na
Total fertilizer consumption (t)	194,087	70,732	179,730	171,805	na	na
Tractors used in agric. (no.)	10,026	13,000	7,818	8,528	na	na

Source: FAOSTAT online database.

Japan

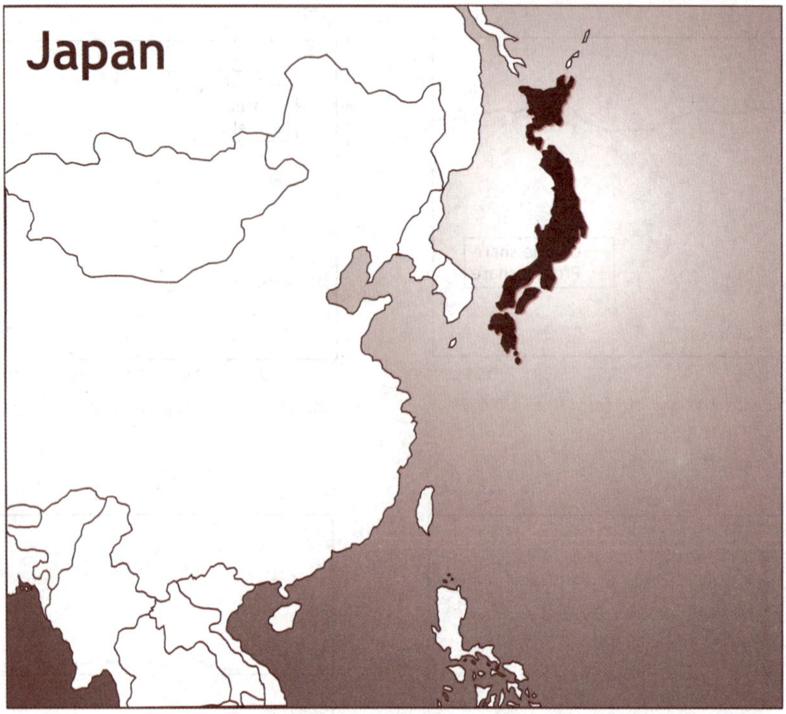

General information

- GNI per capita PPP$, 2000: 27,080
- Internal renewable water resources: 547 km^3
- Main food consumed:
- Rice consumption, 1999: 59.9 kg milled rice per person per year

Production season

	Planting	Harvesting
Main season, North	May-Jun	Aug-Sep
Main season, Central	Apr-May	Aug-Oct
Main season, South	Apr-May	Sep-Nov

The Japanese islands are located between 20° and 45° N latitude and 123° and 146° E longitude. They lie off the eastern coast of Asia, roughly in a crescent shape. The country consists of four main islands with about 4,000 smaller ones. The northern limit of rice cultivation is 44° N. Rice is grown up to 1,400 m altitude in the central region of the main island.

Japan is in AEZ 8, characterized as cool subtropics with summer rainfall. The climate is humid temperate and oceanic with four distinct seasons. Rainfall during the rainy season in June and July is indispensable to rice cultivation. Temperature and solar radiation from April to October are ideal for rice growing.

The population in 1999 was about 127 million. While 21% of the population now lives in rural areas, the agricultural population has declined from about 29 million in 1960 to just 5 million today because of diversification of the economy. Population growth is just 0.2% per annum.

The changing role of rice cultivation in Japan*

Rice cultivation in Japan, which dates back more than 2,000 years, was introduced from China in the late Jomon era. From that time until the present, rice farming has exerted a considerable influence on the social structure and culture of the

*This section was contributed by Dr. Ryoichi Ikeda, Director, Rice Research Division, National Institute of Crop Science, Kannondai, 2-1-18, Tsukuba, Ibaraki, 305-8518, Japan.

country. This is because rice has ever since been the staple food for people in Japan, and it has been the mainstay of Japanese agriculture.

Recently, rice cultivation in Japan has been besieged with a host of problems both within the country and overseas for the following reasons: (1) rice farming is being increasingly considered as a part-time occupation, (2) the rural population is getting older, (3) the scale of operations is too small, (4) the gap between the price of rice in Japan and abroad is increasing, and (5) there is a pressing call for the liberation of the market.

In addition, since the productive capacity of the paddy fields in Japan exceeds the demand for rice, the conversion of paddy fields for the cultivation of other crops has become inevitable. Therefore, the Ministry of Agriculture, Forestry and Fisheries has enacted a policy of wider use of paddy fields in agriculture with a view to increasing the productivity of both paddy rice and other crops, on the basis of the alternative use of fields.

"The paddy fields save the world," as the saying goes. Indeed, under the natural environmental conditions prevailing in Japan, including abundant precipitation of 1,600–1,700 mm per year and steep slopes, the role of reservoirs played by the paddy fields is equivalent to that of a dam with a capacity of 5.1 billion tonnes. In addition, the contribution of paddy fields to land conservation and water purification as well as to the welfare of the population is quite remarkable.

Based on the various inherited circumstances of rice cultivation, Japan was able to raise rice yield by 250% during the past 100 years, in spite of suboptimal solar intensity. The conditions that resulted in such marvelous achievements were (1) the narrow and highly populated country, which demanded highly labor-intensive farming, (2) the respectful and "religious" attitude of the Japanese for rice, and the fact that they had been longing for an abundance of rice, which had never been achieved before, (3) carefully maintained watersheds—hills and mountains covered with trees and grasses—as the sources of irrigation water, (4) well-developed canals, irrigation systems, and water reservoirs covering most of the ricc lands throughout the country, and (5) farmers who were industrious and well prepared for new technol-ogies.

From the Meiji Restoration to the end of World War II, rice cultivation in Japan began to experience initial but substantial changes in fertilizer use and mechanization. These changes led to modernization in the next period: changes from self-supplied manure to purchased fertilizers with higher nitrogen content and other elements and from high inputs of manual labor in tillage, weeding, threshing, and so on to the use of more labor-saving tools and machinery. Such changes affected the characteristics of rice cultivars, resulting in adoption of those more adapted to higher soil fertility or with better resistance to shattering. Together with the pursuit of higher yields, a tendency also appeared for favoring better grain quality, although that ceased temporarily with the outbreak of World War II.

As a result of these improvements, the national average yield increased from about 2.0 t/ha in 1890-1900 to 2.0–2.5 t/ha in the late 1900s, and to 2.5–3.0 t/ha in 1910-20, while the labor needed in rice cultivation decreased from 300 days/ha in 1910-20 to 200 days in the early 1940s.

The progress made in various sectors during the 50 years following World War II was so rapid and revolutionary that Japan attained self-sufficiency in rice in the 1970s. Developments were especially remarkable in breeding for cold tolerance and yield; in controlling diseases, pests, and weeds by means of chemicals; and in mechanization of such arduous tasks as tilling, transplanting, applying fertilizers and other chemicals, harvesting, and threshing/drying. During this period, rice yields increased from about 3.3 to 5.0 t/ha and labor required decreased from 2,200 to 500 hours per hectare. This labor efficiency allowed farmers to leave villages to work in urban industries, causing the average age in rural areas to increase rapidly. Mechanization was required not only in farming but also in various urban industries, and its success resulted in the emigration of workers in their prime of life to urban areas, leaving villages to their wives and to the elderly. However, this tendency is not particular to Japan; it is becoming common also in other rice-growing countries.

As the relative value of the rice industry in the national economy declines, new environmental problems are arising. One example is the abandoned fields in hilly/mountainous areas. Lowland forests, an important source of irrigation water and manure, are also being neglected and sometimes make floods more serious instead of containing them. The pursuit of economic efficiency with minimal sacrifice of natural resources and the provision of favorable environments should be our main goal as a legacy for our descendants.

Although more than 200 leading varieties of nonglutinous rice are cultivated on about 1.8

million ha in Japan, the top ten varieties now occupy about 80% of the cultivated rice area. The major variety, Koshihikari, is grown on more than one-third of the total rice area. The demand in Japan for rice with good eating quality continues to increase.

Recent developments in the rice sector

Agriculture contributes about 3% to GDP. Rice is the most important crop, accounting for 54% of crop area harvested. Because potential rice production now exceeds the demand for rice, the government has encouraged crop diversification. While rice accounted for 47% of calories and 28% of protein in the early 1960s, those shares have declined to just 23% and 12% today. Wheat currently accounts for about 13% of caloric intake and 11% of protein.

Japan is largely self-sufficient in rice, but Japanese consumers today face extremely high rice prices compared with those on the world market. These high prices have made other rice producers eager to compete in this market, but trade restrictions made this impossible until recently. Under the auspices of the World Trade Organization (WTO), Japan now provides market access to imported rice equal to 7–8% of base-period (1986-88) consumption. About 85% of wheat consumption is imported.

Rice environments

Rice ecosystems in Japan occur across a wide range of latitudes, including the subtropical, temperate, and subfrigid zones. Almost all rice is grown under irrigated conditions in the summer. Most rice fields are on the plains of the major river basins. Many rice fields are also found in terraces and valleys.

Rice cultivation in the higher latitudes is a distinguishing characteristic of Japanese rice culture. To cope with cold weather in the northern part of Japan, early maturing varieties resistant to cold weather were developed. Rice growing in the cold area is characterized by the use of good-quality older seedlings for early transplanting, deepwater irrigation to protect the crop from low night temperature, windbreak nets, and the application of organic matter to improve soil fertility.

Production constraints

Rice is occasionally damaged by drought, but the injury is not so serious in irrigated fields. Cold damage experienced in summer is serious in some years in northern Japan. Usually, incidence of rice blast disease is severe in cool summer weather. Heavy rainfall at the end of the rainy season causes local flooding. Rice is sometimes damaged by typhoons, especially in the southern regions. Early season culture has been established to avoid typhoons in southwestern Japan.

Because potential rice production exceeds rice consumption, it is important to find multiple uses for rice fields. However, many rice fields do not have separate irrigation and drainage systems, and this is an important constraint to the use of rice fields for other purposes.

In spite of substantial farm subsidies and price support provided by the government, rice farming cannot compete with other economic activities, and income from it is lower than nonagricultural earnings. Farming operations have been fully mechanized. Nearly all rice seedlings are transplanted using mechanical transplanters. Despite this mechanization, production costs are many times higher than in tropical Asia because of exorbitant land prices and the high opportunity cost of farm labor. Young people are not interested in rice cultivation, which today is carried out mainly by older people. Many Japanese consumers now want environment-friendly rice grown without heavy use of agrochemicals. Some farmers have started producing rice through organic farming to meet this demand.

Production opportunities

Rice research is conducted in research institutes and experiment stations affiliated with prefectural governments as well as with the Ministry of Agriculture, Forestry, and Fisheries (MAFF). [The rice genome research program (RGP), begun in 1991, aims to establish a comprehensive DNA map of rice. Its eventual goal is to isolate genes with useful agronomic traits and use them for human welfare. The RGP is funded by MAFF and the Japan Racing Association.]

As some of the problems encountered in Japan are shared by other Asian countries, cooperation with IRRI is often essential for their solution. Japanese rice research will also help solve problems in neighboring low-income rice-growing countries of South and Southeast Asia.

Reduced production cost, increased productivity through the application of advanced technology, and multipurpose use of rice fields in agriculture are important for sustaining rice cultivation in Japan.

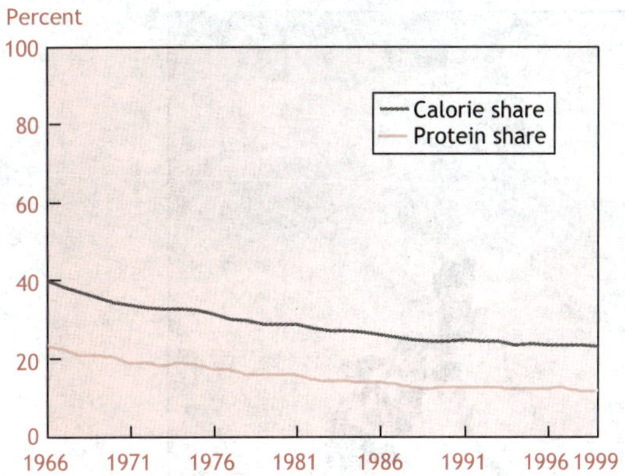

Share of calories and protein from rice, 1966-99

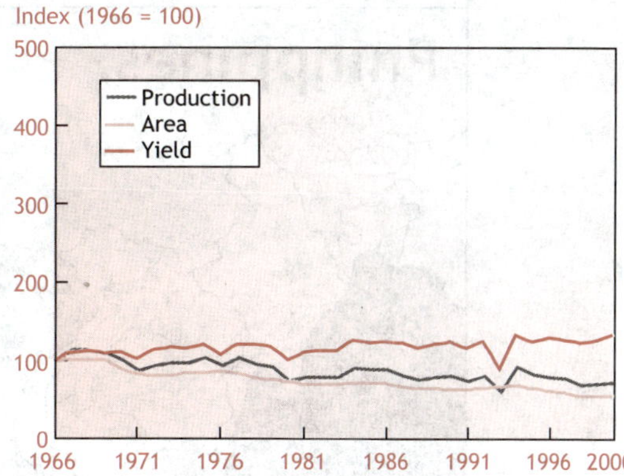

Indices of rice production, area, and yield, 1966-2000

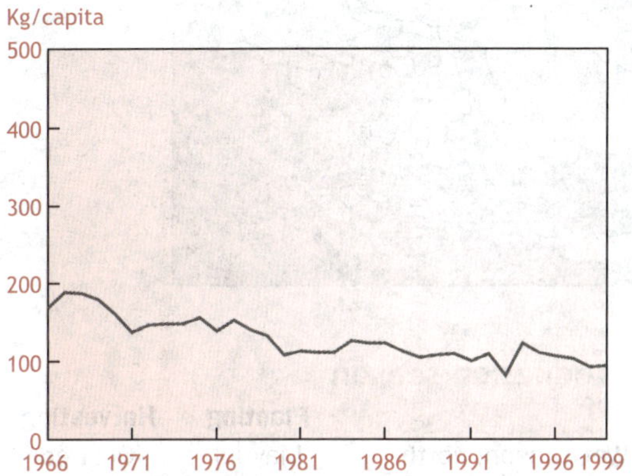

Per capita production (paddy terms), 1966-99

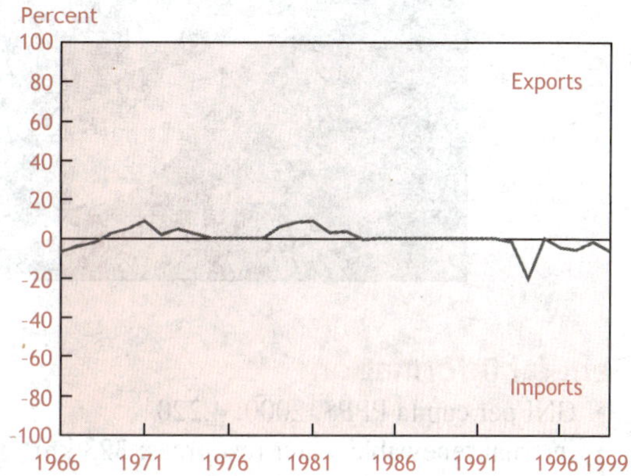

Net trade status, 1966-99

Basic statistics, Japan

	1985	1990	1995	1998	1999	2000
Rice						
Area harvested (ha)	2,342,000	2,074,000	2,118,000	1,801,000	1,788,000	1,770,000
Yield (t/ha)	6.2	6.3	6.3	6.2	6.4	6.7
Production (t)	14,578,000	13,124,000	13,435,000	11,200,000	11,468,800	11,863,000
Rice imports (t)	19,606	18,333	28,971	499,383	664,227	na
Paddy imports (t)	na	na	na	na	na	na
Rice exports (t)	142	16	12,568	358,178	143,953	na
Paddy exports (t)	1	0	1	na	na	na
Others						
Population, total (×10^3)	120,837	123,537	125,472	126,281	126,505	na
Population, agriculture (×10^3)	10,486	8,893	6,736	5,685	5,371	na
Agricultural area (×10^3 ha)	5,879	5,743	5,538	5,405	na	na
Irrigated agricultural area (×10^3 ha)	2,952	2,846	2,745	2,679	na	na
Total fertilizer consumption (t)	2,034,000	1,839,000	1,641,200	1,420,000	na	na
Tractors used in agric. (no.)	1,853,600	2,142,210	2,123,000	2,210,000	na	na

Source: FAOSTAT online database.

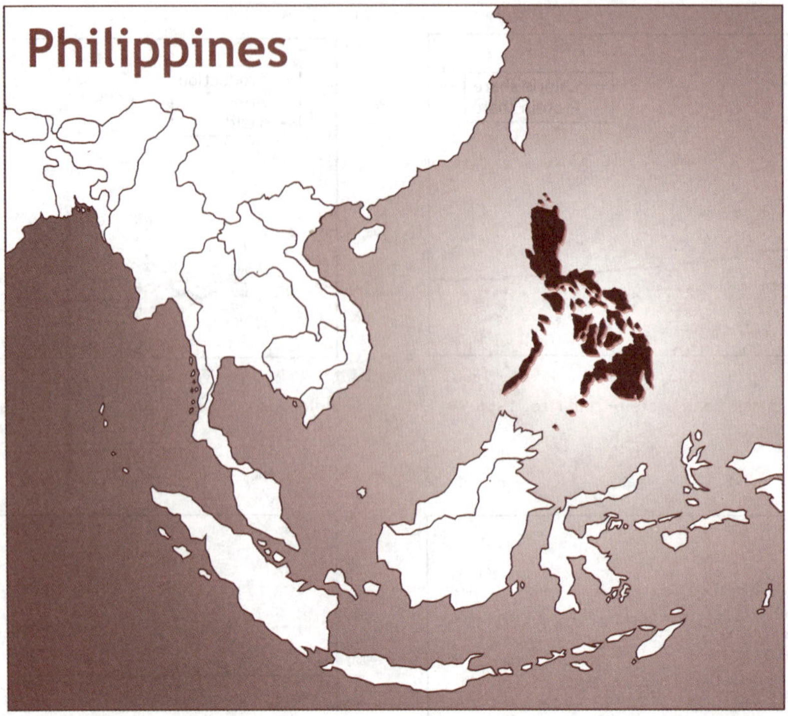

General information

- GNI per capita PPP$, 2000: 4,220
- Internal renewable water resources: 323 km^3
- Incoming water flow: 0 km^3
- Main food consumed: rice, sugar, honey, wheat, roots and tubers, oil and fat
- Rice consumption, 1999: 99.7 kg milled rice per person per year

Production season

	Planting	Harvesting
Wet season, north	May-Jul	Oct-Dec
Dry season, north	Jan-Mar	May-Jun
Wet season, south	Oct-Dec	Mar-May
Dry season, south	May-Jun	Nov-Dec

The Philippines is an archipelago of some 7,100 islands located between 4° and 21° N latitude and 116° and 127° E longitude. The country is bounded by the South China Sea to the west, the Pacific Ocean to the east, the Sulu and Celebes Seas to the south, and the Bashi Channel to the north.

It is in AEZ 3, characterized as the warm humid tropics. The climate is tropical marine, that is, mainly moderated by the surrounding seas, with a November to April northeast monsoon and a May to October southwest monsoon. Climate varies within the country because of a mountainous topography. There are four general climatic types: (1) two pronounced seasons, dry from November to April and wet the rest of the year (Central Luzon, western Visayas), (2) absence of a dry period, but with maximum rains from November to January (eastern Luzon, eastern Visayas, and northeastern Mindanao), (3) dry from November to February and wet the rest of the year (central Visayas, western Bicol, northern Mindanao), and (4) more or less even rainfall distribution thoughout the year (central Mindanao).

The population reached 75 million in 1999. The Philippines (1985-95) has one of the highest population growth rates in Asia at 2.4% per year. However, this growth rate is declining compared with the average of 2.8%/yr in 1975-85. Urbanization has been rapid in recent years and 57% of the population lives in urban areas. Employment in the agricultural sector accounts for 40% of the 31 million-person labor force. Arable land per capita is one of the lowest in developing Asia at just 0.075 ha per capita. The incidence of malnutrition among children age 2–5 (as measured by weight for age) is estimated at 30%.

Recent developments in the rice sector

Agriculture, including forestry and fisheries, contributes about one-sixth of total GDP. Rice constitutes about 30% of total crop area harvested. In the late 1980s and early '90s, maize was the most widely planted crop in the country, but maize area declined to about 2.5 million hectares in the late 1990s. Meanwhile, total harvested rice area reached about 3.8 million ha in 1995 and 4.0 million hectares in 1999 and 2000, its highest level in history despite some loss of agricultural land because of urbanization, land conversion, and industrialization.

Rice is the staple food of Filipinos in most parts of the country, although maize also contributes 20% or more of caloric intake from cereals in parts of Visayas and Mindanao (in Cebu, calories from maize and rice are approximately equal). For the country as a whole, rice accounts for 41% of total caloric intake and 31% of total protein intake.

Because of the Green Revolution package of seeds, fertilizer, and irrigation, national average rice yields doubled from about 1.5 t/ha in the late 1960s to 3.0 t/ha by 1990. Yield from 1985 to 1999 increased at a low rate of 0.9% compared with growth of 2.3% and 2.4% in the late 1960s and early '80s, respectively. These growth rates are attributed to the availability of seeds of high-yielding rice varieties and farm inputs such as fertilizer and irrigation facilities, and the increase in area cultivated among others. Further, the Philippines temporarily achieved self-sufficiency in the early 1980s, but it now imports about 10% of its annual consumption requirements. Despite these imports, rice prices for consumers are the highest in developing Asia (as are farm-gate prices for farmers). The high level of prices is enforced through an import monopoly by the National Food Authority, which also procures paddy from farmers at a support price and stabilizes prices at both farm-gate and retail levels.

Rice environments

In the mid-1990s, about 61% of rice land was irrigated, 35% was rainfed lowland, and the remainder was upland. (As of 2000, of the 4-million-hectare rice area excluding the upland area planted to rice, about 67% is irrigated and the rest is rainfed.) Much of the country's irrigated rice is grown on the central plain of Luzon, the country's ricebowl. Major rice-producing parts of the country are Mindanao (23%), Central Luzon (16%), Cagayan Valley (15%), western Visayas (13%), southern Tagalog (10%), and Ilocos Region (9%). The rest comes mainly from various coastal lowland areas and gently rolling erosional plains, such as in Mindanao and Iloilo. Rainfed rice is found in the Cagayan Valley in northern Luzon, in Iloilo Province, and on the coastal plains of Visayas and Ilocos in northern Luzon. Upland rice is grown in both permanent and shifting cultivation systems scattered throughout the archipelago on rolling to steep lands.

Because of their higher profitability for farmers, modern high-yielding varieties account for the vast majority of rice production, with less than 10% of production coming from traditional varieties. Labor use on rice is relatively low compared with that in many developing Asian countries at about 60 person-days/hectare/crop. Some of the reasons for the relatively low labor use are the widespread use of direct seeding and the mechanization of land preparation and threshing in many parts of the country.

Production constraints

Along with Indonesia, rice production in the Philippines is more severely affected by El Niño events than in other Asian countries, and this led to a large contraction of production in 1998. The major effect of El Niño is a reduction in area harvested, and rainfed areas are the most affected. Irrigated areas are also affected, however, when water availability in dams declines. The Philippines also bears the brunt of many damaging typhoons coming in from the Pacific Ocean. There is also concern about the deterioration of irrigation systems at least partially because of a lack of funding for maintenance. Rainfed lowland rice suffers from uncertain timing of arrival of rains, and drought and submergence—often in the same fields over the course of a single season or in different fields within a farm over the same season. Weeds, drought, diseases (blast), acidic soils, and soil erosion are major problems of upland rice in the Philippines.

Production opportunities

At the forefront of rice research and development (R&D) in the country is the Philippine Rice Research Institute (PhilRice), which was created in 1985. It is a government-owned and controlled corporation attached to the Department of Agriculture (DA) located in Central Luzon and with branch stations in the northern and southern parts of the country.

It seeks to improve the economic conditions of small farmers and, more specifically, to develop and implement a national rice R&D program, sustain gains made in rice production, solve site-specific problems of the rice industry, and direct and coordinate activities of all agencies in the country working on rice in line with its six major programs: direct-seeded rice, transplanted irrigated lowland rice, hybrid rice, rice-based farming systems for adverse environments, policy research and advocacy, and technology promotion and development.

PhilRice pursues these activities in partnership with the member-agencies of the National Rice R&D Network, located in the country's major rice-growing zones. With the PhilRice Central Experiment Station serving as its secretariat, the Network is composed of the PhilRice branch stations, DA-regional research centers, and the premier universities of the country engaged in agricultural research and education, such as the University of the Philippines Los Baños (UPLB), Central Luzon State University (CLSU), Leyte State University (LSU, formerly the Visayas State College of Agriculture or ViSCA), and the Central Mindanao University (CMU). A major activity of PhilRice and the Network is yield and adaptability trials of rice lines developed by Philippine-based public or private institutions working on rice varietal improvement.

Some economic gains at the farm level may be expected through further adoption of direct wet-seeded rice and mechanization of rice production (from seeding/transplanting to postharvest operations). Rainfed lowland rice gains may result from some crop intensification in more favorable areas, further adoption of direct dry-seeded rice, and the development of new rice varieties adapted to rainfed lowland stresses. Upland rice systems in the Philippines are the least sustainable, and future system maintenance will require policies and technologies that encourage a slowing of deforestation rates, the control of soil erosion, and the conversion of cereal production systems into mixed systems with perennial crops on sloping lands.

Rice terraces at Batad, Ifugao Province.

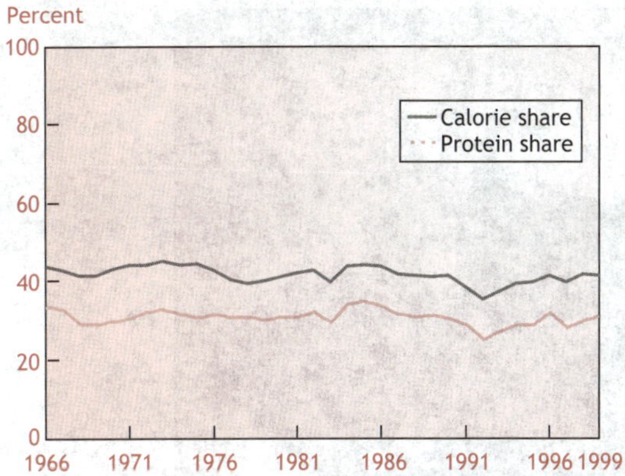

Share of calories and protein from rice, 1966-99

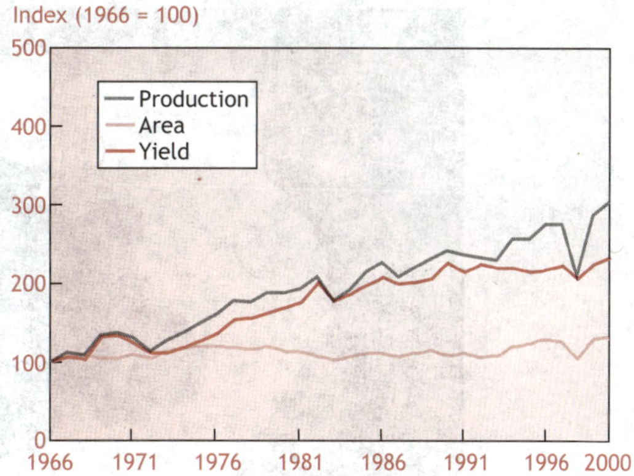

Indices of rice production, area, and yield, 1966-2000

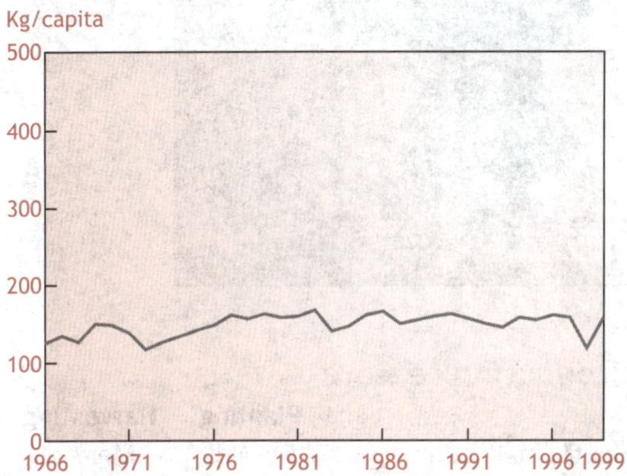

Per capita production (paddy terms), 1966-99

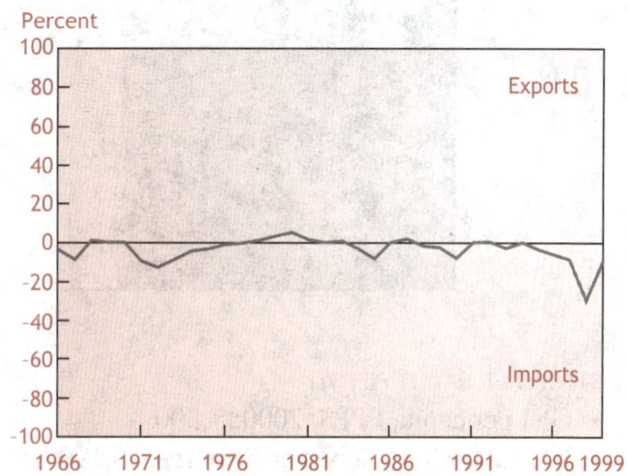

Net trade status, 1966-99

Basic statistics, Philippines

	1985	1990	1995	1998	1999	2000
Rice						
Area harvested (ha)	3,402,610	3,318,720	3,758,700	3,170,042	3,999,839	4,037,085
Yield (t/ha)	2.6	3.0	2.8	3.0	2.9	3.1
Production (t)	8,805,600	9,885,000	10,540,640	8,554,000	11,786,600	12,415,043
Rice imports (t)	538,150	592,727	263,275	2,414,000	834,379	na
Paddy imports (t)	0	0	0	0	1	na
Rice exports (t)	58	2	0	44	294	na
Paddy exports (t)	0	0	0	0	1	na
Others						
Population, total ($\times 10^3$)	54,668	60,687	68,354	72,944	74,454	na
Population, agriculture ($\times 10^3$)	26,655	27,687	28,969	29,577	29,748	na
Agricultural area ($\times 10^3$ ha)	10,910	11,140	11,180	11,280	na	na
Irrigated agricultural area ($\times 10^3$ ha)	1,440	1,550	1,550	1,550	na	na
Total fertilizer consumption (t)	283,181	588,087	598,267	627,930	na	na
Tractors used in agric. (no.)	8,050	10,700	11,500	11,500	na	na

Source: FAOSTAT online database.

General information

- GNI per capita PPP$, 2000: 7,300
- Internal renewable water resources: 5,190 km^3
- Incoming water flow: 1,760 km^3
- Main food consumed: sugar and honey, oil and fat, rice, wheat, meat
- Rice consumption, 1999: 40.3 kg milled rice per person per year

Production season

	Planting	Harvesting
South, main season	Oct-Nov	Mar-Apr
Northeast, main season	Mar-May	Aug-Nov
North, main season	Nov-Dec	Apr-Jun

Brazil extends from 5° N latitude at its frontier with Guyana and Venezuela to as far south as 33° S latitude in the state of Rio Grande do Sul. The longitude limits are from 30° to 74° W.

The country lies within several AEZs: AEZ 1, warm arid and semiarid tropics; AEZ 2, warm subhumid tropics; AEZ 3, warm humid tropics; AEZ 4, cool tropics; and AEZ 8, cool subtropics with summer rainfall. The climate varies from tropical to subtropical, the latter mainly along the southern coast.

Brazil is considered an upper middle income economy. Its per capita gross national income in 2000 was about US$7,320 in purchasing power parity (PPP) terms (PPP adjusts income measures for local price levels), slightly more than one fifth of the level in the U.S. More than 40% of the population in 1995 was below a poverty line of $2 a day (again in PPP terms).

Brazil's population in 1999 was 168 million, making it the fifth most populous country in the world. The current population growth rate is about 1.3% annually, compared with 3% in the early 1960s. Urbanization has proceeded rapidly in the past 40 years, and less than 20% of the people now live in rural areas. Most of these people are employed in agriculture, which accounts for 17% of the economically active population today vis-à-vis 54% in 1961. The agricultural sector contributed 8% of GDP in 1998.

Recent developments in the rice sector

Rice per capita consumption in Brazil was 40.3 kg in 1999, which is slightly less than per capita consumption of cassava. Rice supplied 14% of the calories and 10% of the protein in the diet.

In terms of area harvested, rice is the fifth most important crop in the country, behind soybeans, maize, sugarcane, and dry beans. Coffee and oranges are also important crops, especially in terms of value.

During the past twenty years, the rice area harvested declined steadily from more than 6 million ha in the early 1980s to about 3 million ha in 1998. It then partially recovered to about 3.7 million ha in the subsequent two years. All the decline in area harvested has come from upland rice systems, which accounted for about 80% of total rice area in the mid-1980s. At least part of this reduction in upland rice area was due to increased interest in soybean production in the central part of the country.

National average yields were stagnant from the early 1960s until the mid-1980s, after which they began to grow rapidly from 1.5 t/ha in 1983 to 3.0 t/ha in 2000 (an annual average growth rate of more than 4%). This increase in yields was driven largely by the reduction in upland rice area. Since upland rice has much lower yields than irrigated rice, an increased share of irrigated rice in total area leads to higher national average yields. These opposing trends have largely negated one another in recent years, and annual paddy rice production in the 1990s averaged about 9.5 million t (albeit with substantial fluctuations from year to year). This stagnation in production changed Brazil from a net rice exporter in the 1970s to a consistent net importer in the 1990s.

Rice environments

Brazil is the only country in the world where irrigated and upland rice have similar importance. Upland rice presently accounts for about two-thirds of total rice area (1998-99), but, because yields are much higher in the irrigated system, irrigated rice accounted for about 60% of production in that year.

Irrigated areas are largely planted to modern varieties. Most irrigated rice is concentrated in the two southernmost states of Rio Grande do Sul and Santa Catarina. In 1998-99, irrigated rice achieved an average yield of 5.7 t/ha compared with a national average upland yield of less than 2 t/ha. In Rio Grande do Sul, which accounts for about 75% of the irrigated rice area in Brazil, big farms of about 200 ha are the norm. The cultivation system is highly mechanized, with soil preparation and sowing done in dry soil. However, minimum tillage systems also occupy large areas, and sowing of pregerminated seeds in puddled soil is also gaining in importance. These two systems are alternatives to the conventional system and allow for better control of weeds and red rice. Sowing of pregerminated seeds is dominant in Santa Catarina, where farm sizes are much smaller (average size is about 10 ha).

Much of Brazil's upland rice is planted on the Cerrado in central Brazil. Originally, rice was grown in this area primarily to facilitate deforestation. Now, it is typically done in rotation with pasture (two years of rice, three years of pasture) or soybean. Rotation is essential because of the "yield collapse" that occurs with continuous cultivation of such rice. In the second year of rice cultivation, yields typically decline by about 10%, but in the third year they can fall by as much as 70%. The cause of this problem has been variously attributed to autotoxicity, soil degradation, and soil pests. Thus, crop rotation is essential for profitable farm management.

Production constraints

Most areas planted continuously with rice are infested with weedy rice (red rice) to some degree, significantly affecting yields over time. The sowing of pregerminated seed or the use of no-till systems is helping to reduce the importance of this problem. Another production problem is the occurrence of low temperatures during flowering, primarily in the southern areas of Rio Grande do Sul and Santa Catarina, where irrigated rice is common. This problem can be exacerbated when farmers delay planting because of a rainy spring. Weeds and insect pests (fall armyworm, rice water weevil, rice stem bug) also act as constraints to increased production.

The most important production constraint for upland rice is the abovementioned yield collapse. Because of the fragile structure of the Cerrado soil, soil erosion caused by wind and rain is also an important problem.

Production opportunities

Rice research is organized at both the federal and state levels. Research at the federal level is coordinated by the Empresa Brasileira de Pesquisa Agropecuária (EMBRAPA). The Centro Nacional de Pesquisa de Arroz e Feijão (CNPAF), located in the central region, conducts research on

Irrigated rice research.

all aspects of upland rice and also develops technology for the small irrigated rice areas of central and northern Brazil. The rice program of the Centro de Pesquisa Agropecuária de Terras Baixas de Clima Temperado, also part of the EMBRAPA system, conducts research on all aspects of irrigated rice for the southern states of Rio Grande do Sul and Santa Catarina. At the state level, rice research is organized under state agricultural research organizations. All rice research resources are linked through an EMBRAPA-sponsored national rice network.

The ratio of agricultural research and development expenditures to agricultural GDP in Brazil is high by Latin American standards and approaches that in high-income countries. However, since the mid-1990s, EMBRAPA's funding has declined substantially. The private sector has recently become interested in rice research, both as a donor and as a direct participant, opening up a new source of research resources. However, as of 1996, the private sector accounted for just 2% of agricultural research and development expenditures.

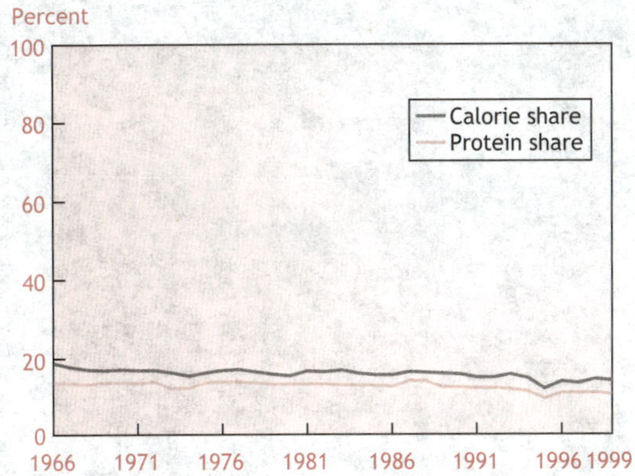

Share of calories and protein from rice, 1966-99

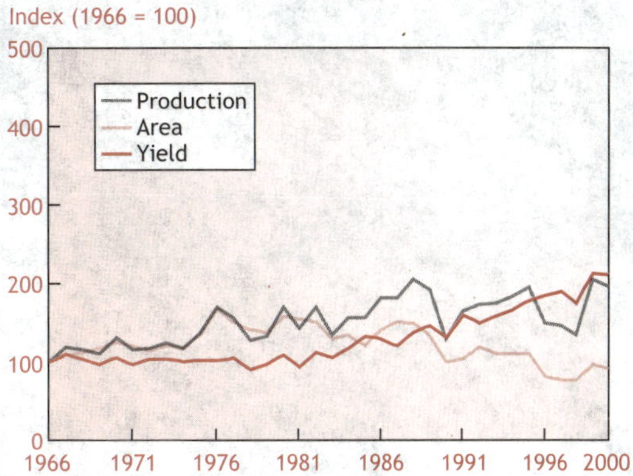

Indices of rice production, area, and yield, 1966-2000

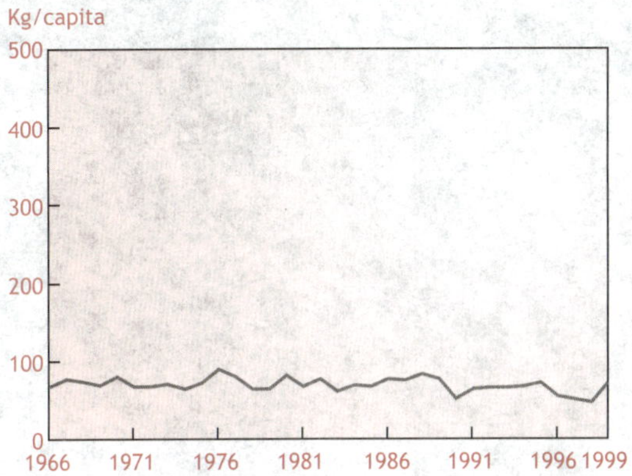

Per capita production (paddy terms), 1966-99

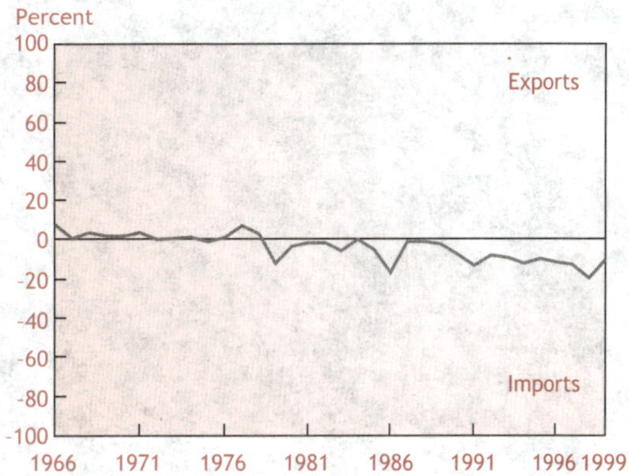

Net trade status, 1966-99

Basic statistics, Brazil

	1985	1990	1995	1998	1999	2000
Rice						
Area harvested (ha)	4,754,692	3,946,690	4,373,540	3,062,200	3,840,040	3,672,090
Yield (t/ha)	1.9	1.9	2.6	2.5	3.1	3.0
Production (t)	9,024,555	7,420,931	11,226,064	7,716,090	11,782,700	11,168,300
Rice imports (t)	339,461	413,825	870,506	1,304,958	984,265	na
Paddy imports (t)	160	70,676	357,658	595,268	637,577	na
Rice exports (t)	3,885	1,248	18,537	6,566	47,639	na
Paddy exports (t)	58	70	79	133	81	na
Others						
Population, total ($\times 10^3$)	135,224	147,940	159,346	165,851	167,988	na
Population, agriculture ($\times 10^3$)	38,744	33,568	30,833	29,167	28,617	na
Agricultural area ($\times 10^3$ ha)	230,081	240,800	250,500	250,200	na	na
Irrigated agricultural area ($\times 10^3$)	2,100	2,700	2,656	2,656	na	na
Total fertilizer consumption (t)	3,197,247	3,164,097	4,205,900	5,736,500	na	na
Tractors used in agric. (no.)	666,309	720,000	790,000	806,000	na	na

Source: FAOSTAT online database.

Rice in 54 other countries

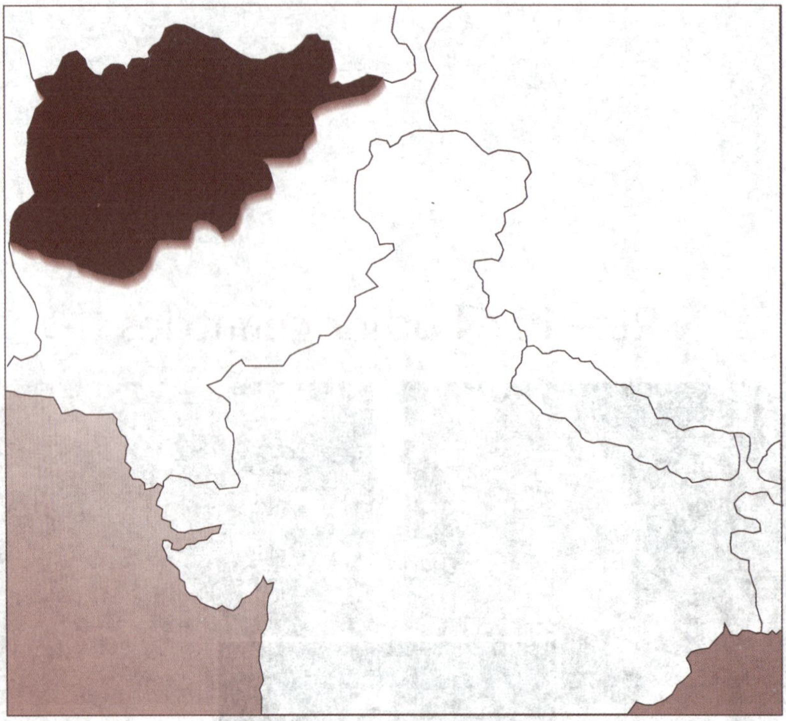

AFGHANISTAN is a landlocked country surrounded by Iran, Pakistan, and the Central Asian Republics. It consists of highland plains separated by mountains. Afghanistan is largely dry and rocky, with some fertile lowlands where rice and other grains, as well as cotton and fruits, are grown. Agricultural activities make up more than half the GDP, although only about 12% of the country is arable. Deforestation has increased rapidly, leading to desertification; potable water is scarce. The main population centers are in eastern valleys.

General information

- Main food consumed: wheat, maize, rice, oil and fat, and meat
- Rice consumption, 1999: 16.7 kg milled rice per person per year

Production season

	Planting	Harvesting
Main season	Apr-Jun	Oct-Nov

Production constraints

Rice production in the country has suffered from the unstable security conditions. Rice yields are still low. If security conditions improve, rice production could be substantially increased by

- Introducing/developing and adopting high-yielding varieties, which are tolerant of low temperature and blast
- Improving input and credit supply systems
- Strengthening research and extension services

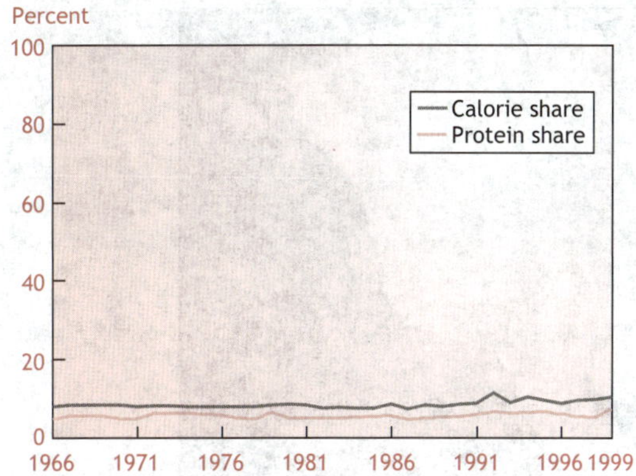

Share of calories and protein from rice, 1966-99

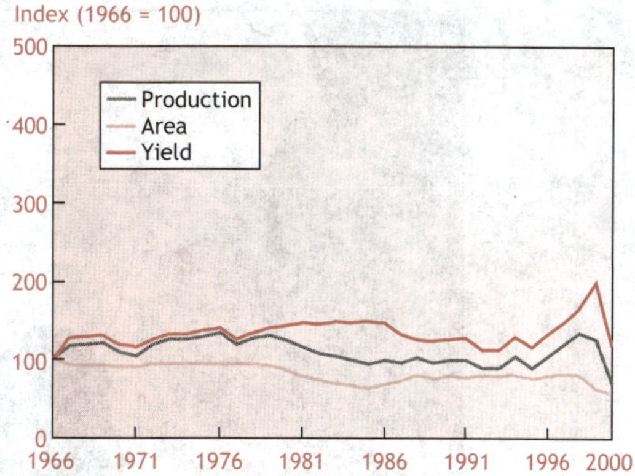

Indices of rice production, area, and yield, 1966-2000

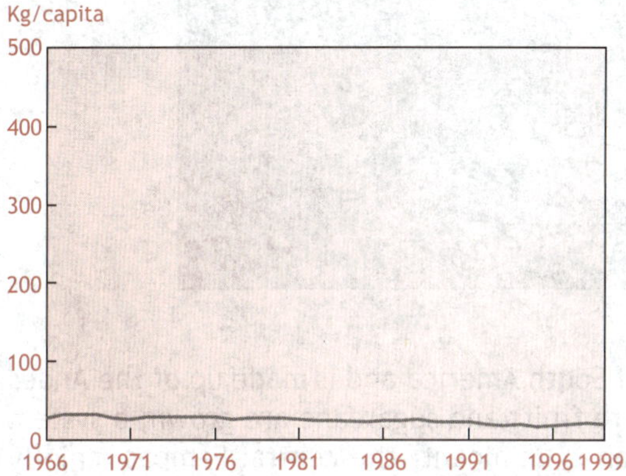

Per capita production (paddy terms), 1966-99

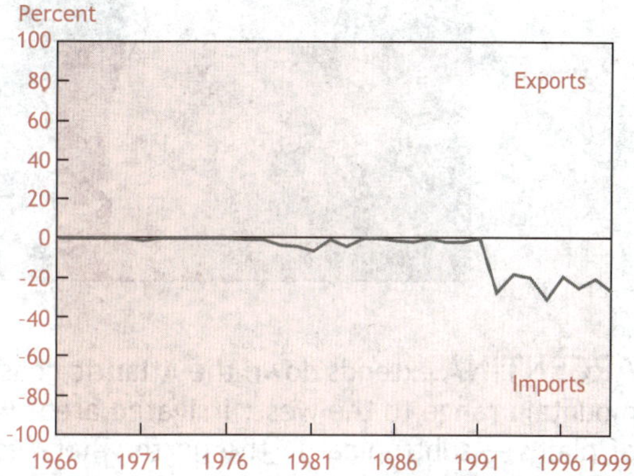

Net trade status, 1966-99

Basic statistics, Afghanistan

	1985	1990	1995	1998	1999	2000
Rice						
Area harvested (ha)	141,000	175,000	170,000	180,000	140,000	130,000
Yield (t/ha)	2.2	1.9	1.8	2.5	3.0	1.8
Production (t)	317,000	333,000	300,000	450,000	420,000	232,800
Rice imports (t)	na	na	na	78,000	106,000	na
Others						
Population, total ($\times 10^3$)	14,519	14,755	19,663	21,354	21,923	na
Population, agriculture ($\times 10^3$)	10,376	10,376	13,507	14,451	14,762	na
Agricultural area ($\times 10^3$ ha)	38,054	38,054	38,054	38,054	na	na
Irrigated agricultural area ($\times 10^3$ ha)	2,586	2,500	2,386	2,386	na	na
Total fertilizer consumption (t)	72,962	44,500	0	5,000	na	na
Tractors used in agric. (no.)	820	850	840	840	na	na

Source: FAOSTAT online database.

ARGENTINA extends down the Atlantic coast of South America and is made up of the Andes mountain range to the west; irrigated areas where fruits and sugarcane are grown; a system of plains—subtropical in the north where cotton is dominant; the central Pampas, chiefly used for grazing; and the cold and arid Patagonian plateau in the south. Arable land is 9% of the total land area, and 7% of that is irrigated. Agriculture makes up 7% of the GDP. The population in 1999 was about 37 million.

General information

- GNI per capita PPP$, 2000: 12,050
- Internal renewable water resources: 694 km^3
- Incoming water flow: 300 km^3
- Main food consumed: meat, oil and fat, sugar and honey, milk, wheat
- Rice consumption, 1999: 3.8 kg milled rice per person per year

Production constraints

The following are constraints to sustainable rice production in Argentina:

- High production costs
- Dependence on external markets for marketing of rice products
- Straighthead, especially on sandy soils in Corrientes
- Weed infestation (*Echinochloa* sp. and *Brachiaria* sp.)
- Low temperature during early and/or reproductive phases
- Yield potential of rice varieties has reached a plateau

Production season

	Planting	Harvesting
Main season	Oct-Nov	Mar-Apr

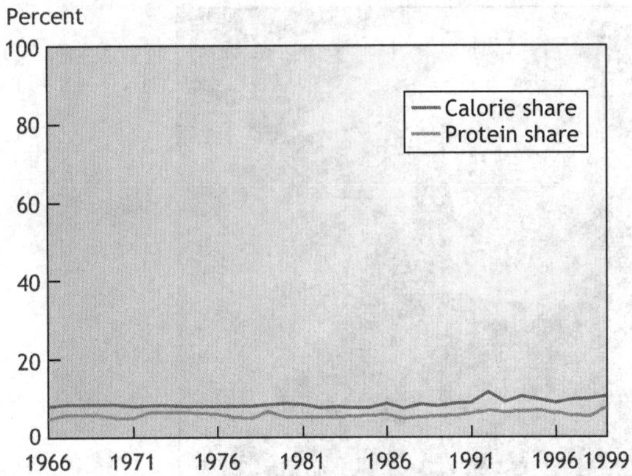

Share of calories and protein from rice, 1966-99

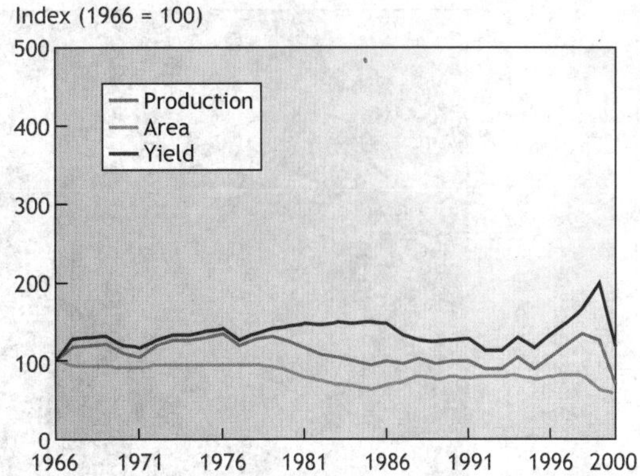

Indices of rice production, area, and yield, 1966-2000

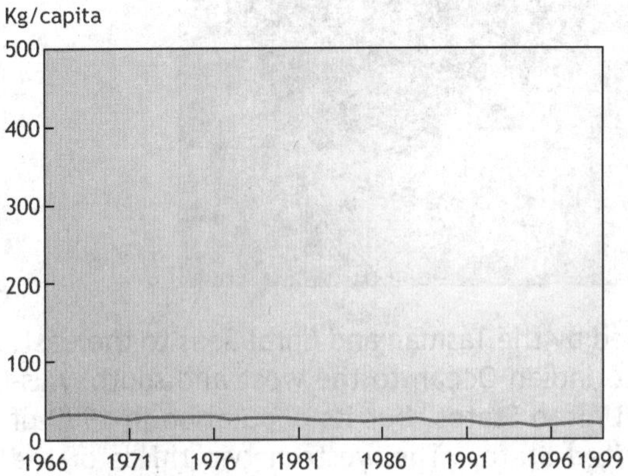

Per capita production (paddy terms), 1966-99

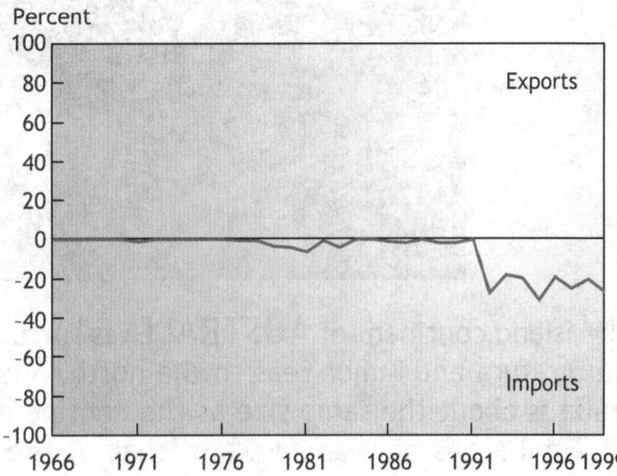

Net trade status, 1966-99

Basic statistics, Afghanistan

	1985	1990	1995	1998	1999	2000
Rice						
Area harvested (ha)	141,000	175,000	170,000	180,000	140,000	130,000
Yield (t/ha)	2.2	1.9	1.8	2.5	3.0	1.8
Production (t)	317,000	333,000	300,000	450,000	420,000	232,800
Rice imports (t)	na	na	na	78,000	106,000	na
Others						
Population, total ($\times 10^3$)	14,519	14,755	19,663	21,354	21,923	na
Population, agriculture ($\times 10^3$)	10,376	10,376	13,507	14,451	14,762	na
Agricultural area ($\times 10^3$ ha)	38,054	38,054	38,054	38,054	na	na
Irrigated agricultural area ($\times 10^3$ ha)	2,586	2,500	2,386	2,386	na	na
Total fertilizer consumption (t)	72,962	44,500	0	5,000	na	na
Tractors used in agric. (no.)	820	850	840	840	na	na

Source: FAOSTAT online database.

The island continent of **AUSTRALIA** is surrounded by the Tasman and Coral Seas to the east, the Arafura and Timor Seas to the north, and the Indian Ocean to the west and south. Australia is about the same size as the continental United States, but its population in 1999 of 18.7 million is only about 7% of that of the United States. The western two-thirds of the continent ranges from hyperarid to semiarid. The majority of the population lives along the relatively well-watered eastern and southeastern coasts.

The climate ranges from the warm humid tropics to cool subtropics. Agriculture contributes about 4% to GDP. The main feature of Australian agriculture is the dominance of large-scale dryland farming and grazing systems. Most irrigated agriculture is found in the Murray-Darling Basin in New South Wales. Irrigated agriculture represents 16% of the value of agricultural production; about 5% of the value of irrigated agriculture is rice.

General information

- GNI per capita PPP$, 2000: 24,970
- Internal renewable water resources: 343 km^3
- Main food consumed: meat, wheat, sugar and honey, oil and fat, milk
- Rice consumption, 1999: 7.9 kg milled rice per person per year

Production season

	Planting	Harvesting
Main season, New South Wales	Oct	Mar-Apr
Main season, West Australia	Nov	Apr-May

Production constraints

Rice yield and production have increased substantially since 1990. The national yield in 2000 was the world's highest. The following are the major constraints to sustainable rice production in Australia:

- Stagnation of the yield potential of rice varieties
- Limited water supply for irrigation
- Increasing salinity of rice soils
- Low temperature
- High production costs
- Increasing concern about environmental degradation

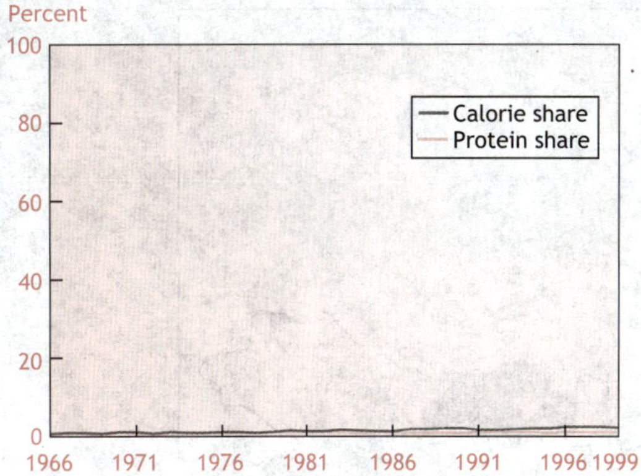

Share of calories and protein from rice, 1966-99

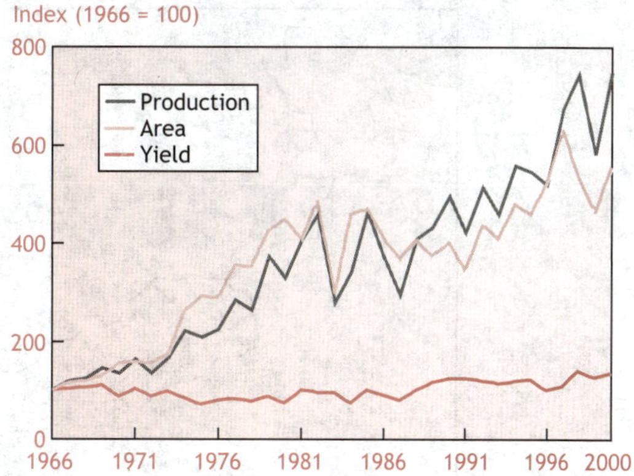

Indices of rice production, area, and yield, 1966-2000

The scale on this graph is not comparable with the scale on production, area, and yield graphs for most other countries.

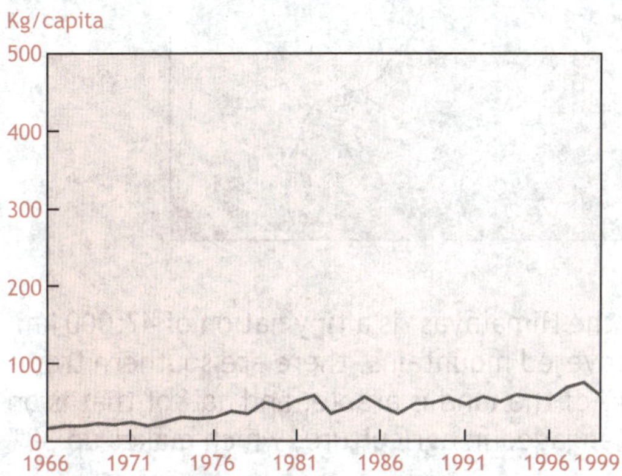

Per capita production (paddy terms), 1966-99

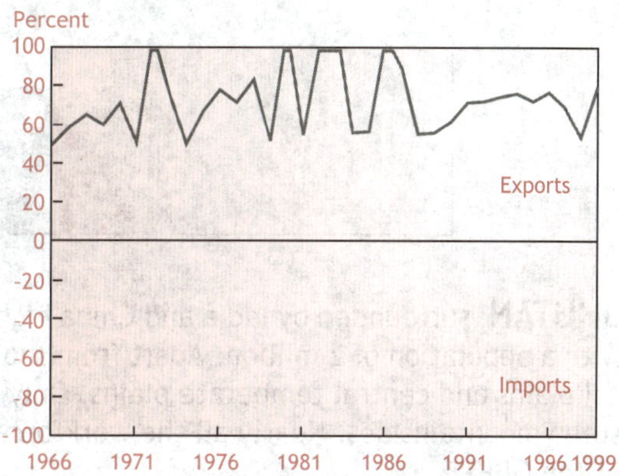

Net trade status, 1966-99

Basic statistics, Australia

	1985	1990	1995	1998	1999	2000
Rice						
Area harvested (ha)	122,049	104,545	118,893	140,000	120,000	145,000
Yield (t/ha)	7.1	8.8	8.5	9.9	9.0	9.7
Production (t)	864,000	924,000	1,015,830	1,390,000	1,084,000	1,400,000
Rice imports (t)	9,562	27,164	31,000	37,156	49,914	na
Paddy imports (t)	1	673	154	40	11	na
Rice exports (t)	341,411	424,287	541,848	551,775	668,591	na
Paddy exports (t)	0	24	431	25,036	71,426	na
Others						
Population, total ($\times10^3$)	15,637	16,884	17,942	18,517	18,701	na
Population, agriculture ($\times10^3$)	937	931	898	878	871	na
Agricultural area ($\times10^3$ ha)	472,960	464,481	463,348	472,000	na	na
Irrigated agricultural area ($\times10^3$ ha)	1,700	1,832	2,400	2,400	na	na
Total fertilizer consumption (t)	1,155,000	1,163,700	1,867,400	2,110,000	na	na
Tractors used in agric. (no.)	322,000	317,000	315,000	315,000	na	na

Source: FAOSTAT online database.

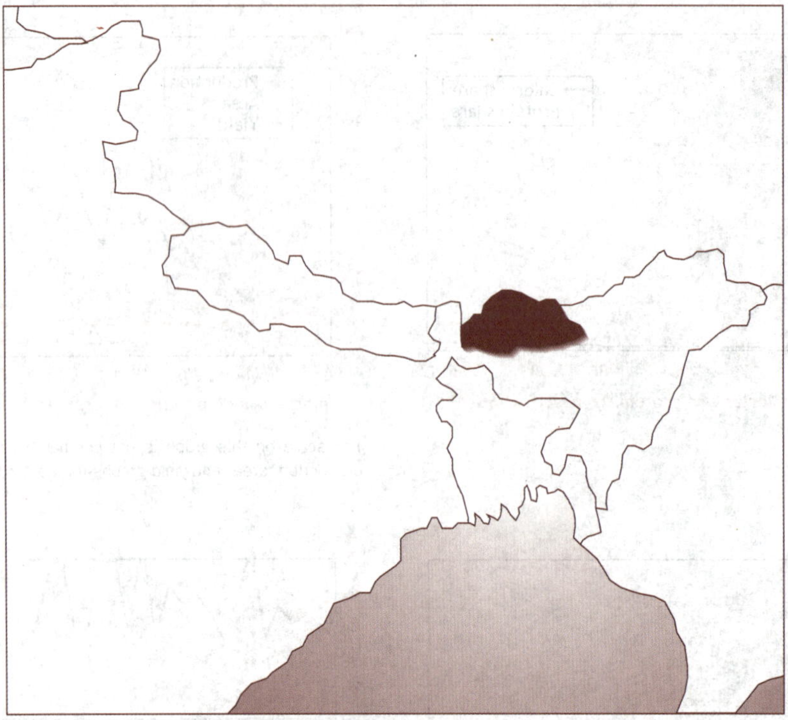

BHUTAN, surrounded by India and China high in the Himalayas, is a tiny nation of 47,000 km² with a population of 2 million. Apart from snow-covered mountains, there are southern tropical plains and central temperate plains. Only 2.4% of the land is arable, and half of that is on steep mountainsides. Nearly all the workforce is engaged in agriculture, which makes up 38% of the GDP.

General information

- GNI per capita PPP$, 2000: 1,440
- Internal renewable water resources: 95 km^3

Production constraints

Both the harvested area and yield of rice have remained more or less unchanged, indicating an urgent need for the creation of dynamic rice research and development programs. There are several constraints to sustainable rice production in Bhutan:

- Low temperature at high altitude
- Most rice soils are sandy loams or loams with low fertility
- Weeds, especially *Potamogeton distinctus*; insects; and diseases, especially blast
- Rural poverty
- Inadequate infrastructure and inadequate credit and input supply
- Poor transportation and distribution systems

Production season

	Planting	Harvesting
Main season	Apr-Jun	Aug-Nov

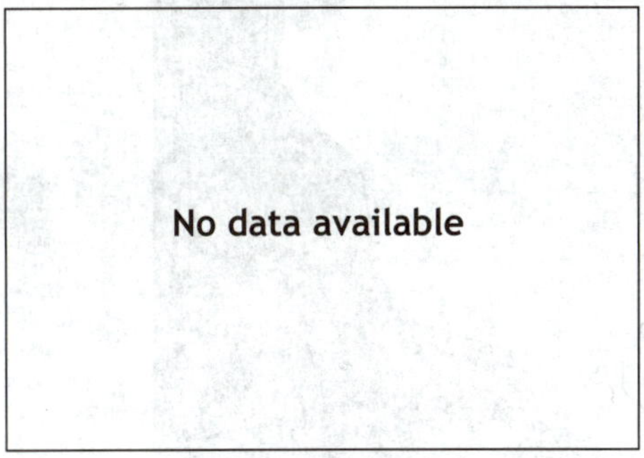

Share of calories and protein from rice, 1966-99

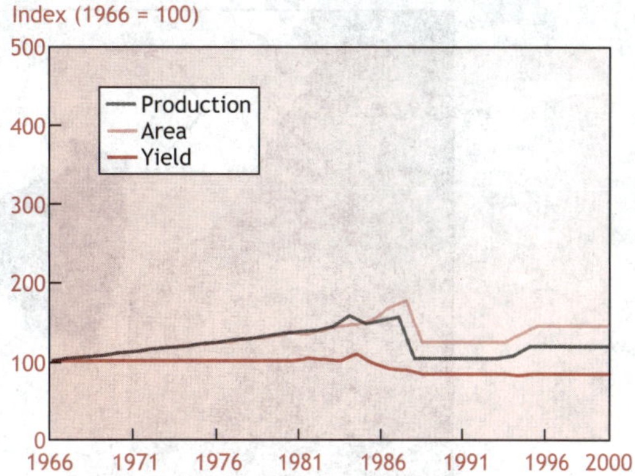

Indices of rice production, area, and yield, 1966-2000

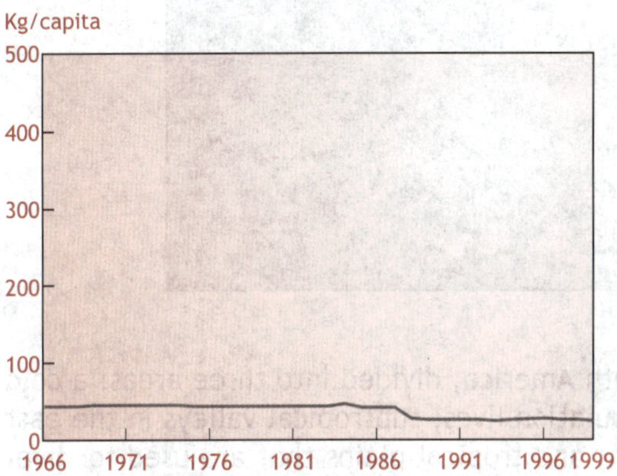

Per capita production (paddy terms), 1966-99

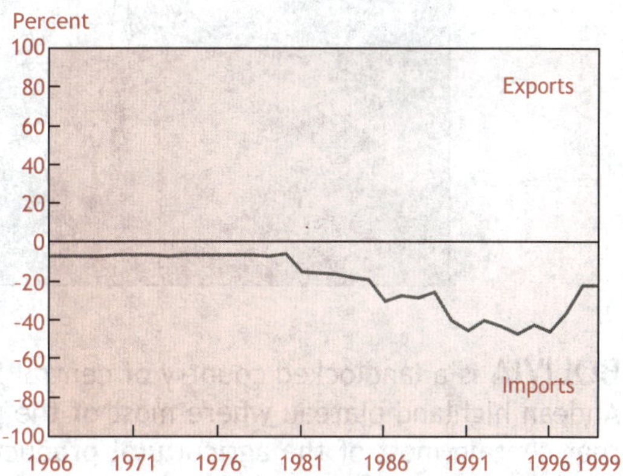

Net trade status, 1966-99

Basic statistics, Bhutan

	1985	1990	1995	1998	1999	2000
Rice						
Area harvested (ha)	32,000	26,000	30,000	30,000	30,000	30,000
Yield (t/ha)	1.9	1.7	1.7	1.7	1.7	1.7
Production (t)	62,000	43,000	50,000	50,000	50,000	50,000
Rice imports (t)	10,000	19,433	24,944	9,700	9,600	na
Rice exports (t)	0	0	5	25	40	na
Others						
Population, total ($\times 10^3$)	1,486	1,696	1,847	2,004	2,064	na
Population, agriculture ($\times 10^3$)	1,400	1,597	1,735	1,880	1,936	na
Agricultural area ($\times 10^3$ ha)	396	403	423	460	na	na
Irrigated agricultural area ($\times 10^3$ ha)	30	39	39	40	na	na
Total fertilizer consumption (t)	100	100	100	100	na	na

Source: FAOSTAT online database.

BOLIVIA is a landlocked country of central South America, divided into three areas: a cold Andean highland plateau where most of the population lives; subtropical valleys in the east that contain most of the agricultural production; and tropical plains that are used for livestock and farming of grains, including rice. Agriculture occupies nearly half the workforce.

General information

- GNI per capita PPP$, 2000: 2,360
- Internal renewable water resources: 300 km^3
- Main food consumed: wheat, sugar and honey, maize, roots and tubers, rice
- Rice consumption, 1999: 14.4 kg milled rice per person per year

Production constraints

There are several constraints to sustainable rice production in Bolivia:

- Drought stress caused by unreliable rainfall, aluminum toxicity, and soil fertility losses in dominant upland systems
- Blast and rice hoja blanca virus
- Weed competition
- Inadequate knowledge and equipment for land preparation for small rice farmers in Santa Cruz, where more than 50% of the rice area is found
- Lack of credit and input supplies to small farmers, especially in Santa Cruz

Production season

	Planting	Harvesting
Main season	Oct-Nov	Feb-Mar

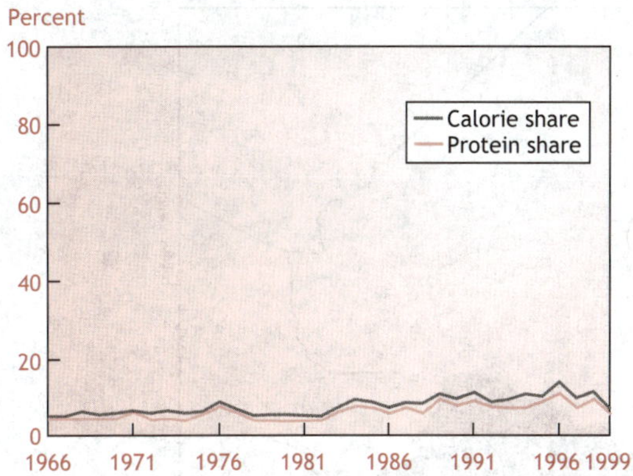

Share of calories and protein from rice, 1966-99

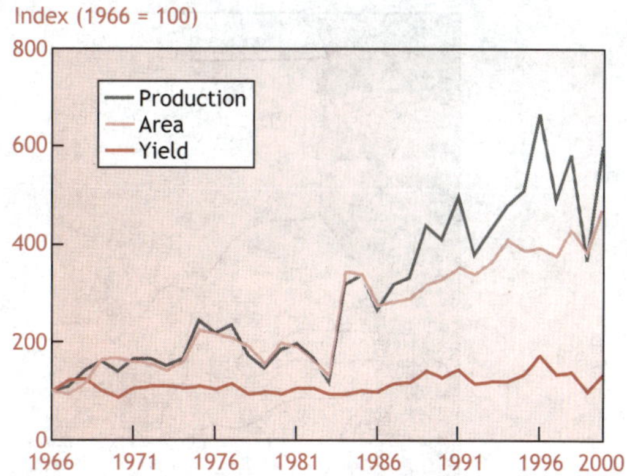

Indices of rice production, area, and yield, 1966-2000

The scale on this graph is not comparable with the scale on production, area, and yield graphs for most other countries.

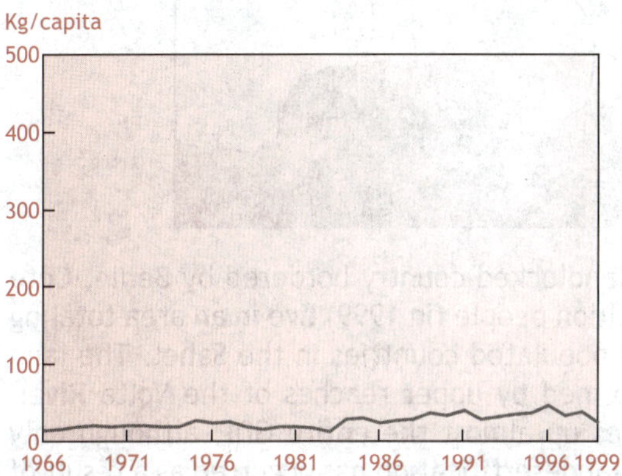

Per capita production (paddy terms), 1966-99

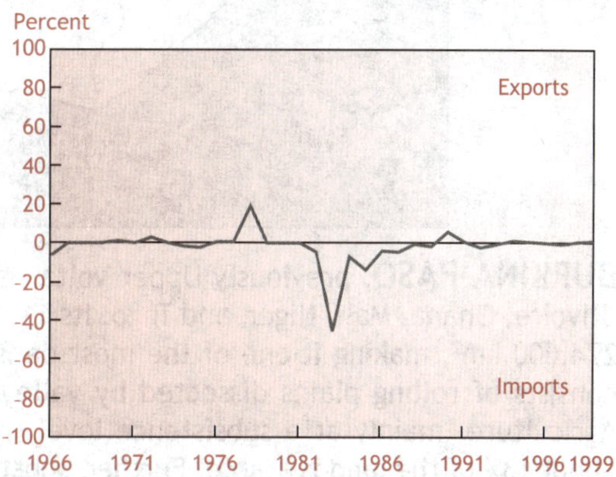

Net trade status, 1966-99

Basic statistics, Bolivia

	1985	1990	1995	1998	1999	2000
Rice						
Area harvested (ha)	112,792	109,381	129,569	143,257	127,740	161,175
Yield (t/ha)	1.5	1.9	2.0	2.1	1.5	1.9
Production (t)	173,156	211,263	263,285	301,341	189,388	310,099
Rice imports (t)	19,000	1,329	274	275	275	na
Paddy imports (t)	0	5	203	na	na	na
Rice exports (t)	na	na	na	1,350	1,350	na
Paddy exports (t)	na	na	na	55	55	na
Others						
Population, total ($\times 10^3$)	5,895	6,573	7,414	7,957	8,142	na
Population, agriculture ($\times 10^3$)	2,833	2,931	3,246	3,445	3,512	na
Agricultural area ($\times 10^3$ ha)	34,697	35,321	35,665	36,034	na	na
Irrigated agricultural area ($\times 10^3$ ha)	125	125	128	128	na	na
Total fertilizer consumption (t)	5,800	5,162	6,769	7,444	na	na
Tractors used in agric. (no.)	4,750	5,200	5,500	5,700	na	na

Source: FAOSTAT online database.

BURKINA FASO, previously Upper Volta, is a landlocked country bordered by Benin, Côte d'Ivoire, Ghana, Mali, Niger, and Togo. Its 11.6 million people (in 1999) live in an area totaling 274,000 km^2, making it one of the most densely populated countries in the Sahel. The land consists of rolling plains dissected by valleys formed by upper reaches of the Volta River. Agriculture, mainly at a subsistence level, makes up almost the entire GDP, although only about 13% of the land is arable. Further, substantial desertification has occurred as a result of previous inappropriate farming techniques used for peanuts and cotton for export. Deforestation has also been rampant, resulting in a lack of fuel wood.

General information

- GNI per capita PPP$, 2000: 970
- Internal renewable water resources: 17.5 km^3
- Main food consumed: millet and sorghum, maize, nuts, oil and fat, meat
- Rice consumption, 1999: 18.5 kg milled rice per person per year

Production constraints

Rice yield has stagnated since 1995. In some irrigated areas, a yield decrease has been reported. The following are the major constraints to sustainable rice production in Burkina Faso:

- Drought in upland areas and drought and flash floods in rainfed lowland (or inland swamp) areas because of irregular weather
- Heat-induced sterility during the off-season in irrigated areas
- Cold-induced sterility in December
- Dominant use of NPK fertilizer formulated for cotton and not adapted to the rice crop
- Inadequate and irregular input supplies: seeds, fertilizer, and credit
- Lack of small farm equipment, especially for postharvest operations
- Lack of effective farmer organizations and cooperatives
- Poor maintenance of irrigation facilities
- Lack of a well-defined rice policy
- Poor road networks and marketing systems
- Labor shortage because of competition from other agricultural activities
- Weak research and extension support

Production season

	Planting	Harvesting
Main season	May-Jun	Oct-Nov
Off-season	Jan-Feb	May-Jun

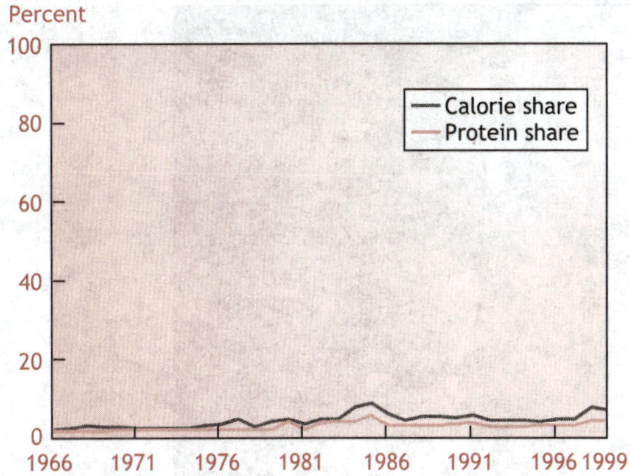

Share of calories and protein from rice, 1966-99

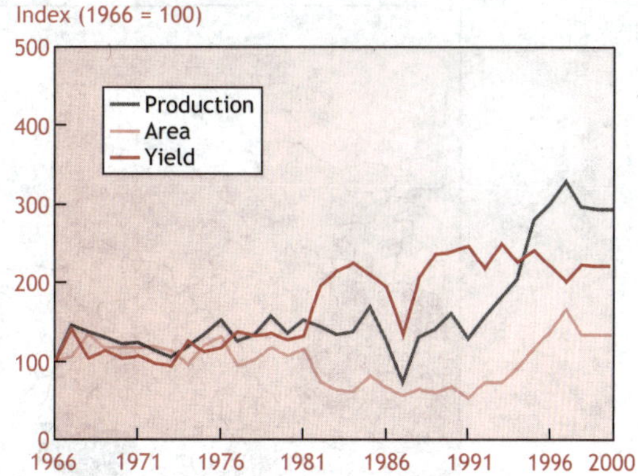

Indices of rice production, area, and yield, 1966-2000

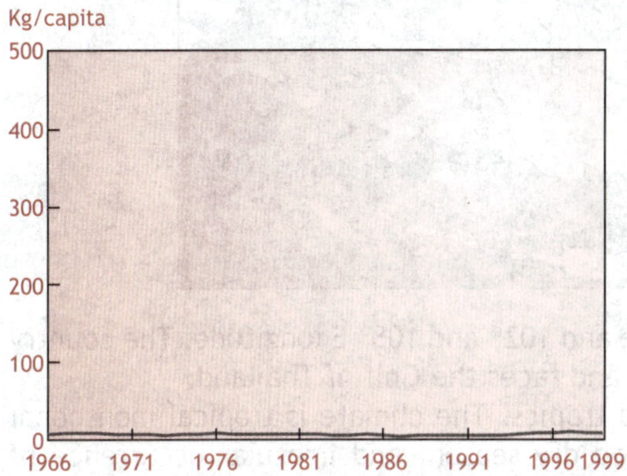

Per capita production (paddy terms), 1966-99

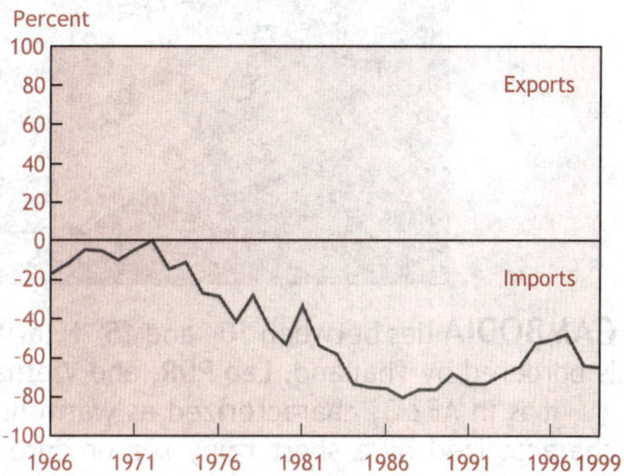

Net trade status, 1966-99

Basic statistics, Burkina Faso

	1985	1990	1995	1998	1999	2000
Rice						
Area harvested (ha)	27,741	23,000	40,000	45,904	46,000	46,000
Yield (t/ha)	1.8	2.1	2.1	1.9	1.9	1.9
Production (t)	50,334	47,800	84,025	88,998	88,000	88,000
Rice imports (t)	104,022	69,654	62,300	107,972	107,972	na
Others						
Population, total ($\times 10^3$)	7,879	9,060	10,415	11,305	11,616	na
Population, agriculture ($\times 10^3$)	7,271	8,370	9,612	10,428	10,713	na
Agricultural area ($\times 10^3$ ha)	9,035	9,575	9,450	9,450	na	na
Irrigated agricultural area ($\times 10^3$ ha)	12	20	25	25	na	na
Total fertilizer consumption (t)	12,136	21,166	24,308	50,232	na	na
Tractors used in agric. (no.)	120	840	1,933	1,993	na	na

Source: FAOSTAT online database.

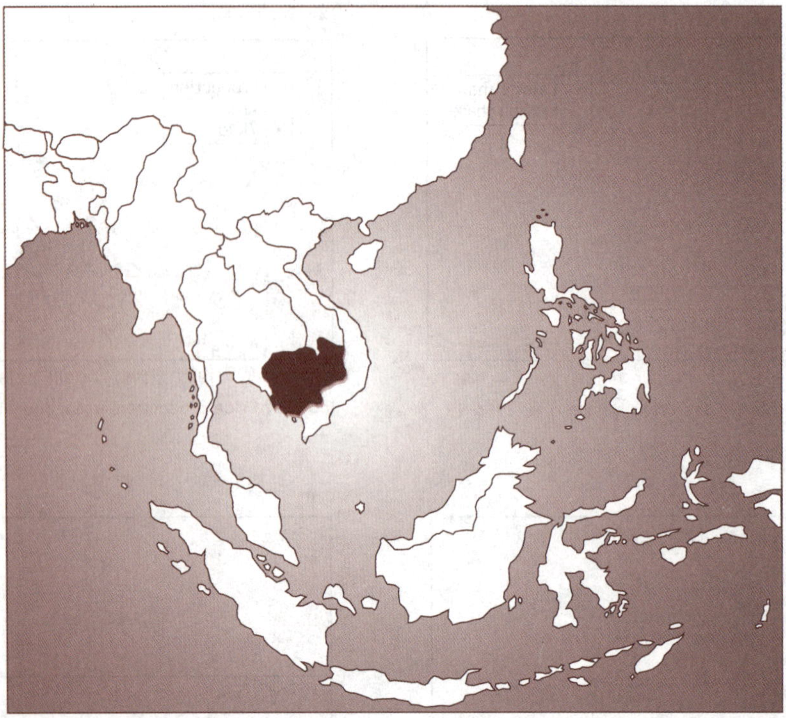

CAMBODIA lies between 10° and 15° N latitude and 102° and 108° E longitude. The country is bordered by Thailand, Lao PDR, and Vietnam, and faces the Gulf of Thailand.

It is in AEZ 3, characterized as warm humid tropics. The climate is tropical monsoonal characterized by a short rainy season, prolonged dry season, and irregular occurrence of rainfall both from year to year and within years. Most rain falls from May to mid-November. Often, a 10- to 15-day dry spell (called the short dry season) occurs in July or August.

The population in 2000 was 12 million. Eighty percent of the population is rural; agriculture makes up 43% of the GDP.

General information

- GNI per capita PPP$, 2000: 1,440
- Internal renewable water resources: 88.1 km^3
- Incoming water flow: 410 km^3
- Main food consumed: rice, meat, maize, fruits, roots and tubers
- Rice consumption, 1999: 165.2 kg milled rice per person per year

Production constraints

There are several constraints to sustainable rice production in Cambodia:

- Drought in rainfed areas, especially during the dry spell during July or August
- Flooding in low-lying areas because of torrential rains
- Sandy and acidic soils with low fertility
- Security situation and land mines
- Weeds, insects (stem borers and gall midge), and diseases
- Shortage of draft power during peak periods
- Rural poverty
- Inadequate infrastructure, inadequate credit and input supply: seed and fertilizers
- Price fluctuation

Production season

	Planting	Harvesting
Main season	Jun-Jul	Nov-Jan
Dry season	Nov-Jan	Apr-May

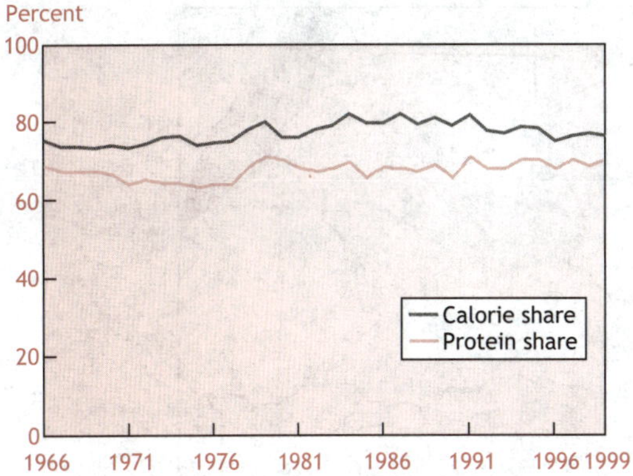

Share of calories and protein from rice, 1966-99

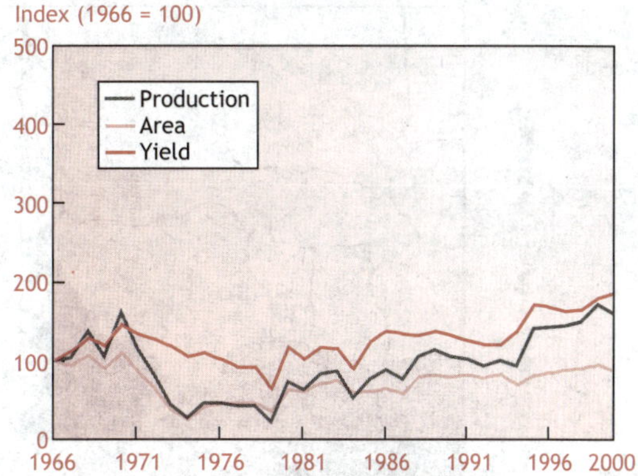

Indices of rice production, area, and yield, 1966-2000

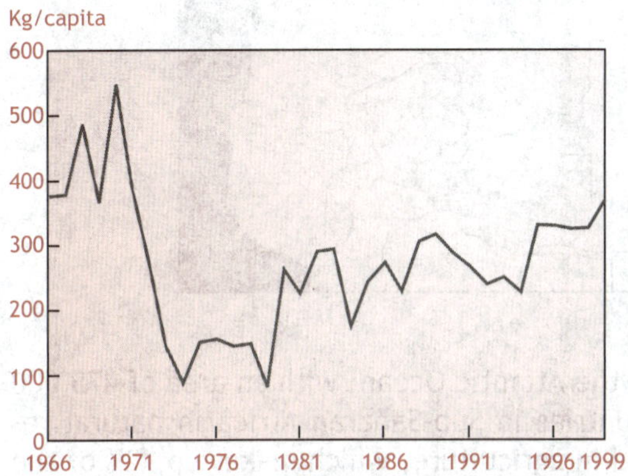

Per capita production (paddy terms), 1966-99

The scale on this graph is not comparable with the scale on per capita production graphs for most other countries.

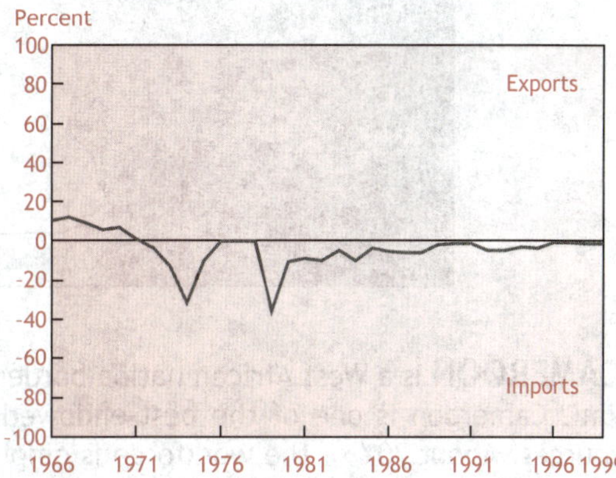

Net trade status, 1966-99

Basic statistics, Cambodia

	1985	1990	1995	1998	1999	2000
Rice						
Area harvested (ha)	1,345,000	1,740,000	1,782,000	1,961,000	2,079,000	1,873,000
Yield (t/ha)	1.3	1.4	1.9	1.8	1.9	2.0
Production (t)	1,812,000	2,500,000	3,318,000	3,515,000	4,041,000	3,762,000
Rice imports (t)	55,000	25,800	81,000	39,200	36,400	na
Rice exports (t)	3,935	0	0	600	2,200	na
Others						
Population, total ($\times10^3$)	7,385	8,652	9,982	10,716	10,945	na
Population, agriculture ($\times10^3$)	5,520	6,388	7,184	7,589	7,710	na
Agricultural area ($\times10^3$ ha)	3,170	5,349	5,307	5,307	na	na
Irrigated agricultural area ($\times10^3$ ha)	180	240	270	270	na	na
Total fertilizer consumption (t)	0	7,800	9,800	12,716	na	na
Tractors used in agric. (no.)	1,233	1,200	1,190	1,190	na	na

Source: FAOSTAT online database.

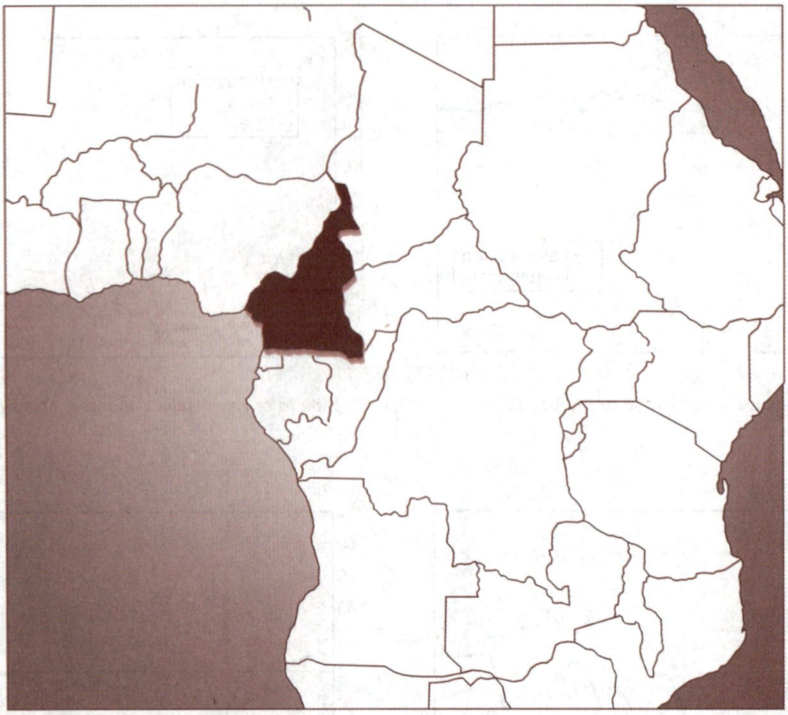

CAMEROON is a West African nation bordering the Atlantic Ocean, with an area of 475,000 km². Cameroon is one of the best-endowed countries in Sub-Saharan Africa in natural resources. About 70% of the workforce is employed in agriculture, which makes up 42% of the GDP. The country has three regions: northern plains, central grasslands, and rich farmlands in the south, where most of the population lives. Rice is mainly grown under irrigation. This area is also under threat of drought and desertification. About 12% of the land is arable.

General information

- GNI per capita PPP$, 2000: 1,590
- Internal renewable water resources: 280 km³
- Main food consumed: roots and tubers, maize, millet and sorghum, fruits, oil and fat
- Rice consumption, 1999: 6.9 kg milled rice per person per year

Production season

	Planting	Harvesting
Main season	May-Jun	Oct-Dec
Off-season	Jan-Feb	May-Jun

Production constraints

Rice yield during 1995-2000 was substantially lower than that during 1985-90. The following are the major constraints to sustainable rice production in Cameroon:

- Low temperature during the off-season in irrigated areas
- Poor marketing systems
- Deteriorating irrigation infrastructure
- Recently, lack of input supply and credit because of reorganization of the public sector
- Weak research support

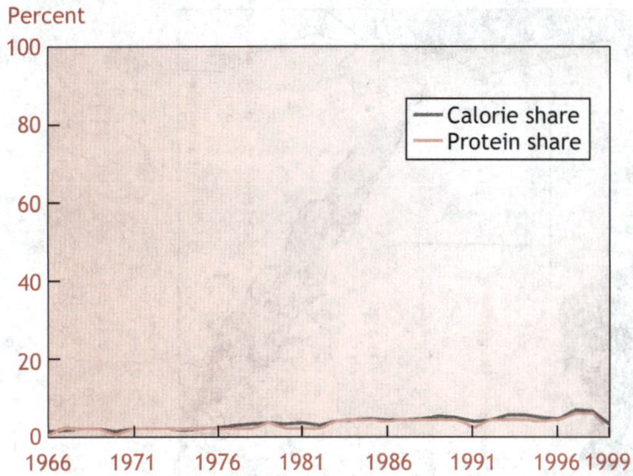

Share of calories and protein from rice, 1966-99

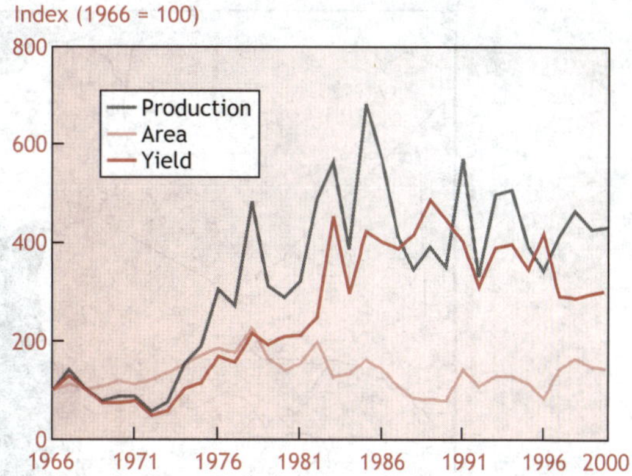

Indices of rice production, area, and yield, 1966-2000

The scale on this graph is not comparable with the scale on production, area, and yield graphs for most other countries.

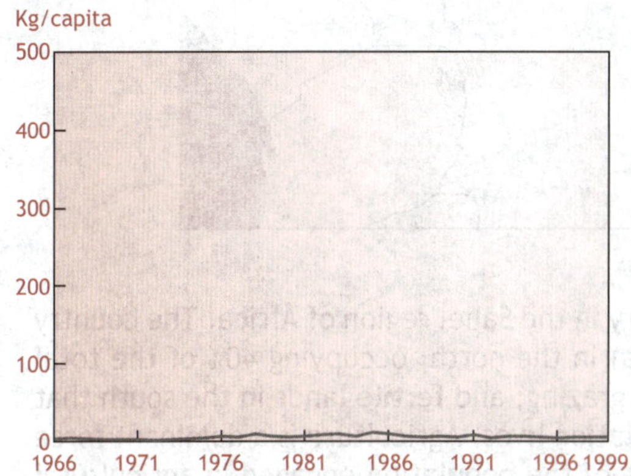

Per capita production (paddy terms), 1966-99

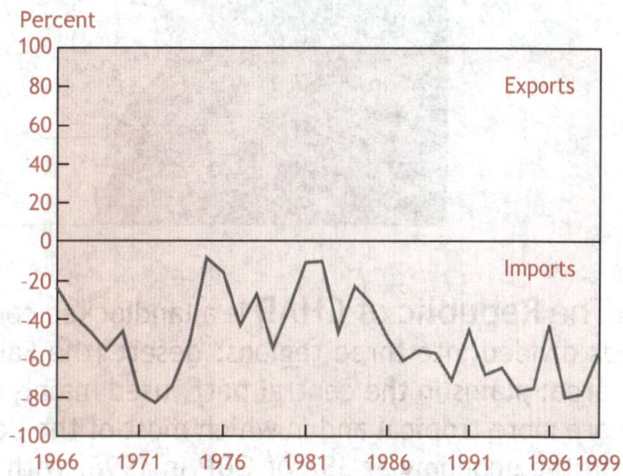

Net trade status, 1966-99

Basic statistics, Cameroon

	1985	1990	1995	1998	1999	2000
Rice						
Area harvested (ha)	22,670	11,040	16,000	22,943	20,387	20,000
Yield (t/ha)	4.7	5.0	3.9	3.2	3.3	3.4
Production (t)	107,399	55,199	62,000	73,288	67,470	68,000
Rice imports (t)	47,756	90,287	124,142	172,312	60,003	na
Rice exports (t)	9,778	311	95	na	na	na
Paddy exports (t)	117	5	0	na	na	na
Others						
Population, total ($\times10^3$)	9,970	11,472	13,182	14,305	14,693	na
Population, agriculture ($\times10^3$)	6,527	7,292	7,626	7,820	7,882	na
Agricultural area ($\times10^3$ ha)	9,160	9,170	9,160	9,160	na	na
Irrigated agricultural area ($\times10^3$ ha)	21	25	33	33	na	na
Total fertilizer consumption (t)	56,500	16,800	30,000	39,533	na	na
Tractors used in agric. (no.)	600	508	500	500	na	na

Source: FAOSTAT online database.

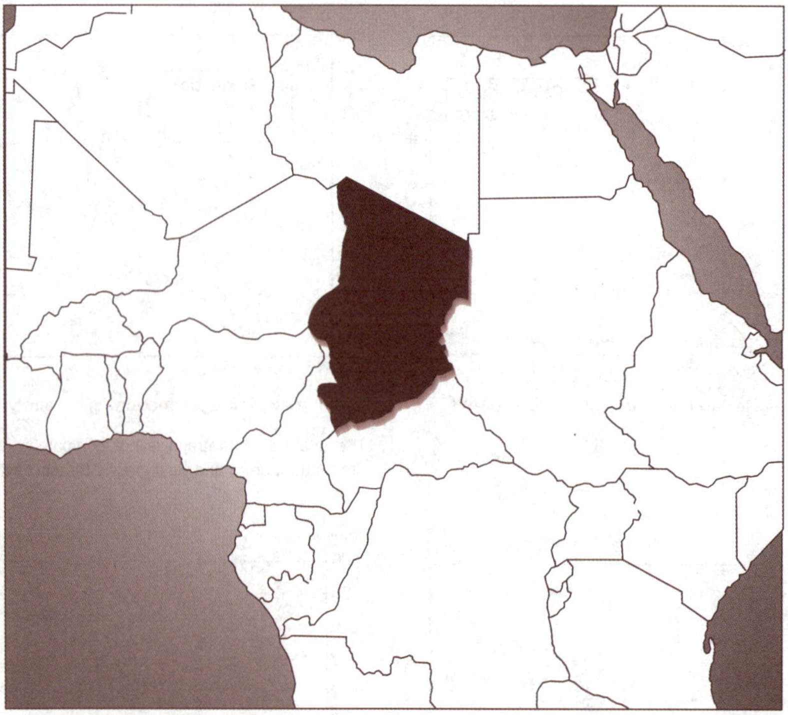

The **Republic of CHAD** is a landlocked country in the Sahel region of Africa. The country is divided into three regions: desert (the Sahara) in the north, occupying 40% of the total area; plains in the central part, used mainly for grazing; and fertile lands in the south that are more tropical and in which most of the population lives. Agriculture is a dominant force in the economy at 38% of GDP in 1998, with 85% of the population engaged in agricultural activities. Arable land, however, is only 3% of the total.

General information

- GNI per capita PPP$, 2000: 870
- Internal renewable water resources: 15 km^3
- Incoming water flow: 28 km^3
- Main food consumed: millet and sorghum, roots and tubers, nuts, oil and fat, sugar and honey
- Rice consumption, 1999: 10.6 kg milled rice per person per year

Production constraints

National rice yields during 1995-2000 were among the world's lowest. The following are the major constraints to sustainable rice production in Chad:

- Low temperature during the off-season in irrigated areas
- Recurrent drought and flood in rainfed lowland areas
- Lack of input supplies: improved seeds, fertilizer, and credit
- Lack of small farm equipment, especially for postharvest operations
- Lack of effective farmer organizations and cooperatives
- Poor maintenance of irrigation facilities
- Poor road networks and marketing systems
- Weak research and extension support

Production season

	Planting	Harvesting
Main season	Jun-Jul	Oct-Dec
Off-season	Jan-Feb	May-Jun

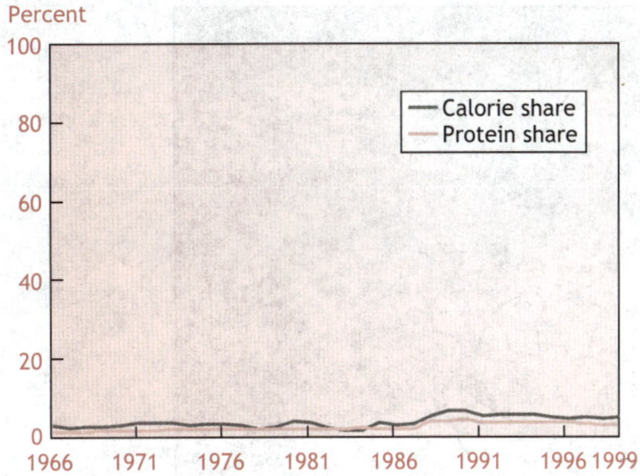

Share of calories and protein from rice, 1966-99

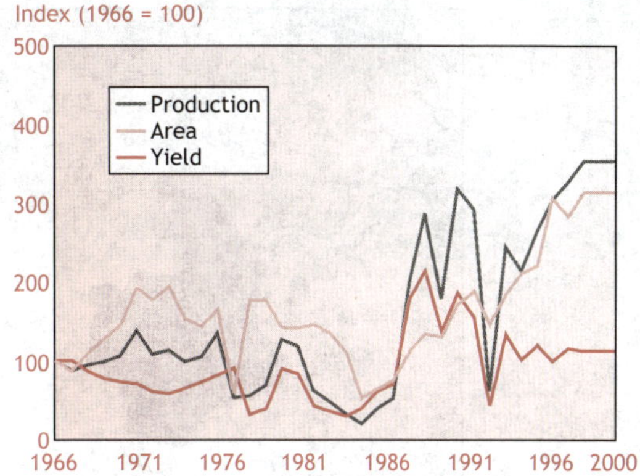

Indices of rice production, area, and yield, 1966-2000

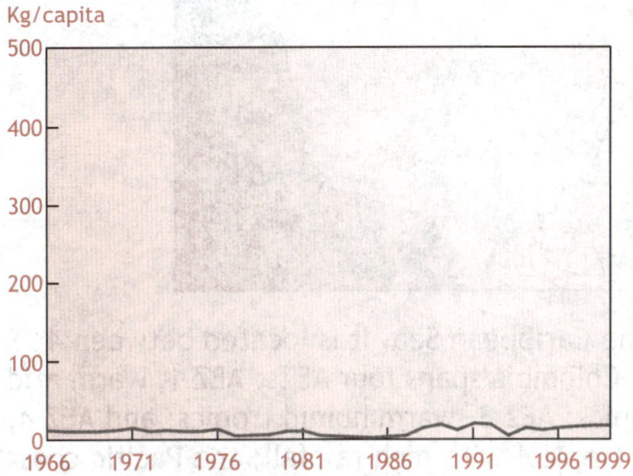

Per capita production (paddy terms), 1966-99

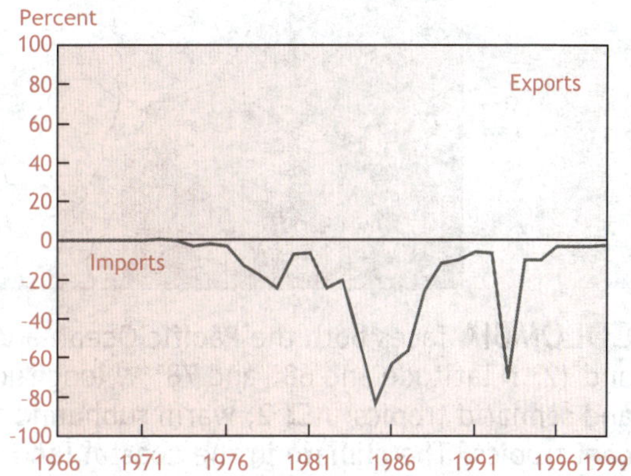

Net trade status, 1966-99

Basic statistics, Republic of Chad

	1985	1990	1995	1998	1999	2000
Rice						
Area harvested (ha)	14,798	36,854	59,633	80,322	88,875	88,875
Yield (t/ha)	0.5	1.8	1.3	1.5	1.5	1.5
Production (t)	7,824	66,027	78,978	120,666	130,521	130,521
Rice imports (t)	213,218	62,530	1,645,837	2,285	2,285	na
Others						
Population, total ($\times10^3$)	5,116	5,746	6,707	7,270	7,458	na
Population, agriculture ($\times10^3$)	4,377	4,781	5,333	5,598	5,676	na
Agricultural area ($\times10^3$ ha)	48,155	48,300	48,450	48,550	na	na
Irrigated agricultural area ($\times10^3$ ha)	14	16	20	20	na	na
Total fertilizer consumption (t)	7,100	5,800	8,558	16,820	na	na
Tractors used in agric. (no.)	160	165	170	170	na	na

Source: FAOSTAT online database.

COLOMBIA faces both the Pacific Ocean and the Caribbean Sea. It is located between 4° S and 12° N latitude and 68° and 78° W longitude. Colombia spans four AEZs: AEZ 1, warm arid and semiarid tropics; AEZ 2, warm subhumid tropics; AEZ 3, warm humid tropics; and AEZ 4, cool tropics. The climate in the coastal lands is tropical with high rainfall; the Pacific coast represents one of the most humid regions in the world. The highlands, where major cities are located, have a rather temperate climate.

The population in 1999 was 41.6 million, 72% of whom live in urban areas. Employment follows the same trend as population distribution, with 70% of the labor force being urban. The major sources of employment are agriculture (30%) and manufacturing (13%). Agriculture contributed 19% to GDP in 1999.

General information

- GNI per capita PPP$, 2000: 6,060
- Internal renewable water resources: 1,070 km^3
- Main food consumed: sugar and honey, rice, maize, oil and fat, meat
- Rice consumption, 1999: 29.3 kg milled rice per person per year

Production constraints

There are several constraints to sustainable rice production in Colombia:

- High production costs and shrinking local market because of a reduction in national consumption
- Blast, rice hoja blanca virus, crinkling or entorchamiento virus, *Rhynchosporium*, *Helminthosporium*
- Weed and red rice competition
- Infertile and aluminum-toxic soil in upland rice areas of eastern plains
- Yield potential of irrigated rice varieties has reached a plateau

Production season

	Planting	Harvesting
Summer season	Mar-Apr	Jul-Aug
Winter season	Aug-Oct	Jan-Mar

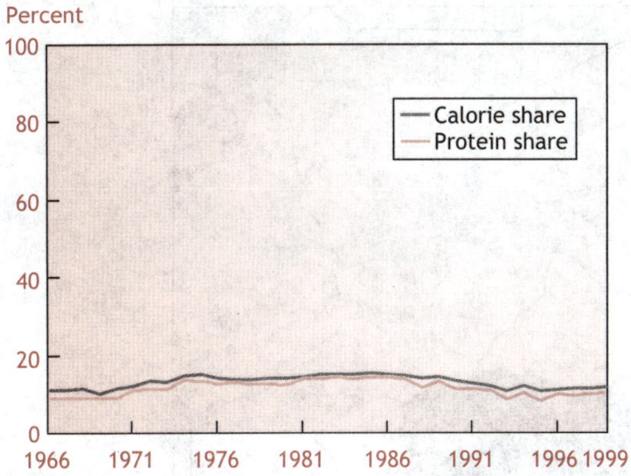

Share of calories and protein from rice, 1966-99

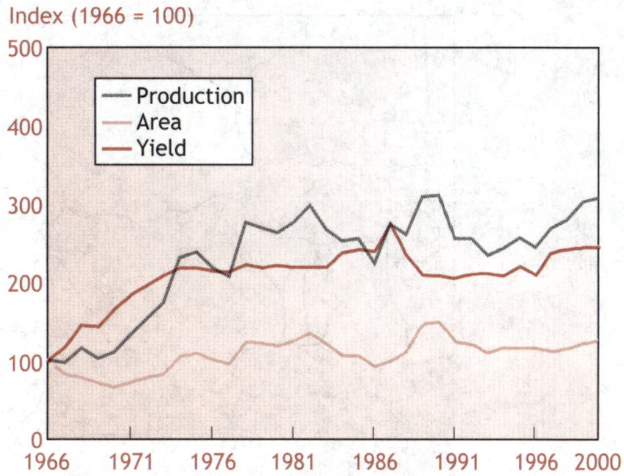

Indices of rice production, area, and yield, 1966-2000

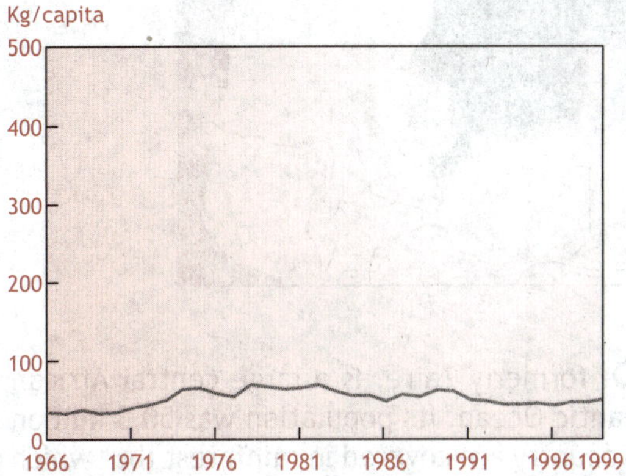

Per capita production (paddy terms), 1966-99

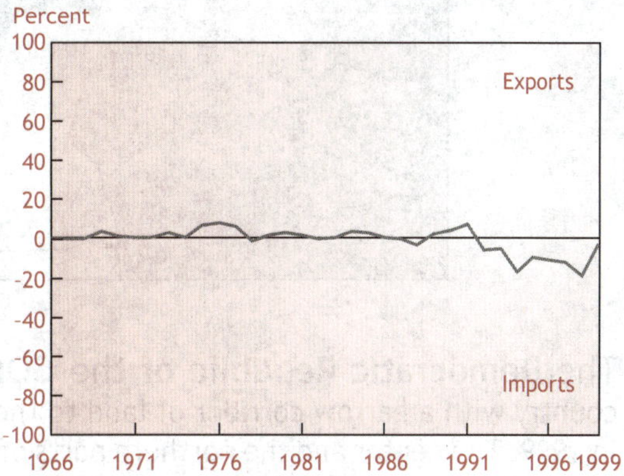

Net trade status, 1966-99

Basic statistics, Colombia

	1985	1990	1995	1998	1999	2000
Rice						
Area harvested (ha)	370,600	521,100	406,761	402,781	431,189	440,000
Yield (t/ha)	4.7	4.1	4.3	4.7	4.8	4.8
Production (t)	1,741,600	2,116,600	1,742,547	1,897,740	2,059,374	2,100,000
Rice imports (t)	70	90	122,548	292,744	37,011	na
Paddy imports (t)	0	0	1,513	308,735	0	na
Rice exports (t)	na	na	na	346	15	na
Paddy exports (t)	160	635	82	83	23	na
Others						
Population, total ($\times10^3$)	31,659	34,970	38,542	40,803	41,564	na
Population, agriculture ($\times10^3$)	10,354	9,314	9,126	9,004	8,958	na
Agricultural area ($\times10^3$ ha)	na	na	na	na	na	na
Irrigated agricultural area ($\times10^3$ ha)	465	650	820	850	na	na
Total fertilizer consumption (t)	364,300	602,500	486,200	627,000	na	na
Tractors used in agric. (no.)	33,450	32,000	23,000	21,000	na	na

Source: FAOSTAT online database.

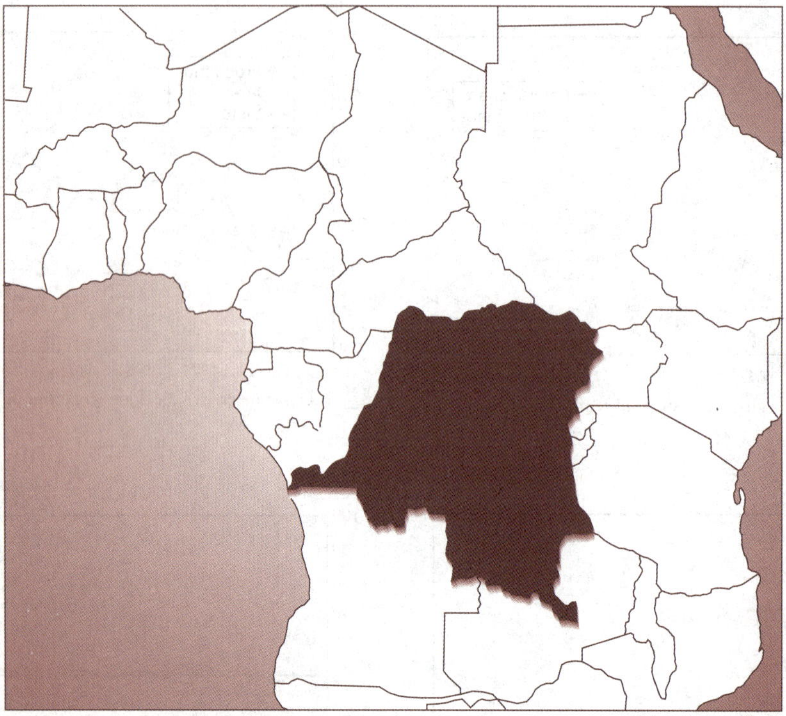

The **Democratic Republic of the CONGO**, formerly Zaire, is a large central African country with a narrow corridor of land to the Atlantic Ocean. Its population was 50.3 million in 1999. The center and the northern parts of the country are covered in rainforest (but which is rapidly disappearing) and are sparsely populated, mainly by subsistence farmers. In the south are extensive grasslands where the bulk of agriculture is carried out. Agriculture constituted 58% of GDP in 1997, occupying some 68% of the workforce.

General information

- GNI per capita PPP$, 2000: 570
- Internal renewable water resources: 935 km^3
- Incoming water flow: 84 km^3
- Main food consumed: roots and tubers, maize, oil and fat, fruits, nuts
- Rice consumption, 1999: 16.5 kg milled rice per person per year

Production constraints

National yield is among the lowest in the world. There are several constraints to sustainable rice production in Congo DR:

- Drought stress, low soil fertility, and erosion of soil and soil fertility in upland areas, which are the dominant rice ecosystem
- Blast and weed competition
- Inadequate and irregular input supplies: seeds, fertilizer, and credit
- Lack of small farm equipment, especially for postharvest operations
- Lack of effective farmers' organizations and cooperatives
- Poor maintenance of irrigation facilities
- Lack of a well-defined rice policy
- Poor road networks and marketing systems
- Limited irrigation facilities and their poor maintenance
- Weak research and extension support
- Recent deterioration of the security situation

Production season

	Planting	Harvesting
Main season, north	Jan-Mar	Jun-Jul
Main season, south	Sep-Oct	Feb-Mar

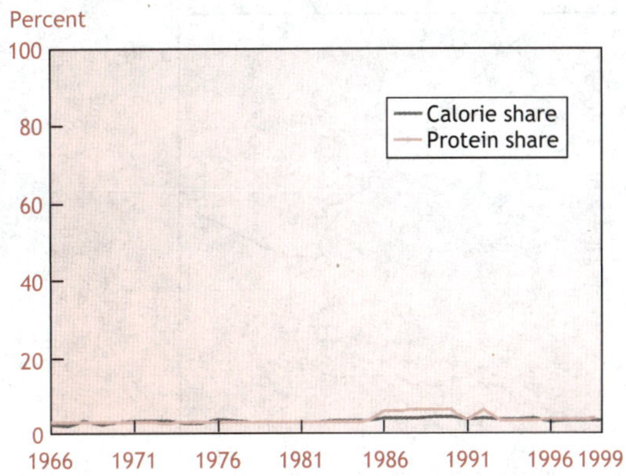

Share of calories and protein from rice, 1966-99

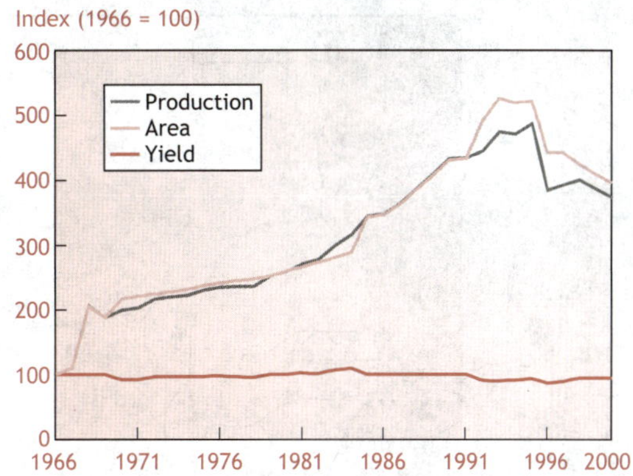

Indices of rice production, area, and yield, 1966-2000

The scale on this graph is not comparable with the scale on production, area, and yield graphs for most other countries.

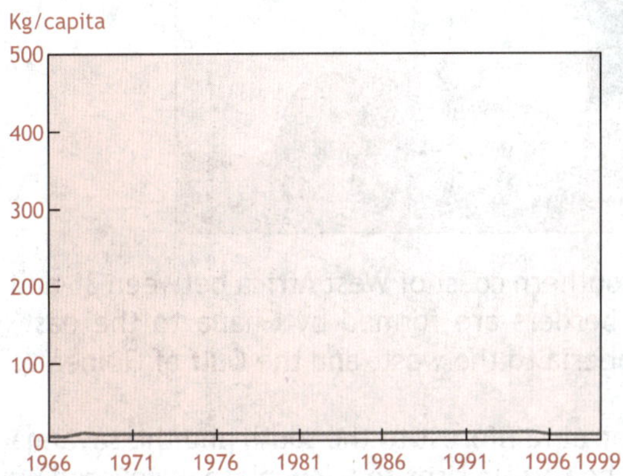

Per capita production (paddy terms), 1966-99

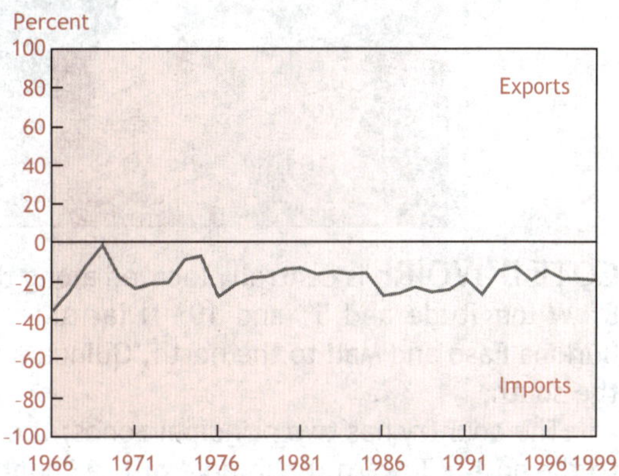

Net trade status, 1966-99

Basic statistics, Democratic Republic of the Congo

	1985	1990	1995	1998	1999	2000
Rice						
Area harvested (ha)	389,030	487,190	590,850	480,340	463,576	447,417
Yield (t/ha)	0.8	0.8	0.7	0.8	0.8	0.8
Production (t)	312,000	392,300	441,000	362,657	350,000	337,800
Rice imports (t)	40,000	85,000	70,741	55,776	55,776	na
Paddy imports (t)	0	0	0	192	192	na
Others						
Population, total ($\times10^3$)	31,669	37,363	45,421	49,139	50,335	na
Population, agriculture ($\times10^3$)	22,078	25,330	29,770	31,528	32,062	na
Agricultural area ($\times10^3$ ha)	22,800	22,860	22,900	22,880	na	na
Irrigated agricultural area ($\times10^3$ ha)	9	10	11	11	na	na
Total fertilizer consumption (t)	6,800	6,200	9,000	0	na	na
Tractors used in agric. (no.)	2,250	2,400	2,430	2,430	na	na

Source: FAOSTAT online database.

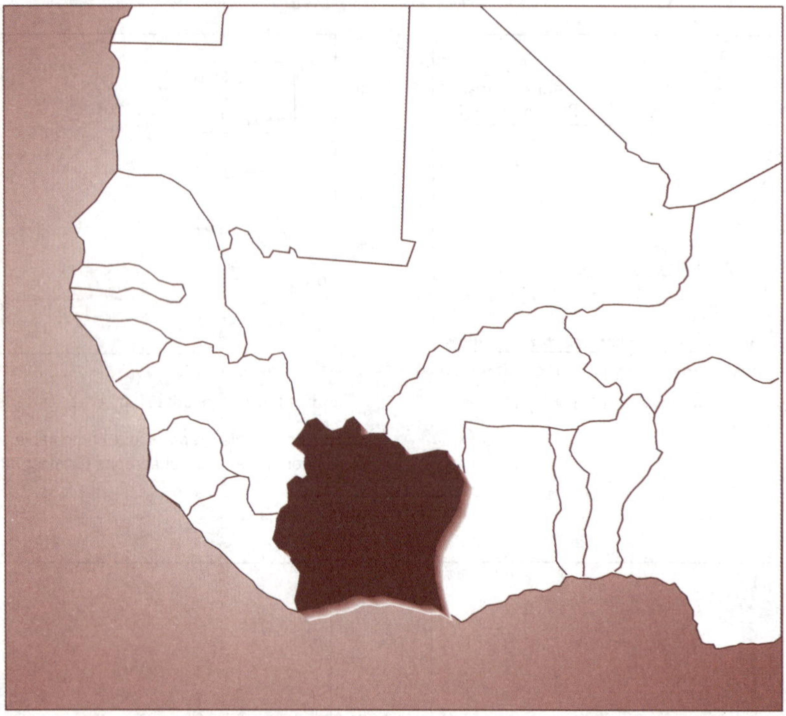

CÔTE D'IVOIRE is centrally located along the southern coast of West Africa between 3° and 8° W longitude and 5° and 10° N latitude. Its borders are formed by Ghana to the east, Burkina Faso and Mali to the north, Guinea and Liberia to the west, and the Gulf of Guinea to the south.

The country has two principal zones: the tropical rainforest to the south and the savanna to the north. Toward the middle of the country, the terrain rises to a low plateau covered by a partially wooded savanna. Here the annual rainfall totals 1,100–1,200 mm during the rainy season from April to October.

The Ivorian population was 14.5 million in 1999, of which at least 3 million are immigrants from neighboring countries. Agriculture supports approximately 60% of the population and produces more than a third of the GDP. Rice is the principal food crop grown in many areas and is one of the most important staple foods for the country's urban population.

General information

- GNI per capita PPP$, 2000: 1,500
- Internal renewable water resources: 76.7 km^3
- Incoming water flow: 1.0 km^3
- Main food consumed: roots and tubers, rice, oil and fat, maize, fruit
- Rice consumption, 1999: 74.3 kg milled rice per person per year

Production season

	Planting	Harvesting
Main season, north	May-Jun	Oct-Dec
Main season, south	Apr-May	Sep-Nov
Off-season	Dec-Feb	Apr-Jun

Production constraints

The major constraints to sustainable rice production in Côte d'Ivoire are

- Dominance of upland shifting cultivation
- Drought stress, low soil fertility, and erosion of soil and soil fertility in upland areas, which are the dominant rice ecosystem
- Blast and weed competition
- Inadequate and irregular input supplies: seeds, fertilizer, and credit
- Lack of well-defined rice policies: rice has been given low priority in national policy
- Weak research and extension support
- Poor road networks and marketing systems

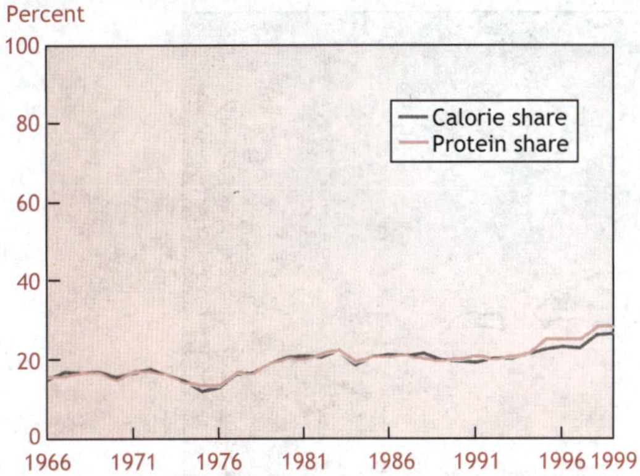

Share of calories and protein from rice, 1966-99

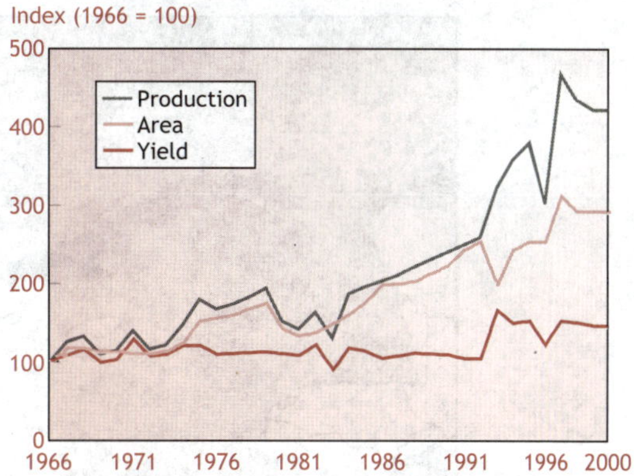

Indices of rice production, area, and yield, 1966-2000

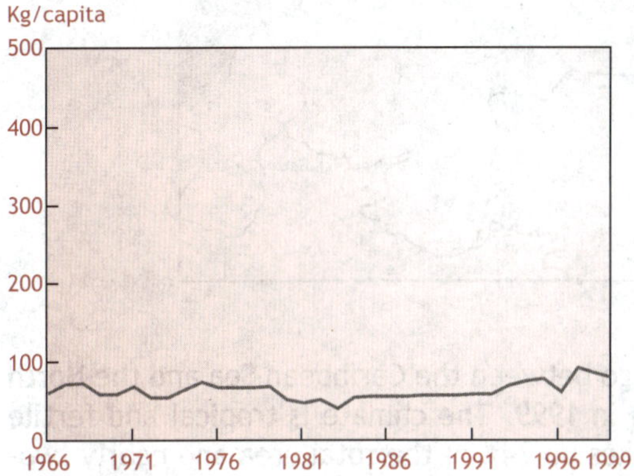

Per capita production (paddy terms), 1966-99

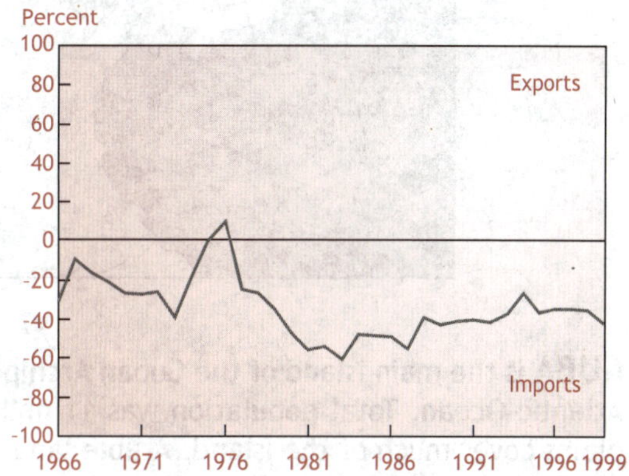

Net trade status, 1966-99

Basic statistics, Côte d'Ivoire

	1985	1990	1995	1998	1999	2000
Rice						
Area harvested (ha)	450,000	572,000	650,000	750,000	750,000	750,000
Yield (t/ha)	1.2	1.2	1.6	1.6	1.5	1.5
Production (t)	540,000	660,400	1,045,452	1,197,000	1,161,518	1,161,518
Rice imports (t)	242,358	235,000	350,000	440,000	573,400	na
Paddy imports (t)	0	0	1	na	na	na
Rice exports (t)	1,233	0	67	na	na	na
Paddy exports (t)	0	0	99	na	na	na
Others						
Population, total (×10³)	9,878	11,635	13,528	14,292	14,526	na
Population, agriculture (×10³)	6,163	6,976	7,397	7,352	7,314	na
Agricultural area (×10³ ha)	18,180	18,930	20,020	20,350	na	na
Irrigated agricultural area (×10³ ha)	54	66	73	73	na	na
Total fertilizer consumption (t)	41,500	35,700	66,000	113,400	na	na
Tractors used in agric. (no.)	3,300	3,550	3,800	3,800	na	na

Source: FAOSTAT online database.

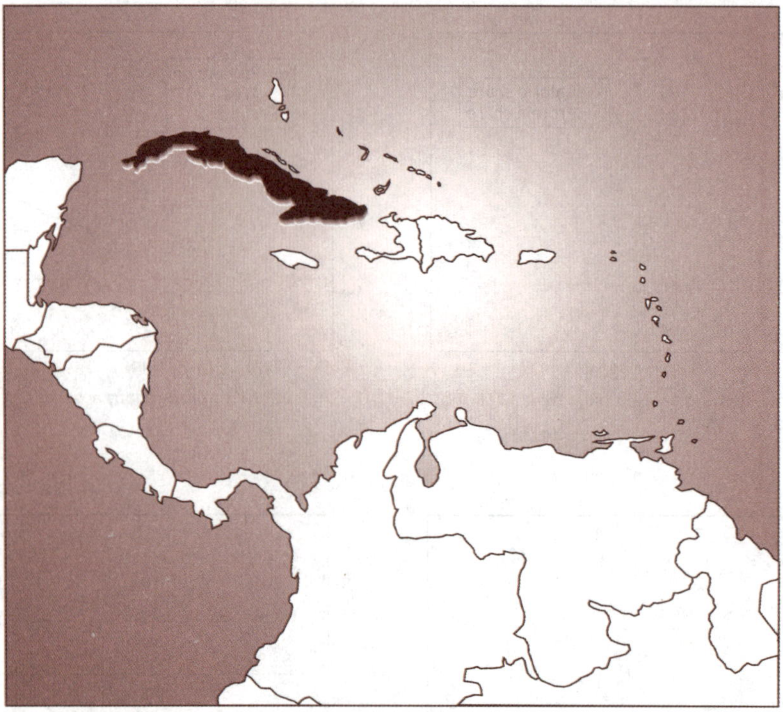

CUBA is the main island of the Cuban Archipelago between the Caribbean Sea and the North Atlantic Ocean. Total population was 11 million in 1999. The climate is tropical and fertile plains cover much of the island. Arable land makes up 24% of the total area and nearly two-thirds is planted to sugarcane, the main crop. Agriculture overall, however, forms only 7% of the GDP, while employing a quarter of the workforce.

General information

- Internal renewable water resources: 34.5 km^3
- Main food consumed: sugar and honey, wheat, rice, oil and fat, meat
- Rice consumption, 1999: 46.2 kg milled rice per person per year

Production season

	Planting	Harvesting
Main season	Mar-Apr	Jul-Aug

Production constraints

The following are the major constraints to sustainable rice production in Cuba:

- Rice hoja blanca virus and blast
- Weed and red rice competition
- Degradation of irrigation systems
- Limited irrigation water supply
- Salinity
- Inadequate credit and input supply
- Low temperature during the flowering stage

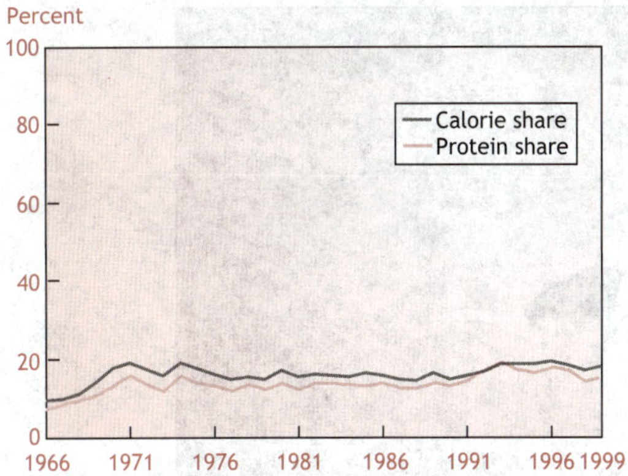

Share of calories and protein from rice, 1966-99

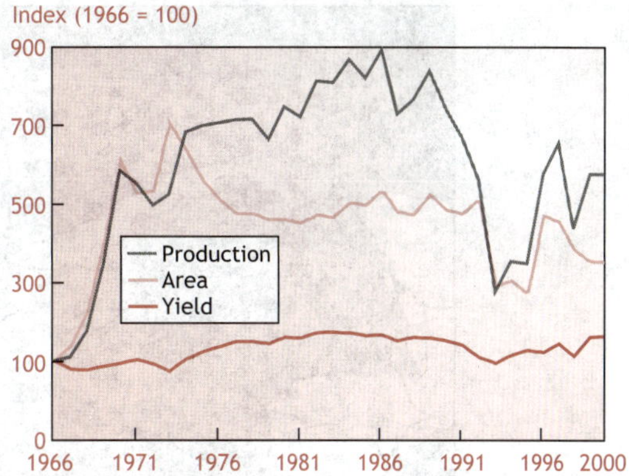

Indices of rice production, area, and yield, 1966-2000

The scale on this graph is not comparable with the scale on production, area, and yield graphs for most other countries.

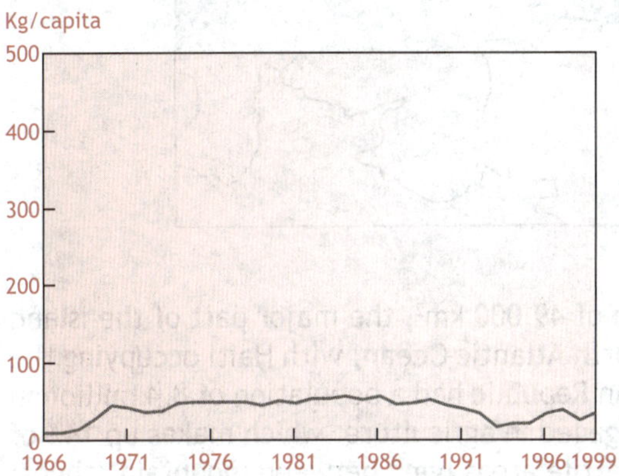

Per capita production (paddy terms), 1966-99

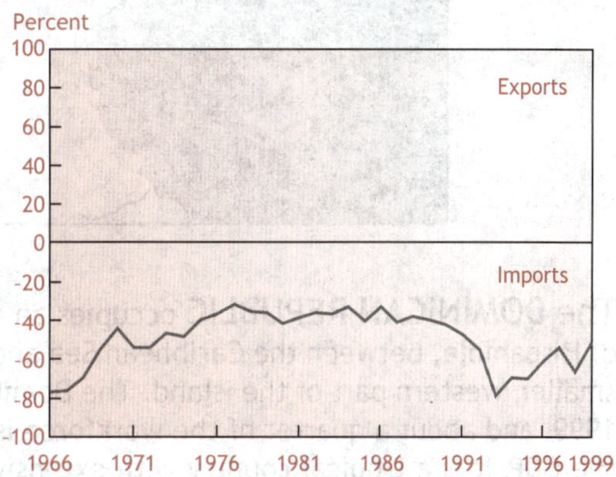

Net trade status, 1966-99

Basic statistics, Cuba

	1985	1990	1995	1998	1999	2000
Rice						
Area harvested (ha)	159,200	154,896	87,034	123,137	112,824	112,824
Yield (t/ha)	3.3	3.1	2.6	2.3	3.3	3.3
Production (t)	524,320	473,673	222,846	280,412	368,770	368,770
Rice imports (t)	342,246	308,432	404,248	375,000	267,200	na
Others						
Population, total ($\times10^3$)	10,115	10,628	10,964	11,116	11,160	na
Population, agriculture ($\times10^3$)	2,423	2,217	2,020	1,900	1,861	na
Agricultural area ($\times10^3$ ha)	6,222	6,960	6,920	6,665	na	na
Irrigated agricultural area ($\times10^3$ ha)	861	890	870	870	na	na
Total fertilizer consumption (t)	585,300	580,000	244,000	169,708	na	na
Tractors used in agric. (no.)	68,585	77,800	78,000	78,000	na	na

Source: FAOSTAT online database.

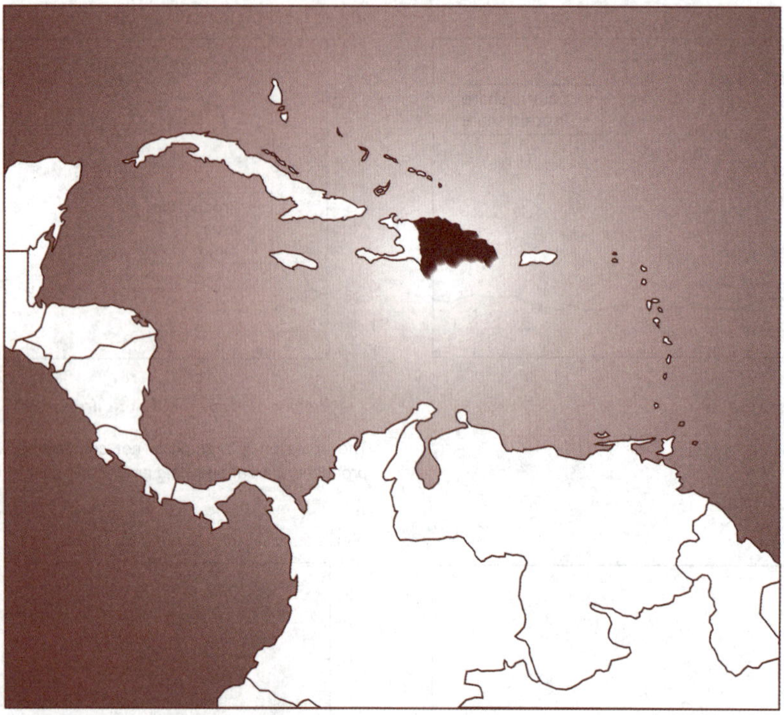

The **DOMINICAN REPUBLIC** occupies an area of 49,000 km^2, the major part of the island of Hispaniola, between the Caribbean Sea and North Atlantic Ocean, with Haiti occupying the smaller, western part of the island. The Dominican Republic had a population of 8.4 million in 1999, and about a quarter of the workforce is engaged in agriculture, which makes up 14% of the GDP. It is a tropical country with extensive fertile areas lying between mountain ranges; about 20% is arable land.

General information

- GNI per capita PPP$, 2000: 5,710
- Internal renewable water resources: 20 km^3
- Main food consumed: rice, oil and fat, sugar and honey, fruits, wheat
- Rice consumption, 1999: 42.9 kg milled rice per person per year

Production constraints

There are several constraints to sustainable rice production in the Dominican Republic:

- Limited irrigation water supply and poor maintenance of irrigation infrastructure
- Poor drainage in the Yuma and Camu river basins and salinity problems in valleys and basins of the Yaque River in Dajabón
- Yield potentials of irrigated rice varieties have reached a plateau
- Weed competition: *Heteranthera reniformis, Echinochloa colona, Ischaemum rugosum*
- Blast, rice hoja blanca virus, *Helminthosporium oryzae*, *Corticium sasaki*, *Cercospora oryzae*
- Insects: *Lissorhoptrus orizophilus*, thrips, *Spodoptera frijiperda*, *Solubea* sp.
- High production costs, lack of credit supply
- National policy of maintaining low food (rice) prices for urban consumers

Production season

	Planting	Harvesting
Main season	Mar-Apr	Jul-Aug

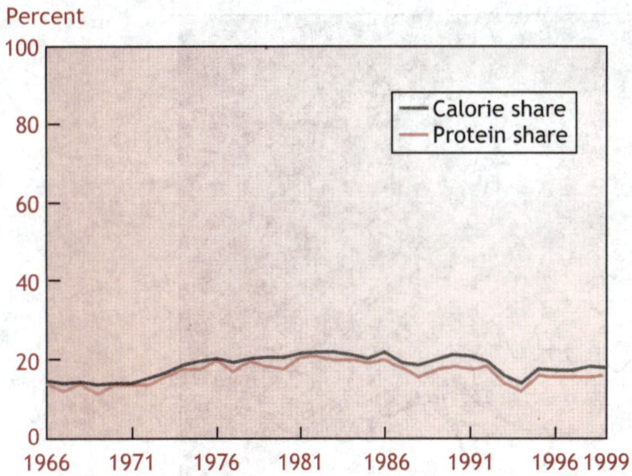

Share of calories and protein from rice, 1966-99

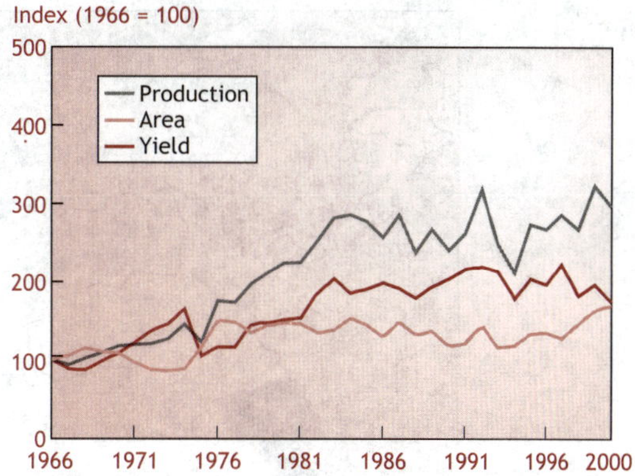

Indices of rice production, area, and yield, 1966-2000

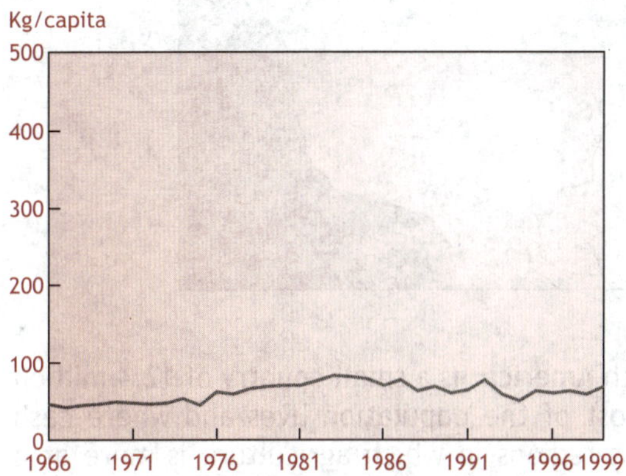

Per capita production (paddy terms), 1966-99

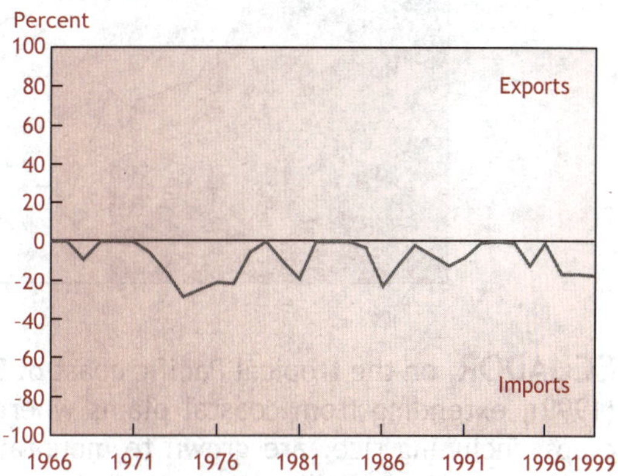

Net trade status, 1966-99

Basic statistics, Dominican Republic

	1985	1990	1995	1998	1999	2000
Rice						
Area harvested (ha)	110,301	89,361	101,988	111,123	124,719	129,290
Yield (t/ha)	4.5	4.8	4.8	4.3	4.6	4.1
Production (t)	493,756	427,597	486,741	474,595	574,357	527,014
Rice imports (t)	12,455	41,000	47,900	66,000	82,100	na
Others						
Population, total ($\times10^3$)	6,376	7,110	7,823	8,232	8,364	na
Population, agriculture ($\times10^3$)	1,907	1,866	1,686	1,576	1,540	na
Agricultural area ($\times10^3$ ha)	3,522	3,590	3,609	3,639	na	na
Irrigated agricultural area ($\times10^3$ ha)	198	225	259	259	na	na
Total fertilizer consumption (t)	61,000	88,790	94,000	95,500	na	na
Tractors used in agric. (no.)	2,250	2,330	2,350	2,350	na	na

Source: FAOSTAT online database.

ECUADOR, on the tropical Pacific coast of South America, is a small country of 12.4 million (1999), extending from coastal plains where most of the population lives and where cash crops, including rice, are grown to mountainous regions in which agriculture is more at a subsistence level. Fully one-third of the workforce is agricultural; arable land constitutes about 6% of the land area and agriculture accounts for 14% of the GDP.

General information

- GNI per capita PPP$, 2000: 2,910
- Internal renewable water resources: 314 km^3
- Main food consumed: oil and fat, rice, sugar and honey, maize, wheat
- Rice consumption, 1999: 50.5 kg milled rice per person per year

Production constraints

There are several constraints to sustainable rice production in Ecuador:

- Irrigation infrastructure is still weak
- Inadequate equipment for land preparation and harvesting
- Insufficient control of water in rainfed areas
- Blast, hoja blanca virus
- Weed competition and insects
- Reduction in public support for rice research and extension activities
- Underdeveloped private seed industry

Production season

	Planting	Harvesting
Winter season	Dec-Feb	Apr-Jun
Summer season	May-Jul	Sep-Dec

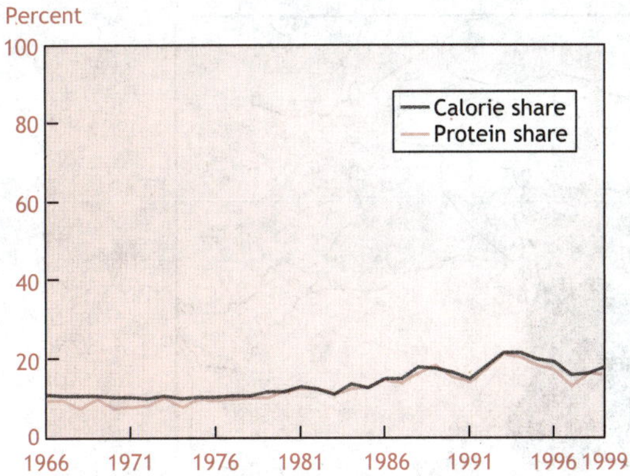

Share of calories and protein from rice, 1966-99

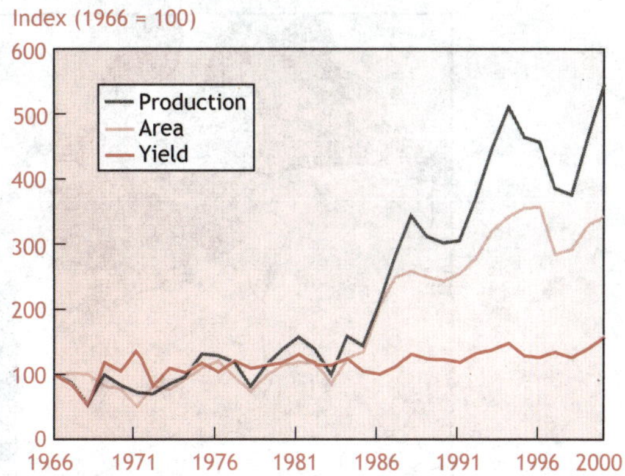

Indices of rice production, area, and yield, 1966-2000

The scale on this graph is not comparable with the scale on production, area, and yield graphs for most other countries.

Ecuador

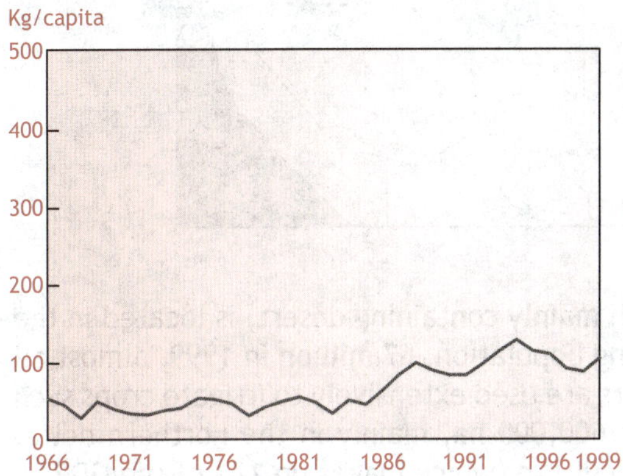

Per capita production (paddy terms), 1966-99

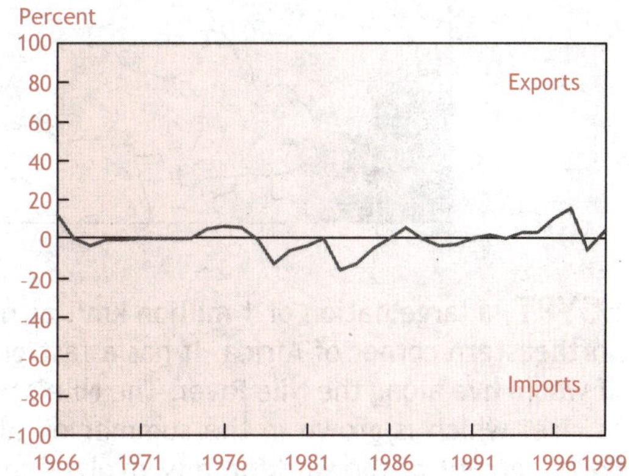

Net trade status, 1966-99

Basic statistics, Ecuador

	1985	1990	1995	1998	1999	2000
Rice						
Area harvested (ha)	149,897	269,190	395,600	325,329	366,130	380,000
Yield (t/ha)	2.7	3.1	3.3	3.2	3.5	4.0
Production (t)	397,354	840,361	1,290,518	1,042,992	1,289,684	1,520,000
Rice imports (t)	14,267	19,219	928	96,973	2,793	na
Paddy imports (t)	0	0	524	121,568	1,002	na
Rice exports (t)	0	0	26,153	54,156	30,076	na
Paddy exports (t)	0	0	2,000	1	0	na
Others						
Population, total ($\times10^3$)	9,099	10,264	11,460	12,175	12,411	na
Population, agriculture ($\times10^3$)	3,474	3,585	3,557	3,521	3,506	na
Agricultural area ($\times10^3$ ha)	7,230	7,846	8,108	8,108	na	na
Irrigated agricultural area ($\times10^3$ ha)	720	820	860	865	na	na
Total fertilizer consumption (t)	72,338	67,218	101,000	172,511	na	na
Tractors used in agric. (no.)	7,800	8,700	8,900	8,900	na	na

Source: FAOSTAT online database.

EGYPT, a large nation of 1 million km^2, although mainly containing desert, is located in the northeastern corner of Africa. It has a fast-growing population, 67 million in 1999, almost all of whom live along the Nile River. The Nile's waters are used extensively to irrigate crops such as rice, which is grown in the summer on about 600,000 ha, mainly in the northern delta. Forty percent of the workforce is involved in agriculture, which makes up 17% of the GDP.

General information

- GNI per capita PPP$, 2000: 3,670
- Internal renewable water resources: 1.8 km^3
- Incoming water flow: 66.7 km^3
- Main food consumed: wheat, maize, sugar and honey, oil and fat, rice
- Rice consumption, 1999: 40.6 kg milled rice per person per year

Production constraints

Yields during 1995-2000 were among the world's highest. The following are constraints to sustainable rice production in Egypt:

- Shortage of water
- Salinity
- Pests and disease (blast)
- Yield potentials of current varieties have reached a plateau

Production season

	Planting	Harvesting
Main season	May	Oct

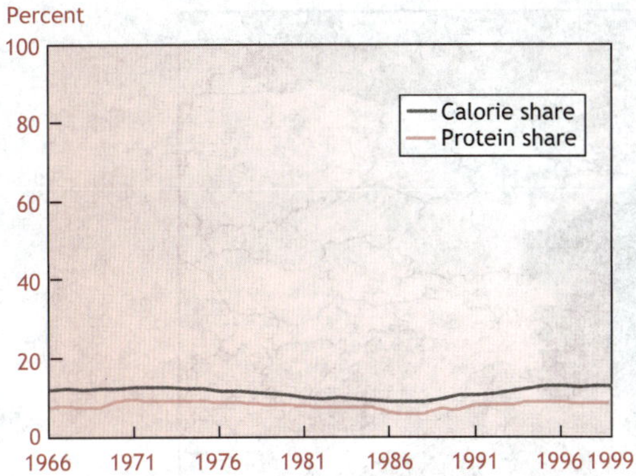

Share of calories and protein from rice, 1966-99

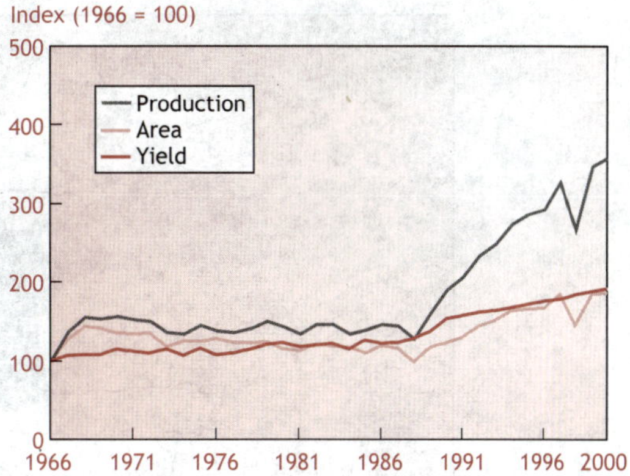

Indices of rice production, area, and yield, 1966-2000

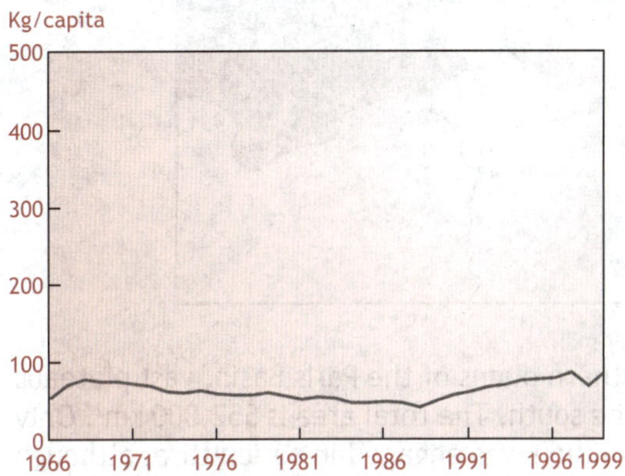

Per capita production (paddy terms), 1966-99

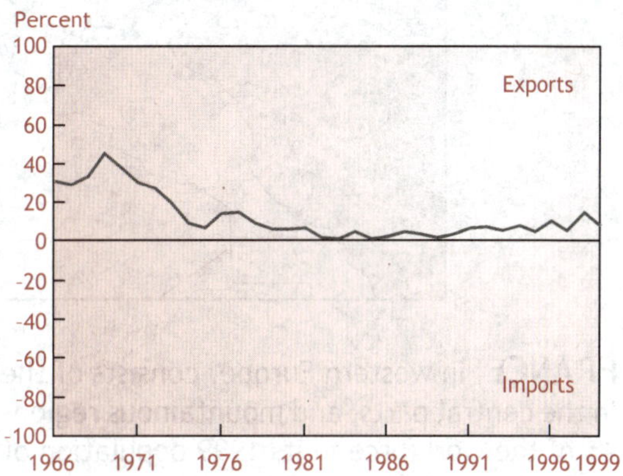

Net trade status, 1966-99

Basic statistics, Egypt

	1985	1990	1995	1998	1999	2000
Rice						
Area harvested (ha)	388,600	435,908	588,538	514,687	655,082	660,000
Yield (t/ha)	5.9	7.3	8.1	8.7	8.9	9.1
Production (t)	2,311,300	3,167,421	4,788,097	4,474,110	5,816,960	5,996,830
Rice imports (t)	16	2,443	795	698	6,627	na
Paddy imports (t)	25	0	0	na	na	na
Rice exports (t)	16,352	75,718	156,787	428,925	306,977	na
Paddy exports (t)	800	0	0	1,000	0	na
Others						
Population, total ($\times10^3$)	49,748	56,333	62,282	65,978	67,226	na
Population, agriculture ($\times10^3$)	25,665	24,708	24,956	25,041	25,057	na
Agricultural area ($\times10^3$ ha)	2,497	2,648	3,283	3,300	n	na
Irrigated agricultural area ($\times10^3$ ha)	2,497	2,648	3,283	3,300	n	na
Total fertilizer consumption (t)	863,500	964,815	1,113,195	1,112,652	na	na
Tractors used in agric. (no.)	51,856	57,000	89,080	90,000	na	na

Source: FAOSTAT online database.

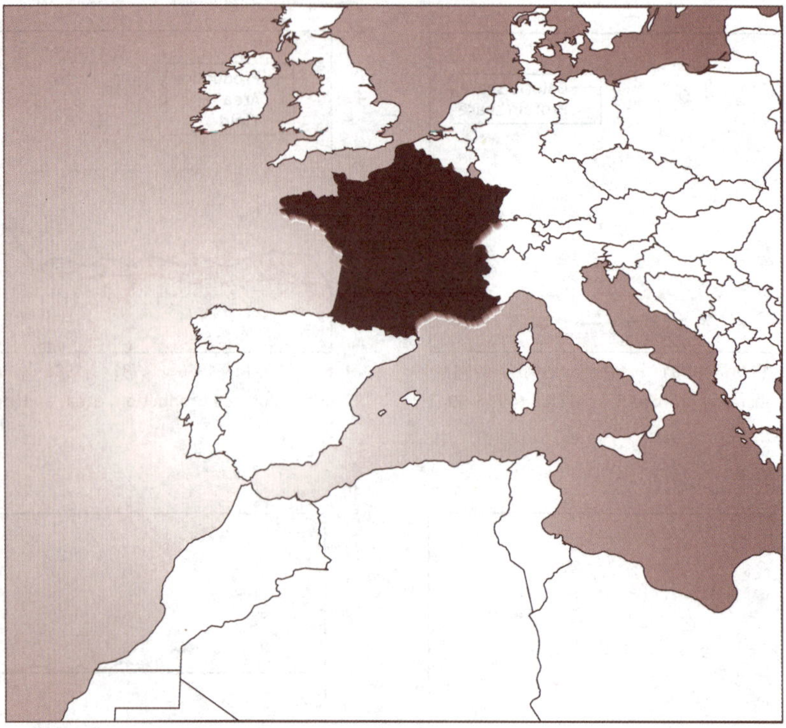

FRANCE, in western Europe, consists of the northern plains of the Paris Basin, vast plateaus in the central parts, and mountainous regions in the south. The total area is 552,000 km^2. Only 4% of the workforce in its 1999 population of 59 million was engaged in agriculture, although arable land covers one-third of the country. Agriculture accounts for only 3% of the GDP. There are extensive rice-growing areas in the Camargue region on the Mediterranean coast of France.

General information

- GNI per capita PPP$, 2000: 24,420
- Internal renewable water resources: 170 km^3
- Incoming water flow: 15 km^3
- Main food consumed: wheat, meat, oil and fat, milk, sugar and honey
- Rice consumption, 1999: 4.7 kg milled rice per person per year

Production season

	Planting	Harvesting
Main season	Apr-May	Sep-Oct

Production constraints

Rice yield has stagnated since 1990. There are several constraints to sustainable rice production:

- Low temperature
- Stagnation of the yield potential of rice varieties
- High production costs
- Blast
- Red rice

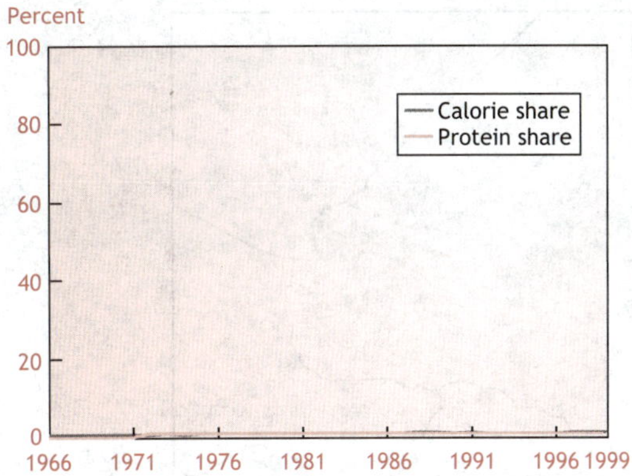

Share of calories and protein from rice, 1966-99

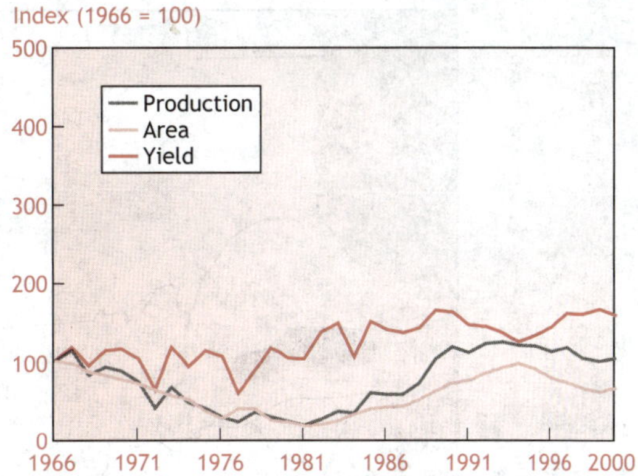

Indices of rice production, area, and yield, 1966-2000

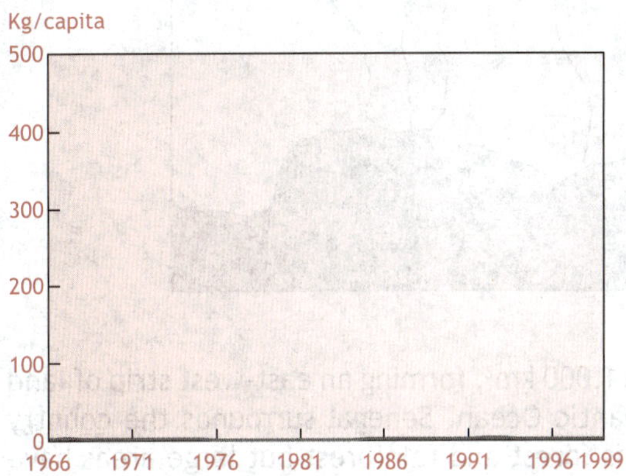

Per capita production (paddy terms), 1966-99

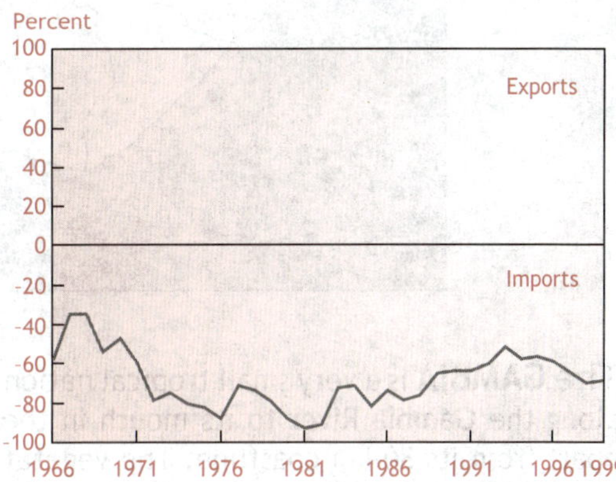

Net trade status, 1966-99

Basic statistics, France

	1985	1990	1995	1998	1999	2000
Rice						
Area harvested (ha)	11,200	20,400	25,200	18,420	16,970	18,565
Yield (t/ha)	5.5	5.9	4.9	5.8	6.1	5.8
Production (t)	61,560	121,300	122,400	107,300	103,250	107,000
Rice imports (t)	283,169	265,758	280,993	379,973	394,092	na
Paddy imports (t)	31,985	8,431	8,663	7,787	7,494	na
Rice exports (t)	17,897	43,831	74,373	61,455	74,942	na
Paddy exports (t)	1,832	5,977	31,888	32,385	32,212	na
Others						
Population, total ($\times10^3$)	55,170	56,718	58,020	58,683	58,886	na
Population, agriculture ($\times10^3$)	3,794	3,115	2,493	2,173	2,075	na
Agricultural area ($\times10^3$ ha)	31,442	30,570	30,059	29,944	na	na
Irrigated agricultural area ($\times10^3$ ha)	1,050	1,300	1,630	2,000	na	na
Total fertilizer consumption (t)	5,694,700	5,683,000	4,914,500	4,831,000	na	na
Tractors used in agric. (no.)	1,491,200	1,440,000	1,311,700	1,270,000	na	na

Source: FAOSTAT online database.

The GAMBIA is a very small tropical nation of 11,000 km^2, forming an east-west strip of land along the Gambia River to its mouth in the Atlantic Ocean. Senegal surrounds the country apart from its 80-km coastline. The vegetation is forest and rainforest but large areas have been deforested for firewood and agriculture. Only 1% of the land is arable. Agriculture occupies 52% of the workforce in the population of 1.3 million (1999) and is 23% of the GDP.

General information

- GNI per capita PPP$, 2000: 1,620
- Internal renewable water resources: 3 km^3
- Incoming water flow: 5 km^3
- Main food consumed: rice, millet and sorghum, oil and fat, sugar and honey, wheat
- Rice consumption, 1999: 89.7 kg milled rice per person per year

Production constraints

There are several major constraints to sustainable rice production in The Gambia:

- Periodic drought in upland areas and drought and flash flood in rainfed lowland (or inland swamp) and tidal wetland (or mangrove) areas because of irregular weather
- Inadequate and irregular input supplies: seeds, fertilizer, and credit
- Lack of small farm equipment, especially for postharvest operations
- Poor maintenance of developed swamps
- Poor drainage, iron toxicity in undeveloped swamps
- Lack of effective farmers' organizations and cooperatives
- Lack of well-defined rice policies
- Poor road networks and marketing systems
- Weak research and extension support

Production season

	Planting	Harvesting
Main season	May-Jun	Oct-Nov
Off-season	Jan-Feb	May-Jun

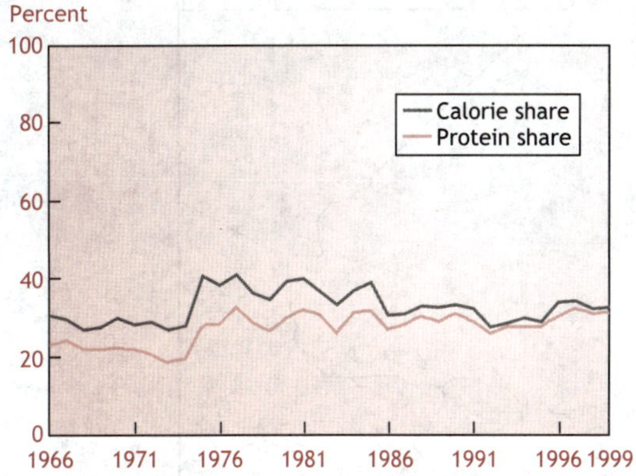

Share of calories and protein from rice, 1966-99

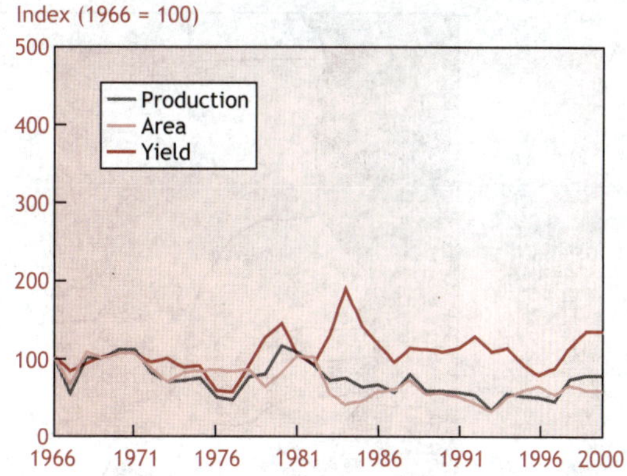

Indices of rice production, area, and yield, 1966-2000

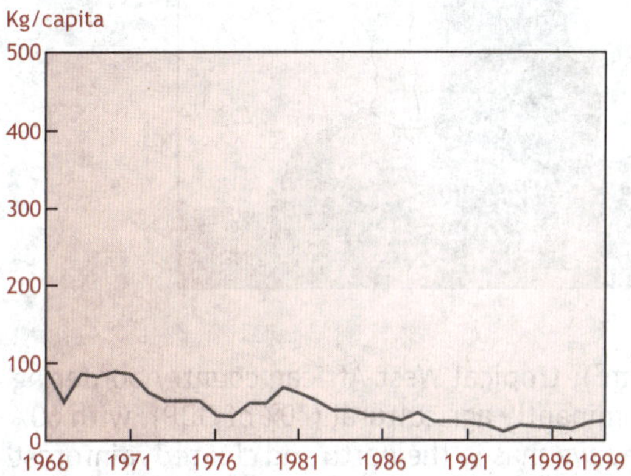

Per capita production (paddy terms), 1966-99

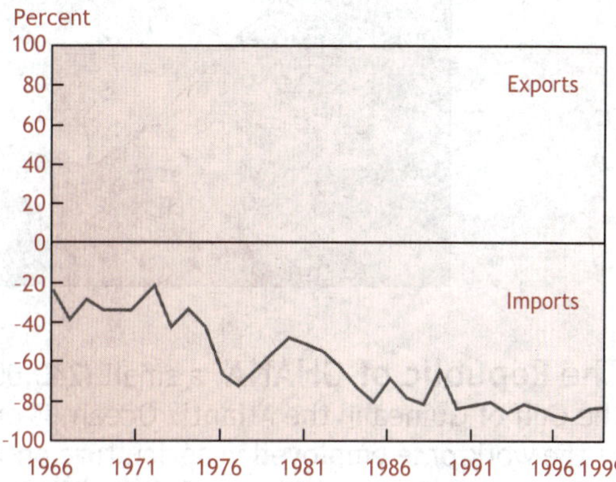

Net trade status, 1966-99

Basic statistics, The Gambia

	1985	1990	1995	1998	1999	2000
Rice						
Area harvested (ha)	12,100	14,608	15,403	17,338	15,786	15,786
Yield (t/ha)	1.9	1.5	1.2	1.5	1.8	1.8
Production (t)	23,100	21,431	18,952	26,636	28,873	28,873
Rice imports (t)	64,783	75,055	68,207	94,265	94,265	na
Rice exports (t)	0	0	5	5	5	na
Others						
Population, total ($\times10^3$)	745	920	1,111	1,229	1,268	na
Population, agriculture ($\times10^3$)	619	754	894	978	1,005	na
Agricultural area ($\times10^3$ ha)	359	375	380	395	na	na
Irrigated agricultural area ($\times10^3$ ha)	1	1	2	2	na	na
Total fertilizer consumption (t)	3,900	600	944	1,500	na	na
Tractors used in agric. (no.)	43	43	45	45	na	na

Source: FAOSTAT online database.

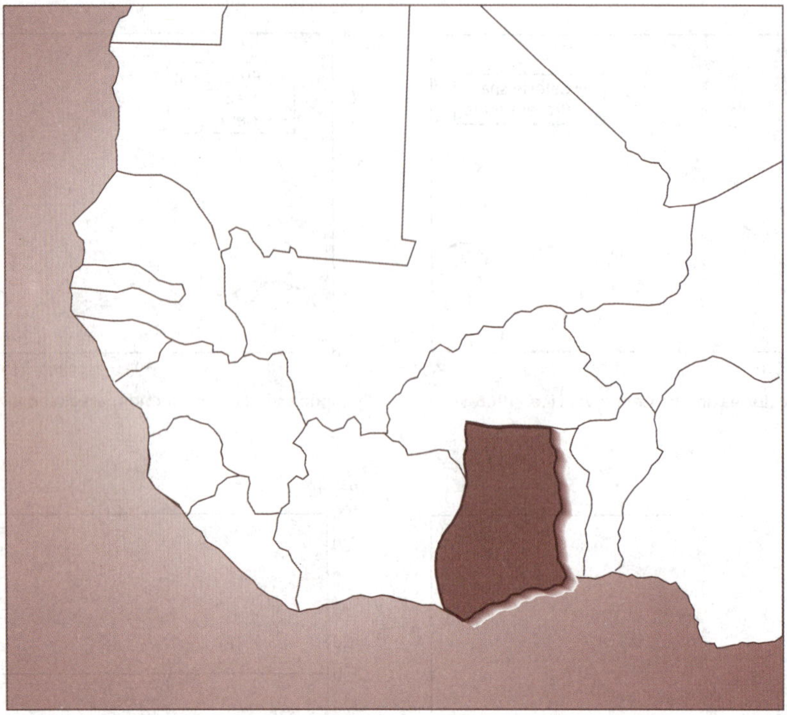

The **Republic of GHANA**, a small (240,000 km^2), tropical West African country bordering the Gulf of Guinea in the Atlantic Ocean, is predominantly agricultural (40% of GDP), with 60% of the workforce employed in agriculture on wide savannas in the north and cleared rainforest in the south. Deforestation and desertification are of increasing concern. The population in 1999 was 19.7 million.

General information

- GNI per capita PPP$, 2000: 1,910
- Internal renewable water resources: 30.3 km^3
- Incoming water flow: 22.9 km^3
- Main food consumed: roots and tubers, maize, oil and fat, fruits, millet and sorghum
- Rice consumption, 1999: 8.8 kg milled rice per person per year

Production constraints

The major constraints to sustainable rice production in Ghana are

- Drought in upland areas and drought and flash flood in rainfed lowland (or inland swamp) areas because of irregular weather
- Lack of suitable varieties for irrigated and rainfed lowland (or swamp) areas
- Inadequate and irregular input supplies: seeds, fertilizer, and credit
- Lack of small farm equipment, especially for postharvest operations
- Lack of effective farmers' organizations and cooperatives
- Poor maintenance of irrigation facilities
- Lack of well-defined rice policies
- Poor road networks and marketing systems
- Labor shortage because of competition from other crops
- Weak research and extension support

Production season

	Planting	Harvesting
Main season	May-Jun	Oct-Nov
Off-season	Jan-Feb	May-Jun

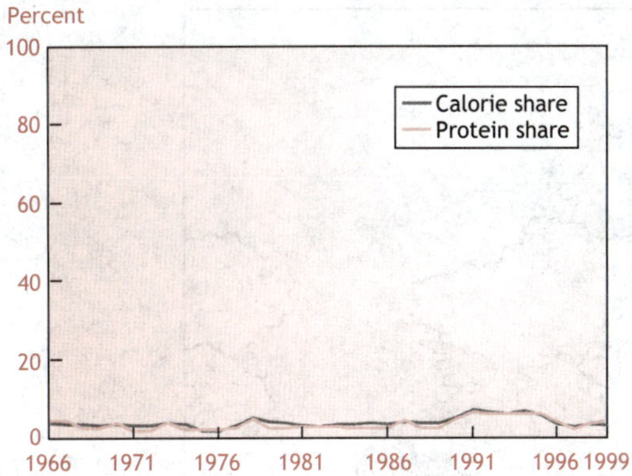

Share of calories and protein from rice, 1966-99

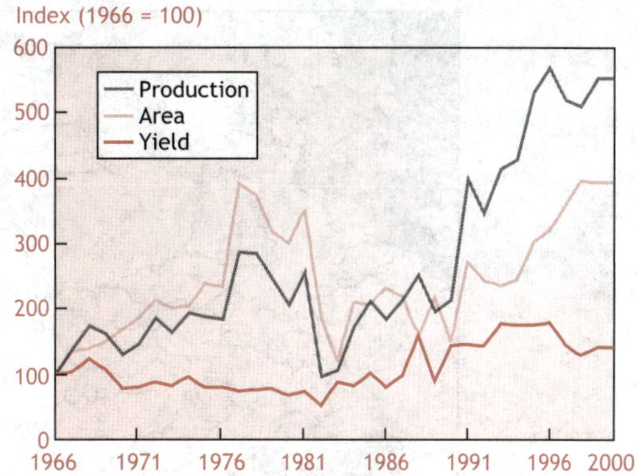

Indices of rice production, area, and yield, 1966-2000

The scale on this graph is not comparable with the scale on production, area, and yield graphs for most other countries.

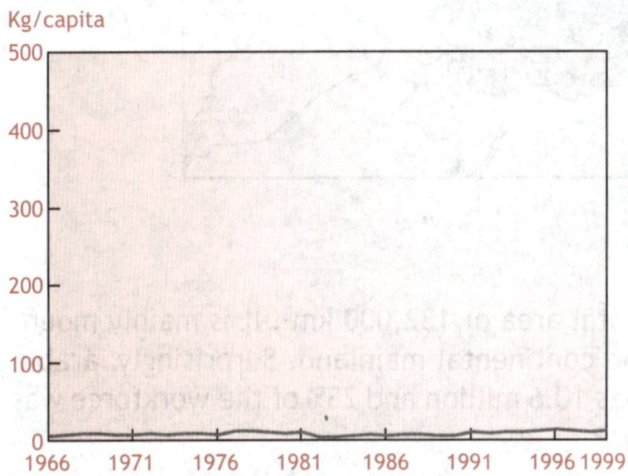

Per capita production (paddy terms), 1966-99

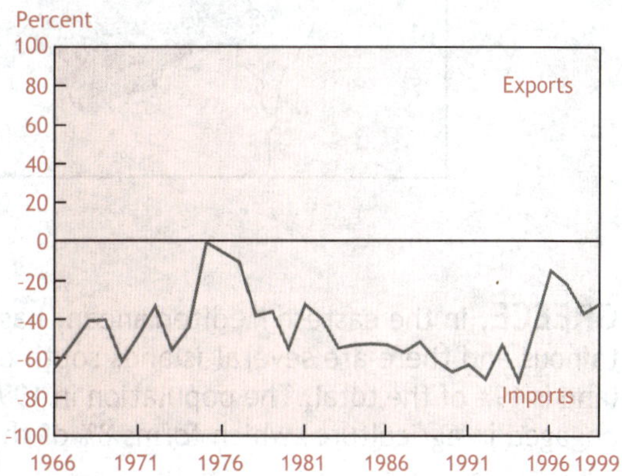

Net trade status, 1966-99

Basic statistics, Republic of Ghana

	1985	1990	1995	1998	1999	2000
Rice						
Area harvested (ha)	68,000	49,000	99,900	130,400	130,000	130,000
Yield (t/ha)	1.2	1.7	2.0	1.5	1.6	1.6
Production (t)	80,000	80,900	201,720	193,600	209,750	209,750
Rice imports (t)	60,000	113,000	106,000	77,986	69,131	na
Paddy imports (t)	0	0	0	1,718	31	na
Rice exports (t)	0	0	0	108	734	na
Paddy exports (t)	0	0	0	166	67	na
Others						
Population, total ($\times10^3$)	12,933	15,128	17,649	19,162	19,678	na
Population, agriculture ($\times10^3$)	7,723	8,870	10,130	10,858	11,104	na
Agricultural area ($\times10^3$ ha)	12,400	12,605	12,900	13,628	na	na
Irrigated agricultural area ($\times10^3$ ha)	7	8	11	11	na	na
Total fertilizer consumption (t)	12,500	13,000	9,700	15,140	na	na
Tractors used in agric. (no.)	4,120	4,120	3,700	3,570	na	na

Source: FAOSTAT online database.

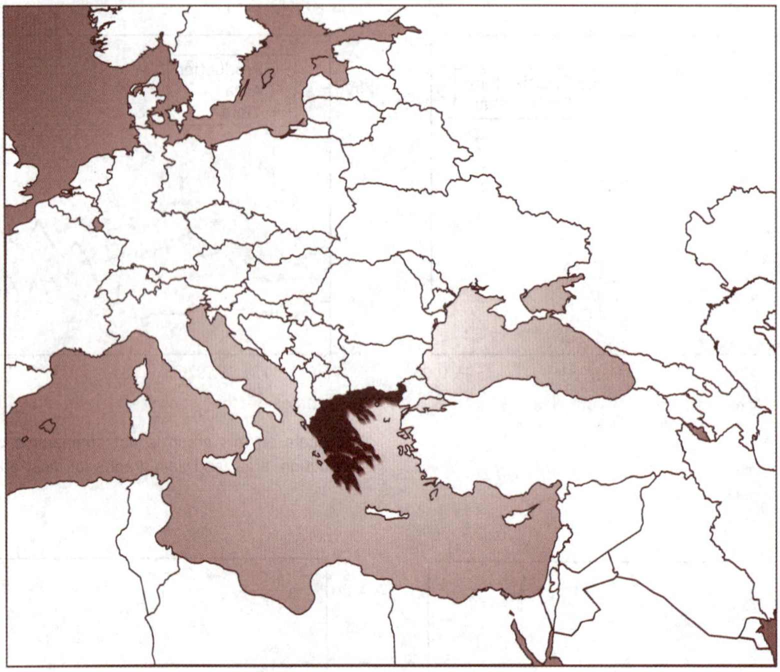

GREECE, in the eastern Mediterranean, has a total area of 132,000 km². It is mainly mountainous and there are several islands south of the continental mainland. Surprisingly, arable land is 19% of the total. The population in 1999 was 10.6 million and 23% of the workforce was engaged in agriculture, which forms 8% of the GDP.

General information

- GNI per capita PPP$, 2000: 16,860
- Internal renewable water resources: 45.1 km^3
- Incoming water flow: 13.5 km^3
- Main food consumed: wheat, oil and fat, meat, milk, sugar and honey
- Rice consumption, 1999: 7.5 kg milled rice per person per year

Production season

	Planting	Harvesting
Main season	Apr-May	Oct-Nov

Production constraints

Rice yield increased substantially during the last five years. There are several constraints to sustainable rice production in Greece:

- Low temperature
- Stagnation of the yield potential of rice varieties
- High production costs
- Blast
- Dependence on external markets for the product

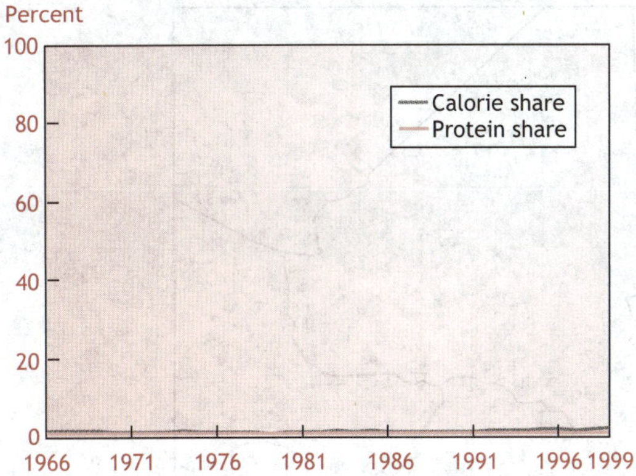

Share of calories and protein from rice, 1966-99

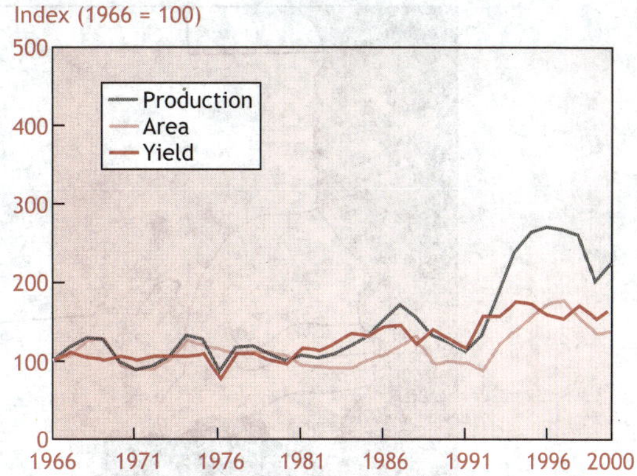

Indices of rice production, area, and yield, 1966-2000

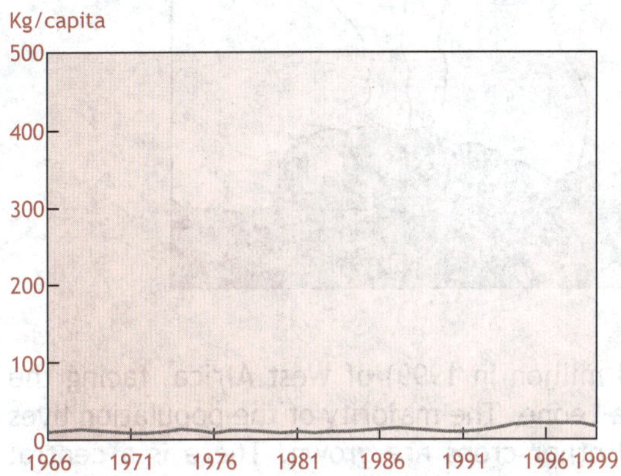

Per capita production (paddy terms), 1966-99

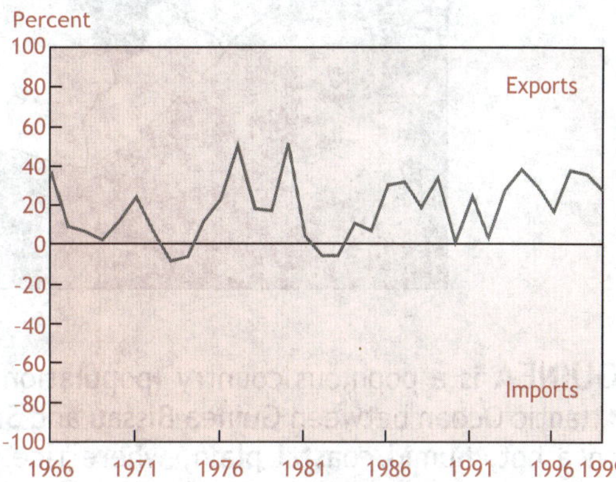

Net trade status, 1966-99

Basic statistics, Greece

	1985	1990	1995	1998	1999	2000
Rice						
Area harvested (ha)	16,859	16,500	26,105	25,948	22,400	23,000
Yield (t/ha)	6.2	6.0	8.1	8.1	7.2	7.8
Production (t)	103,765	99,000	211,599	208,975	161,435	180,000
Rice imports (t)	7,410	4,673	6,433	6,793	10,067	na
Paddy imports (t)	125	675	1,599	950	984	na
Rice exports (t)	12,534	5,592	49,608	59,094	41,126	na
Paddy exports (t)	15,464	2	64,747	53,906	6,215	na
Others						
Population, total ($\times 10^3$)	9,934	10,220	10,489	10,600	10,626	na
Population, agriculture ($\times 10^3$)	2,189	1,912	1,664	1,523	1,478	na
Agricultural area ($\times 10^3$ ha)	9,195	9,160	9,111	9,091	na	na
Irrigated agricultural area ($\times 10^3$ ha)	1,099	1,195	1,328	1,422	na	na
Total fertilizer consumption (t)	685,000	685,105	505,000	486,000	na	na
Tractors used in agric. (no.)	183,410	215,755	236,197	241,000	na	na

Source: FAOSTAT online database.

GUINEA is a populous country (population 7.4 million in 1999) of West Africa, facing the Atlantic Ocean between Guinea Bissau and Sierra Leone. The majority of the population lives on a hot, humid coastal plain, where rice and other crops are grown. There is a central mountainous region and a drier northeastern area where maize and similar crops are cultivated. Arable land is 2.5% of the total (246,000 km^2). Nearly 90% of the workforce is agricultural, yet agriculture makes up only 24% of the GDP.

General information

- GNI per capita PPP$, 2000: 1,930
- Internal renewable water resources: 226 km^3
- Incoming water flow: none
- Main food consumed: rice, roots and tubers, oil and fat, fruits, vegetables
- Rice consumption, 1999: 62.1 kg milled rice per person per year

Production constraints

The following are the major constraints to sustainable rice production in Guinea:

- Drought in upland areas and drought and flash flood in rainfed lowland (or inland swamp) and tidal wetland (or mangrove) areas because of irregular weather
- Inadequate and irregular input supplies: seeds, fertilizer, and credit
- Lack of small farm equipment, especially for postharvest operations
- Poor maintenance of developed swamps
- Poor drainage, iron toxicity in undeveloped swamps
- Lack of effective farmers' organizations and cooperatives
- Poor road networks and marketing systems
- Inadequate land tenure status

Production season

	Planting	Harvesting
Main season	Apr-Jun	Sep-Nov
Off-season	Dec-Feb	Apr-Jun

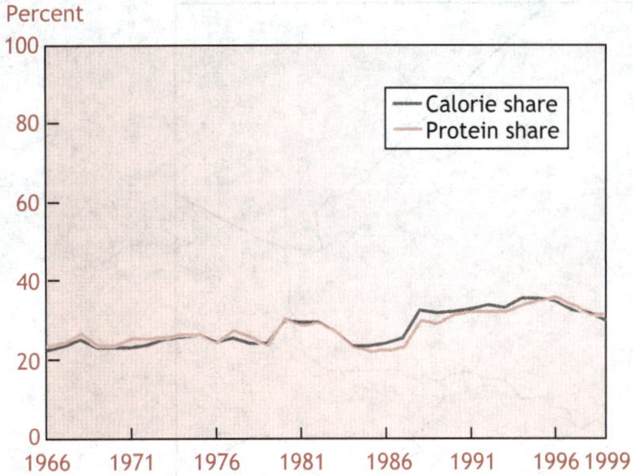

Share of calories and protein from rice, 1966-99

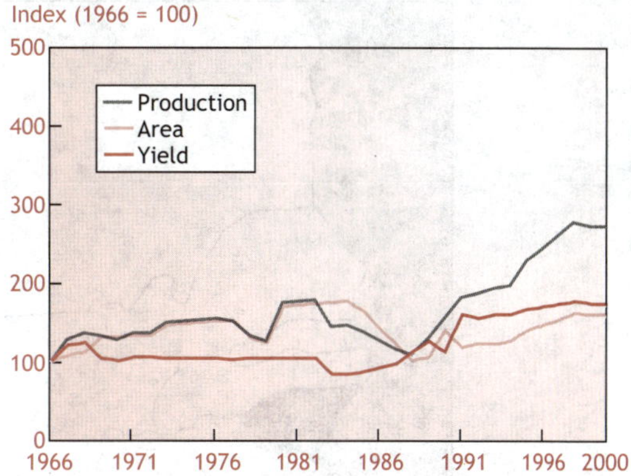

Indices of rice production, area, and yield, 1966-2000

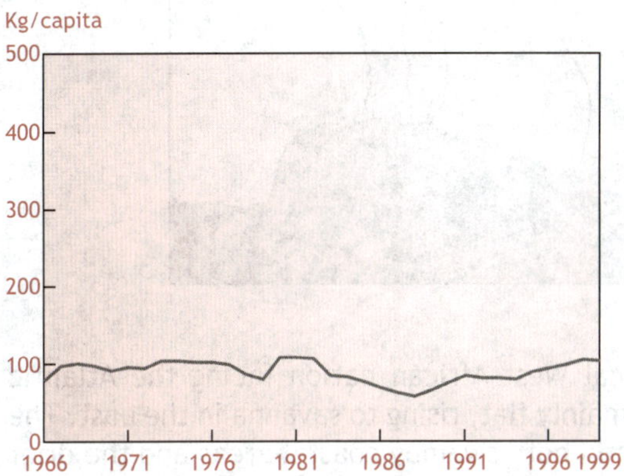

Per capita production (paddy terms), 1966-99

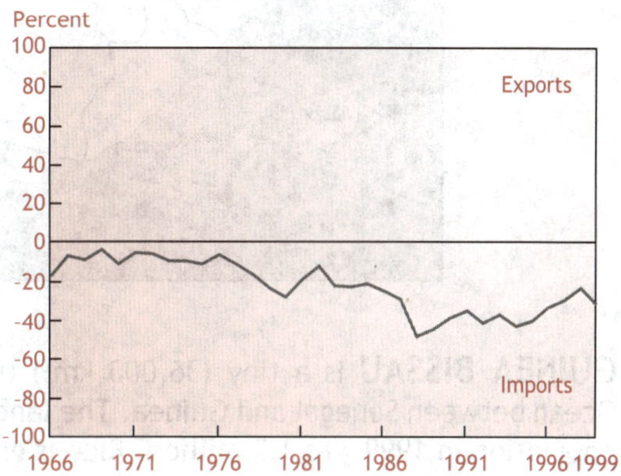

Net trade status, 1966-99

Basic statistics, Guinea

	1985	1990	1995	1998	1999	2000
Rice						
Area harvested (ha)	510,000	436,209	438,403	503,017	500,000	500,000
Yield (t/ha)	0.7	1.0	1.4	1.5	1.5	1.5
Production (t)	380,000	423,821	630,511	763,955	750,000	750,000
Rice imports (t)	70,000	182,158	290,750	159,000	240,000	na
Others						
Population, total ($\times10^3$)	4,987	5,755	7,153	7,337	7,360	na
Population, agriculture ($\times10^3$)	4,441	5,018	6,123	6,205	6,198	na
Agricultural area ($\times10^3$ ha)	11,895	12,016	12,183	8,233	na	na
Irrigated agricultural area ($\times10^3$ ha)	90	90	93	95	na	na
Total fertilizer consumption (t)	360	1,161	5,106	3,284	na	na
Tractors used in agric. (no.)	220	350	500	542	na	na

Source: FAOSTAT online database.

GUINEA BISSAU is a tiny (36,000 km²) tropical West African nation facing the Atlantic Ocean between Senegal and Guinea. The land is mainly flat, rising to savanna in the east. The population in 1999 was 1.2 million. Rice is grown in both swampy coastal areas and the drier savanna region. Eleven percent of the land is arable and agriculture, which occupies 85% of the workforce, constitutes 54% of the GDP.

General information

- GNI per capita PPP$, 2000: 710
- Internal renewable water resources: 16 km³
- Incoming water flow: 9 km³
- Main food consumed: rice, oil and fat, nuts, millet and sorghum, roots and tubers
- Rice consumption, 1999: 87.3 kg milled rice per person per year

Production constraints

The major constraints to sustainable rice production in Guinea Bissau are as follows:

- Drought in upland areas and drought and flash flood in rainfed lowland (or inland swamp) and tidal wetland (or mangrove) areas because of irregular weather
- Acidity and salinity in tidal wetland (or mangrove) areas
- Inadequate and irregular input supplies: seeds, fertilizer, and credit
- Lack of small farm equipment, especially for postharvest operations
- Poor maintenance of developed swamps
- Poor drainage, iron toxicity in undeveloped swamps
- Poor maintenance of dikes, saltwater intrusion
- Lack of effective farmers' organizations and cooperatives
- Lack of a well-defined rice policy
- Poor road networks and marketing systems
- Weak research and extension support

Production season

	Planting	Harvesting
Main season	May-Aug	Oct-Jan

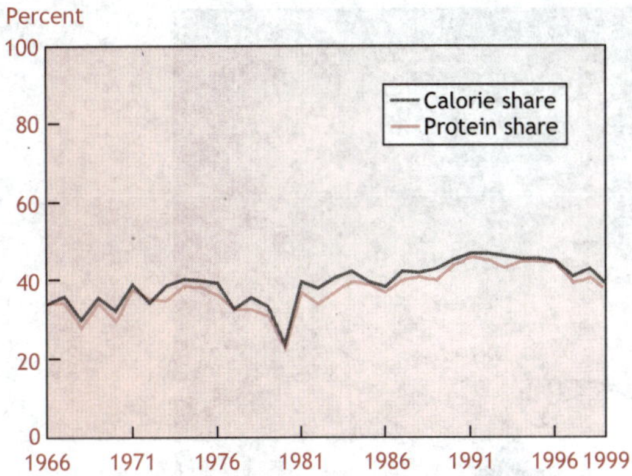

Share of calories and protein from rice, 1966-99

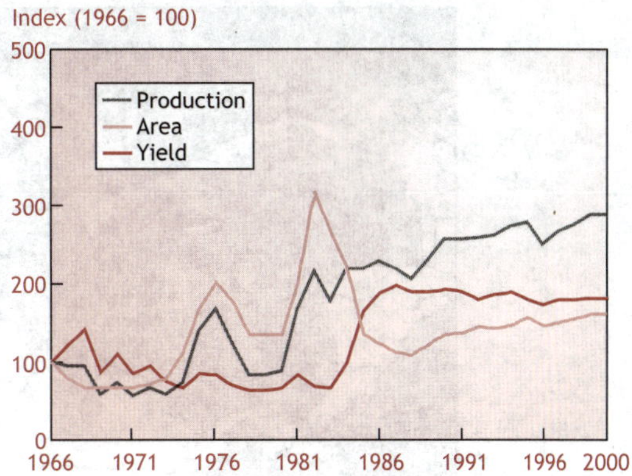

Indices of rice production, area, and yield, 1966-2000

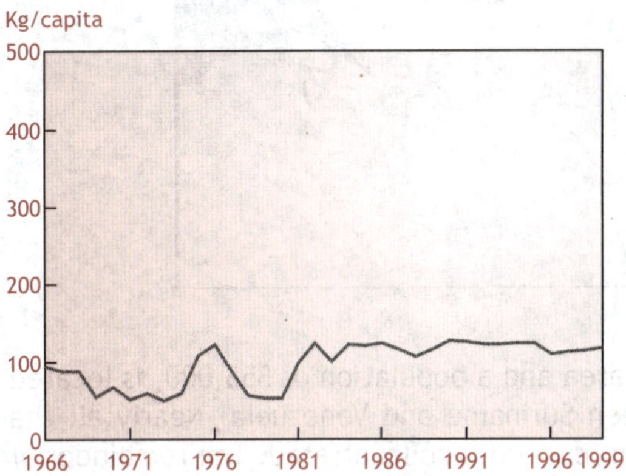

Per capita production (paddy terms), 1966-99

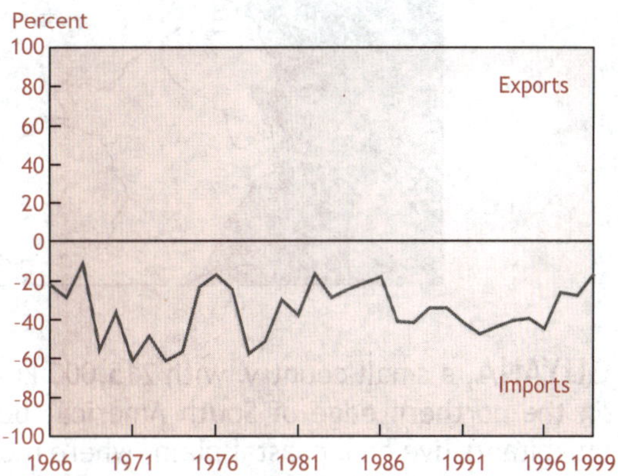

Net trade status, 1966-99

Basic statistics, Guinea Bissau

	1985	1990	1995	1998	1999	2000
Rice						
Area harvested (ha)	60,000	60,100	69,881	69,000	72,000	72,000
Yield (t/ha)	1.8	2.1	1.9	1.9	1.9	1.9
Production (t)	105,000	123,314	133,266	132,000	138,282	138,282
Rice imports (t)	19,169	43,268	58,955	34,000	18,200	na
Others						
Population, total ($\times 10^3$)	877	973	1,086	1,161	1,187	na
Population, agriculture ($\times 10^3$)	758	830	914	968	986	na
Agricultural area ($\times 10^3$ ha)	1,400	1,420	1,425	1,430	na	na
Irrigated agricultural area ($\times 10^3$ ha)	17	17	17	17	na	na
Total fertilizer consumption (t)	0	586	300	600	na	na
Tractors used in agric. (no.)	18	19	19	19	na	na

Source: FAOSTAT online database.

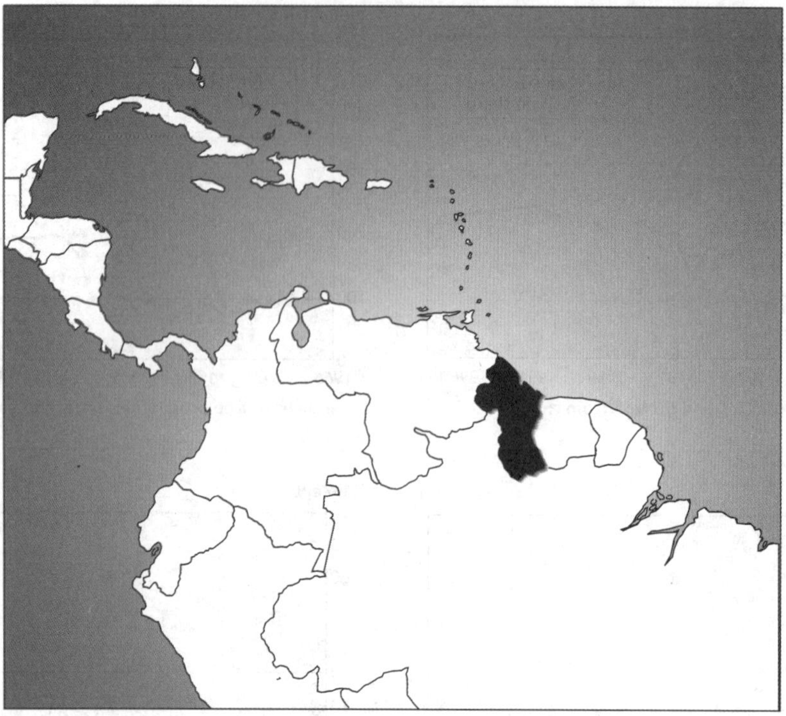

GUYANA, a small country, with 215,000 km^2 in area and a population of 855,000, is located on the northern edge of South America, between Suriname and Venezuela. Nearly all the inhabitants live on a coastal plain, where most agriculture is concentrated. The remainder of the country consists of a rainforest belt backed by an extensive mountain range. Rice and sugarcane are the major crops; only 2.2% of the land is arable. Agriculture forms 35% of the GDP and occupies 22% of the workforce.

General information

- GNI per capita PPP$, 2000: 3,670
- Internal renewable water resources: 241 km^3
- Main food consumed: rice, wheat, sugar and honey, fruits, meat
- Rice consumption, 1999: 85.3 kg milled rice per person per year

Production constraints

There are several constraints to sustainable rice production in Guyana:

- Inadequate development of irrigation and drainage infrastructure
- Weed competition
- Blast, hoja blanca virus
- Lack of credit to farmers
- Inadequate input supply, especially improved seeds and fertilizer
- Inadequate drying and storage facilities
- Lack of export facilities, especially bulk loading

Production season

	Planting	Harvesting
Main season	Jan-Feb	May-Jun

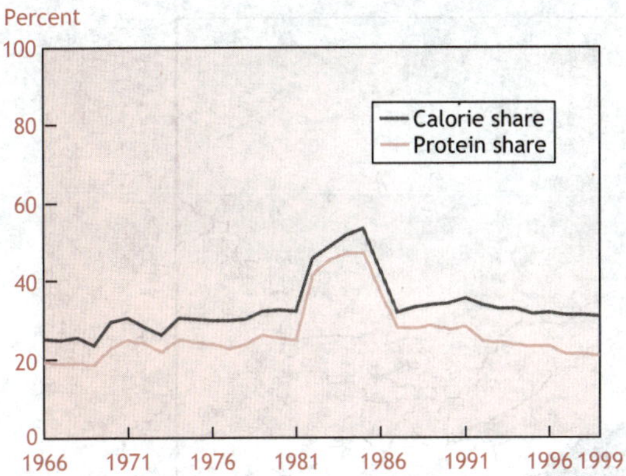

Share of calories and protein from rice, 1966-99

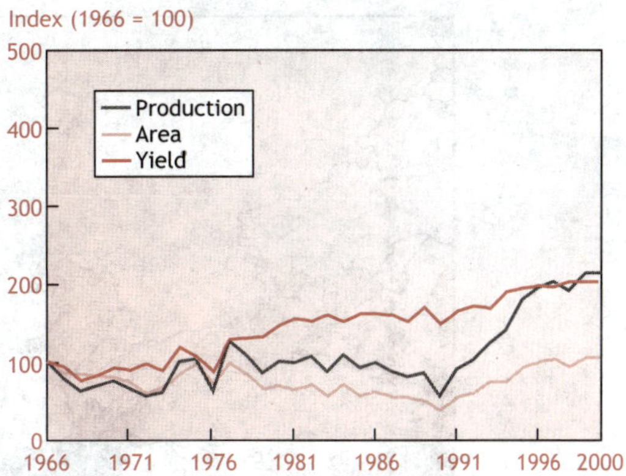

Indices of rice production, area, and yield, 1966-2000

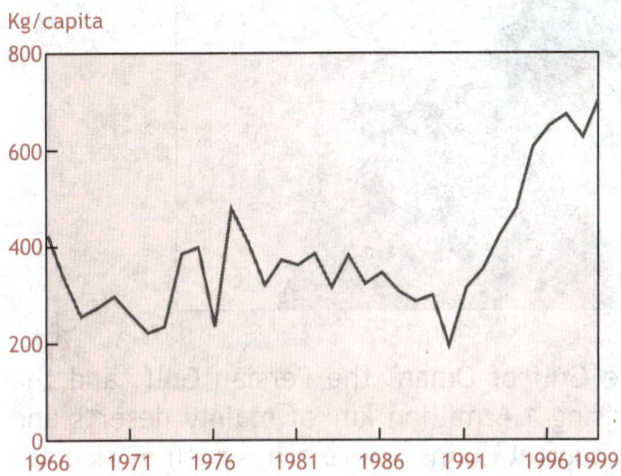

Per capita production (paddy terms), 1966-99

The scale on this graph is not comparable with the scale on per capita production graphs for most other countries.

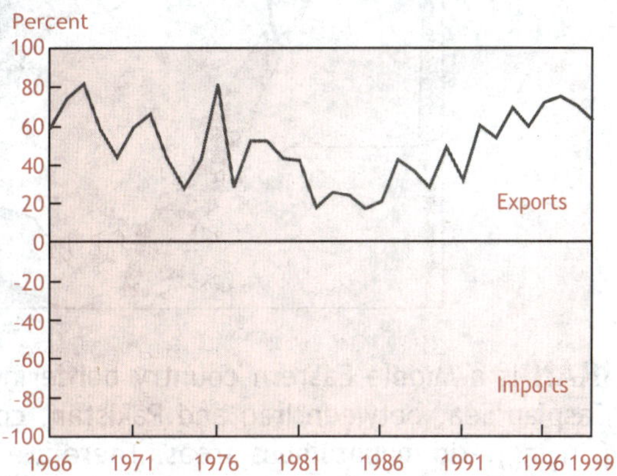

Net trade status, 1966-99

Basic statistics, Guyana

	1985	1990	1995	1998	1999	2000
Rice						
Area harvested (ha)	77,779	51,345	126,777	129,419	145,000	145,000
Yield (t/ha)	3.3	3.0	4.0	4.1	4.1	4.1
Production (t)	256,167	155,740	503,278	532,000	600,000	600,000
Rice exports (t)	29,339	50,943	200,544	249,760	249,760	na
Others						
Population, total ($\times10^3$)	793	795	830	850	855	na
Population, agriculture ($\times10^3$)	192	174	164	158	156	na
Agricultural area ($\times10^3$ ha)	1,725	1,725	1,726	1,726	na	na
Irrigated agricultural area ($\times10^3$ ha)	127	135	150	150	na	na
Total fertilizer consumption (t)	12,333	12,000	15,000	16,200	na	na
Tractors used in agric. (no.)	3,550	3,600	3,630	3,630	na	na

Source: FAOSTAT online database.

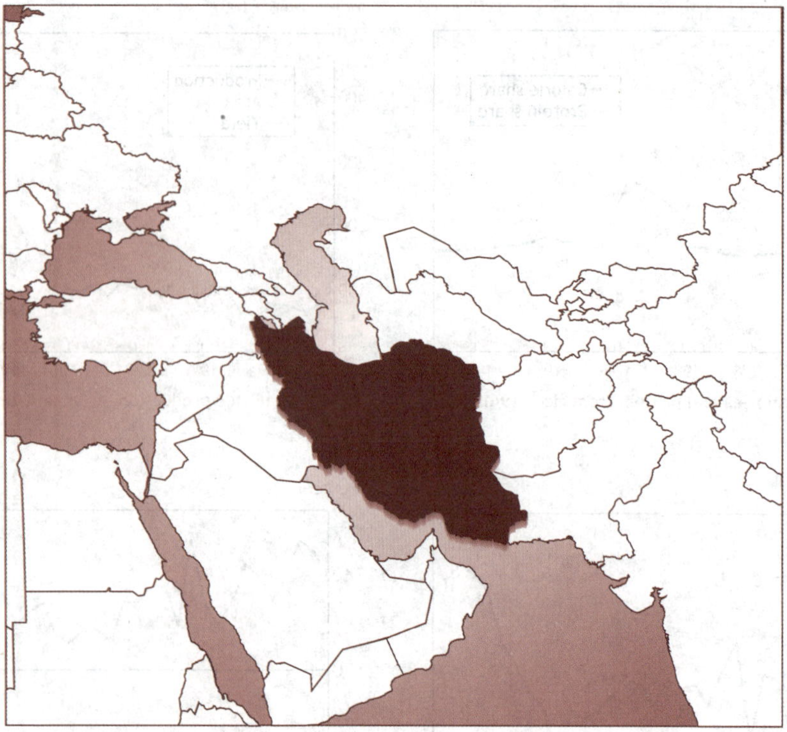

IRAN is a Middle Eastern country bordering the Gulf of Oman, the Persian Gulf, and the Caspian Sea, between Iraq and Pakistan, comprising 1.6 million km^2 of mainly deserts and fringing, arid mountainous areas. There are also coastal plains where crops such as rice are grown. Ten percent of the land is arable. A third of the workforce in the population of 66.8 million is occupied in agriculture, which forms 21% of the GDP.

General information

- GNI per capita PPP$, 2000: 5,910
- Internal renewable water resources: 117.5 km^3
- Main food consumed: wheat, oil and fat, rice, sugar and honey, fruits
- Rice consumption, 1999: 28.7 kg milled rice per person per year

Production season

	Planting	Harvesting
Main season	May-Jun	Sep-Oct

Production constraints

Rice is grown mostly at low altitudes and is irrigated. Both the harvested area and yield have remained more or less unchanged during the last five years. There are several constraints to sustainable rice production:

- High rainfall during September, causing difficulty in harvest and postharvest operations
- Increasing drought stress because of shortage of irrigation water
- Popular preference for high grain quality and aromatic rice
- Disease (blast) and insects (stem borers)

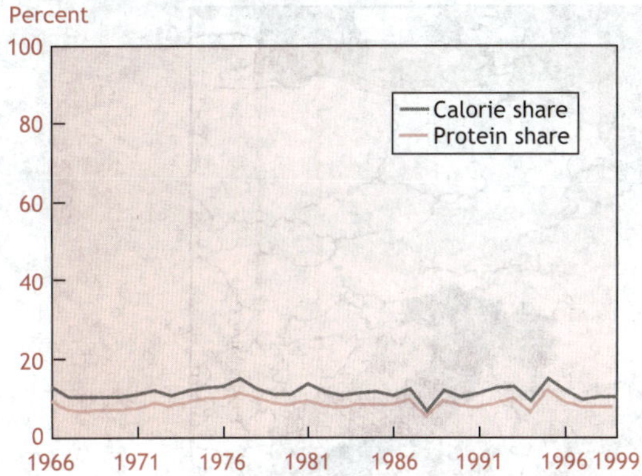

Share of calories and protein from rice, 1966-99

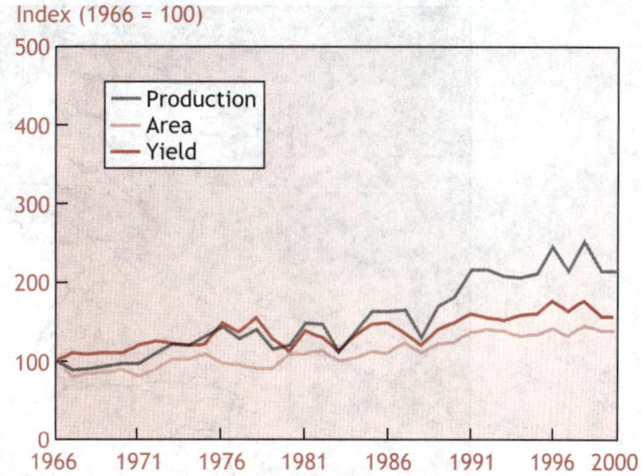

Indices of rice production, area, and yield, 1966-2000

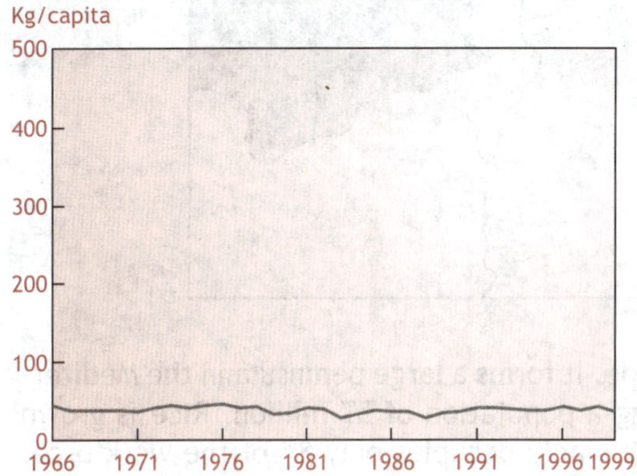

Per capita production (paddy terms), 1966-99

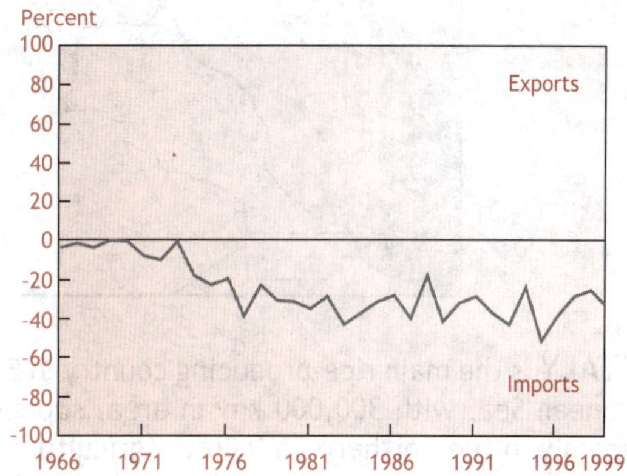

Net trade status, 1966-99

Basic statistics, Iran

	1985	1990	1995	1998	1999	2000
Rice						
Area harvested (ha)	478,592	524,266	565,575	614,963	587,150	587,150
Yield (t/ha)	3.7	3.8	4.1	4.5	4.0	4.0
Production (t)	1,776,085	1,981,019	2,300,901	2,770,574	2,348,241	2,348,241
Rice imports (t)	539,321	620,000	1,633,000	631,292	852,000	na
Paddy imports (t)	30,433	0	0	na	na	na
Rice exports (t)	0	0	0	12	425	na
Others						
Population, total (×10^3)	47,622	56,309	62,324	65,758	66,796	na
Population, agriculture (×10^3)	17,050	18,531	18,607	18,518	18,447	na
Agricultural area (×10^3 ha)	59,870	60,500	63,018	62,803	na	na
Irrigated agricultural area (×10^3 ha)	6,800	7,000	7,264	7,562	na	na
Total fertilizer consumption (t)	902,557	1,161,000	1,009,023	1,252,500	na	na
Tractors used in agric. (no.)	150,000	215,000	228,450	228,200	na	na

Source: FAOSTAT online database.

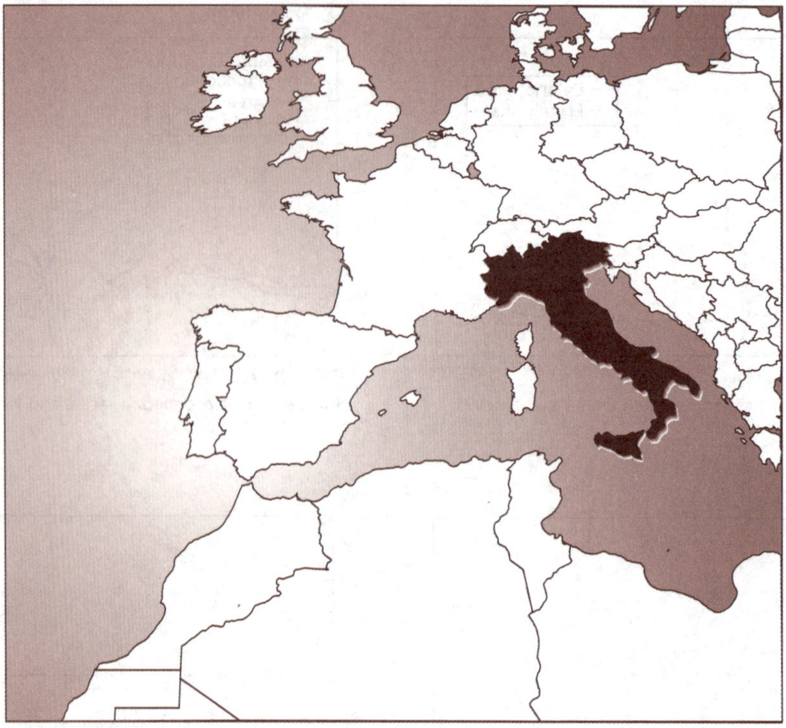

ITALY is the main rice-producing country in Europe. It forms a large peninsula in the Mediterranean Sea, with 300,000 km^2 in area, supporting a population of 57 million. Rice is grown mainly in the northern Po Valley. Agriculture as a whole occupies only 8% of the workforce. Arable land makes up 31% of the total area, but agriculture contributes only 2.6% to the GDP.

General information

- GNI per capita PPP$, 2000: 23,470
- Internal renewable water resources: 179.4 km^3
- Incoming water flow: 7.6 km^3
- Main food consumed: wheat, oil and fat, meat, sugar and honey, milk
- Rice consumption, 1999: 5.9 kg milled rice per person per year

Production season

	Planting	Harvesting
Main season	Apr-May	Sep-Oct

Production constraints

Rice yield has stagnated during the last decade. The main constraints to sustainable rice production in Italy are

- Low temperature
- Stagnation of the yield potential of rice varieties
- High production costs
- Blast
- Red rice
- Dependence on external market for the product

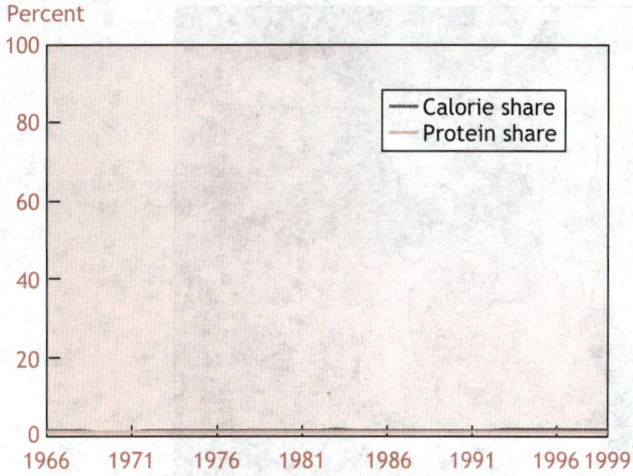

Share of calories and protein from rice, 1966-99

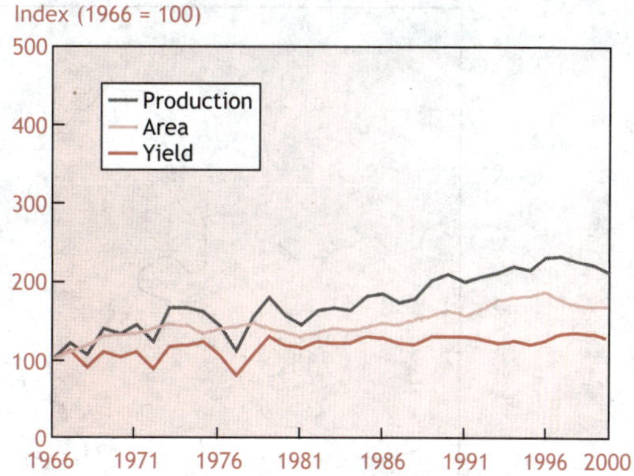

Indices of rice production, area, and yield, 1966-2000

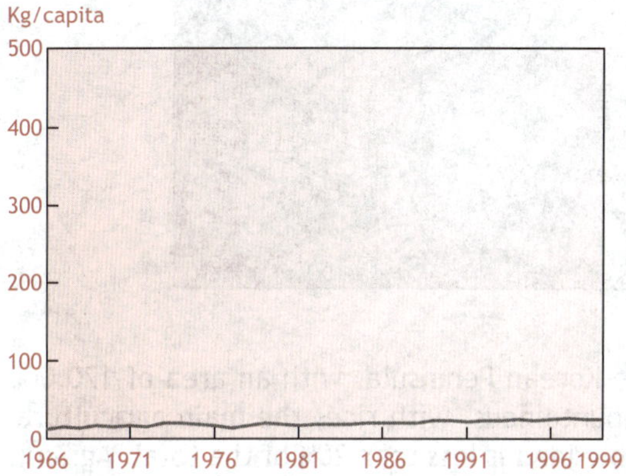

Per capita production (paddy terms), 1966-99

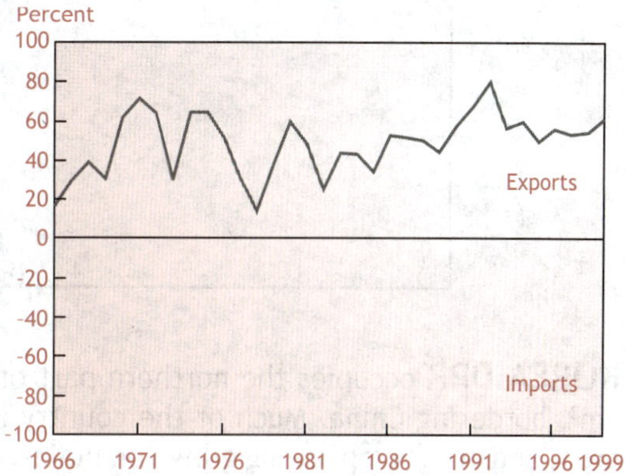

Net trade status, 1966-99

Basic statistics, Italy

	1985	1990	1995	1998	1999	2000
Rice						
Area harvested (ha)	185,510	214,124	239,259	222,705	220,795	220,850
Yield (t/ha)	6.1	6.0	5.5	6.3	6.2	5.9
Production (t)	1,122,700	1,290,700	1,320,850	1,393,524	1,362,452	1,300,000
Rice imports (t)	352,201	51,065	58,383	60,908	66,212	na
Paddy imports (t)	131,892	1,928	36,726	30,466	11,548	na
Rice exports (t)	727,432	576,947	523,898	601,597	667,367	na
Others						
Paddy exports (t)	25,276	41,369	24,511	13,677	13,105	na
Population, total (×10^3)	56,771	57,023	57,338	57,369	57,343	na
Population, agriculture (×10^3)	6,030	4,904	3,885	3,362	3,201	na
Agricultural area (×10^3 ha)	17,095	16,840	15,333	15,377	na	na
Irrigated agricultural area (×10^3 ha)	2,425	2,711	2,698	2,698	na	na
Total fertilizer consumption (t)	2,102,149	1,944,380	1,865,900	1,742,000	na	na
Tractors used in agric. (no.)	1,227,134	1,429,756	1,460,000	1,475,000	na	na

Source: FAOSTAT online database.

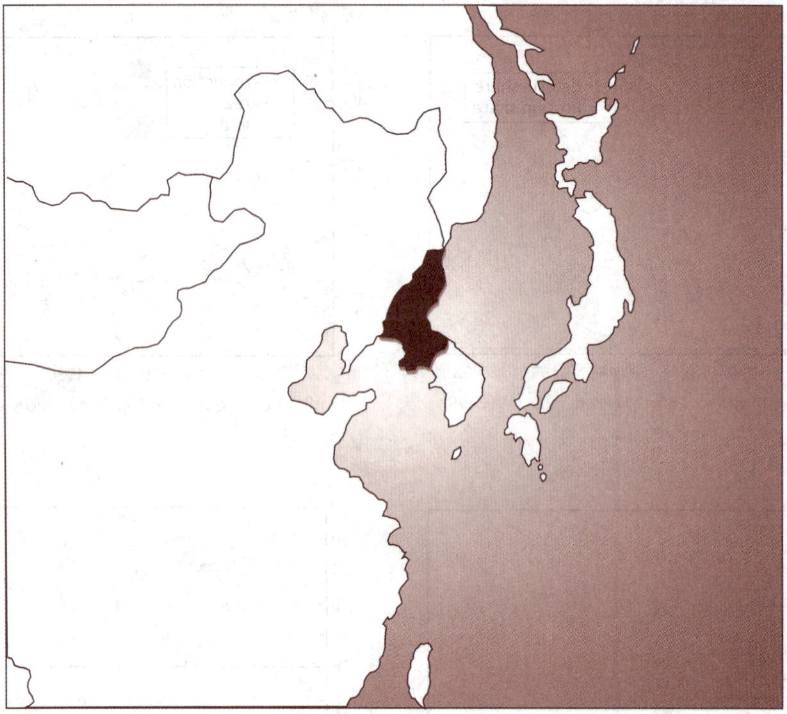

KOREA DPR occupies the northern part of the Korean Peninsula, with an area of 120,000 km^2, bordering China. Much of the country is mountainous, with rice, the main agricultural crop, and other crops being grown on plains. Arable land is less than 20% of the total. Agriculture makes up 30% of the GDP.

General information

- Internal renewable water resources: 67 km^3
- Main food consumed: rice, maize, roots and tubers, wheat, pulses
- Rice consumption, 1999: 75.4 kg milled rice per person per year

Production season

	Planting	Harvesting
Main season	Apr-May	Aug-Oct

Production constraints

During recent years, rice production has declined because of a substantial decrease in rice yield. The main constraints to sustainable rice production in Korea DPR are

- Uneven rainfall distribution during the rainy season—heavy rains during September and October
- Soils with low fertility
- Lack of production inputs
- Weeds, insects, and diseases, especially blast
- Unsuitable system of farm management (collective cooperative)

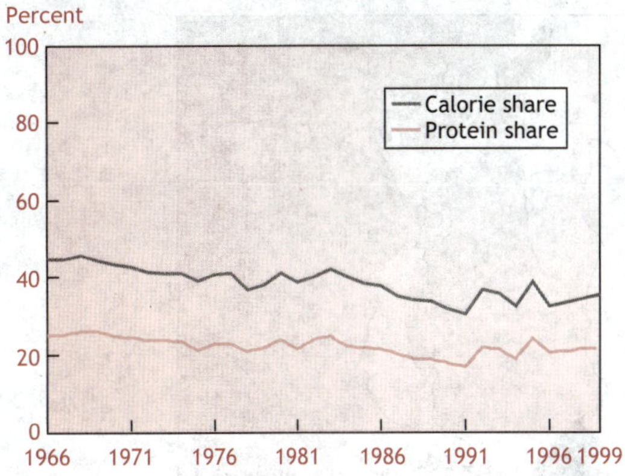

Share of calories and protein from rice, 1966-99

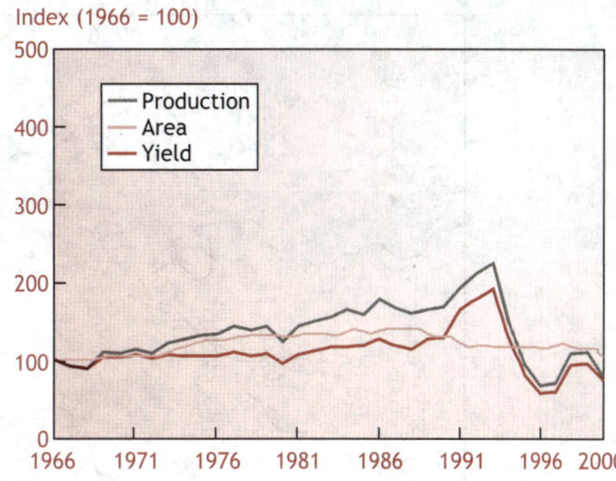

Indices of rice production, area, and yield, 1966-2000

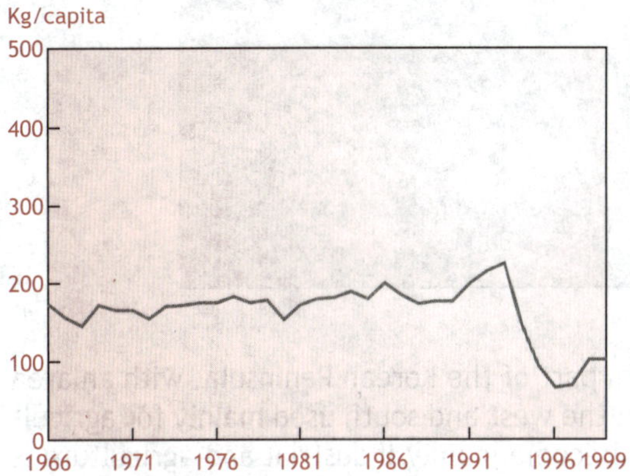

Per capita production (paddy terms), 1966-99

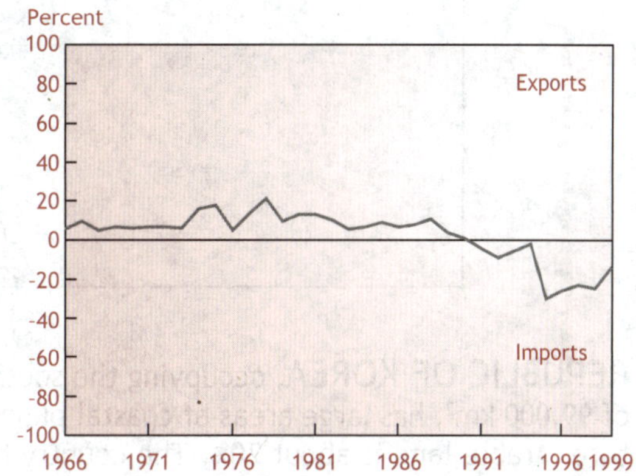

Net trade status, 1966-99

Basic statistics, Korea DPR

	1985	1990	1995	1998	1999	2000
Rice						
Area harvested (ha)	670,000	650,000	582,000	580,000	580,000	535,000
Yield (t/ha)	5.0	5.5	3.5	4.0	4.0	3.2
Production (t)	3,368,500	3,570,000	2,016,000	2,307,000	2,343,000	1,690,000
Rice imports (t)	0	27,000	587,000	507,500	250,000	na
Rice exports (t)	200,000	43,000	0	na	na	na
Others						
Population, total ($\times10^3$)	18,945	20,461	22,239	23,348	23,702	na
Population, agriculture ($\times10^3$)	7,844	7,778	7,550	7,382	7,315	na
Agricultural area ($\times10^3$ ha)	2,005	2,050	2,050	2,050	na	na
Irrigated agricultural area ($\times10^3$ ha)	1,270	1,420	1,460	1,460	na	na
Total fertilizer consumption (t)	844,400	832,400	104,200	157,205	na	na
Tractors used in agric. (no.)	68,000	73,000	75,000	75,000	na	na

Source: FAOSTAT online database.

REPUBLIC OF KOREA, occupying the southern part of the Korean Peninsula, with an area of 99,000 km^2, has large areas of coastal plains in the west and south used mainly for agriculture. Arable land is about 20%. The country has become mainly industrial and agriculture is only 5% of the GDP.

General information

- GNI per capita PPP$, 2000: 17,300
- Internal renewable water resources: 66.1 km^3
- Main food consumed: rice, wheat, sugar and honey, oil and fat, alcohol beverages
- Rice consumption, 1999: 94.2 kg milled rice per person per year

Production constraints

There are several constraints to sustainable rice production in the Republic of Korea:

- High costs of production
- Low return from rice production and aging of rice farmers
- Increasing preference of the population for japonica rice
- Because of already high national yield, a further increase in rice yield requires the development and use of novel technologies
- High possibility of water shortage during early stages of crop
- Flood caused by heavy rains during July-September
- Salinity caused by tidal influence in coastal areas
- Micronutrient deficiency, especially Mg, Zn, and Bo
- Weeds, insects, and disease (blast)

Production season

	Planting	Harvesting
Main season	May-Jun	Oct-Nov

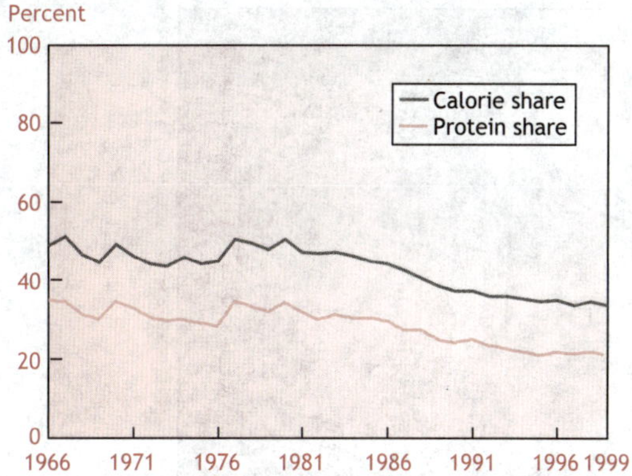

Share of calories and protein from rice, 1966-99

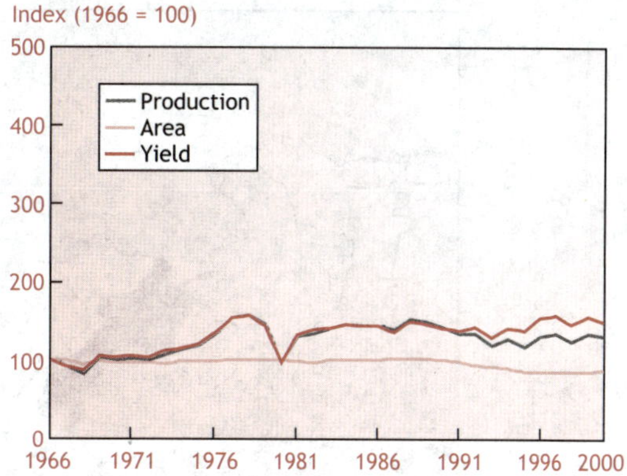

Indices of rice production, area, and yield, 1966-2000

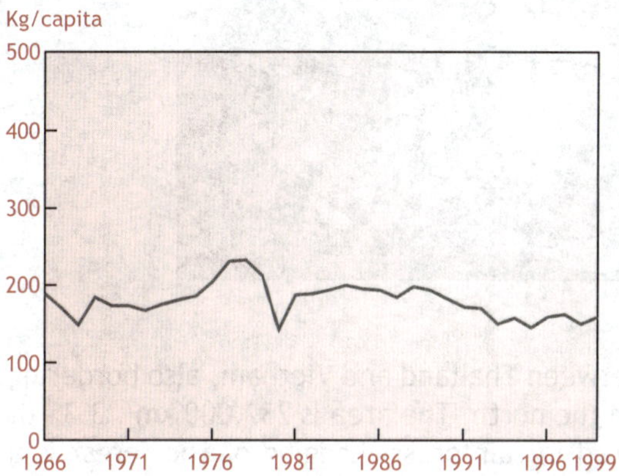

Per capita production (paddy terms), 1966-99

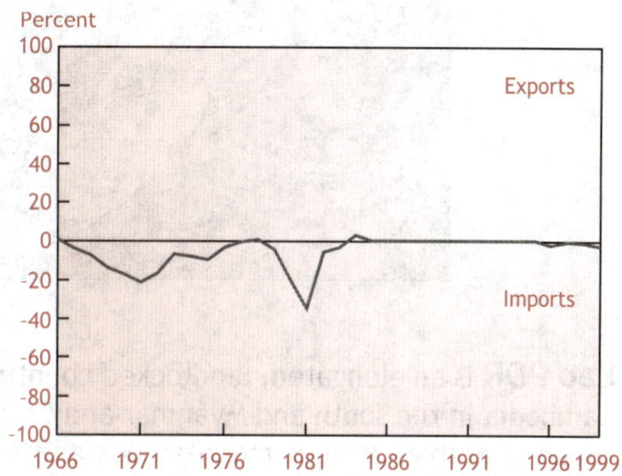

Net trade status, 1966-99

Basic statistics, Republic of Korea

	1985	1990	1995	1998	1999	2000
Rice						
Area harvested (ha)	1,236,768	1,244,341	1,055,337	1,056,483	1,058,600	1,072,000
Yield (t/ha)	6.4	6.2	6.1	6.4	6.9	6.6
Production (t)	7,855,262	7,721,968	6,387,301	6,779,290	7,270,500	7,067,000
Rice imports (t)	306	1,611	731	61,618	155,818	na
Paddy imports (t)	6	0	0	39	20	na
Rice exports (t)	155	544	493	47	6	na
Paddy exports (t)	0	0	0	68	4	na
Others						
Population, total ($\times 10^3$)	40,806	42,869	44,949	46,109	46,480	na
Population, agriculture ($\times 10^3$)	9,547	6,917	5,348	4,570	4,334	na
Agricultural area ($\times 10^3$ ha)	2,220	2,179	2,048	1,969	na	na
Irrigated agricultural area ($\times 10^3$ ha)	1,325	1,345	1,206	1,159	na	na
Total fertilizer consumption (t)	807,000	915,941	972,595	874,000	na	na
Tractors used in agric. (no.)	12,389	41,203	100,412	157,888	na	

Source: FAOSTAT online database.

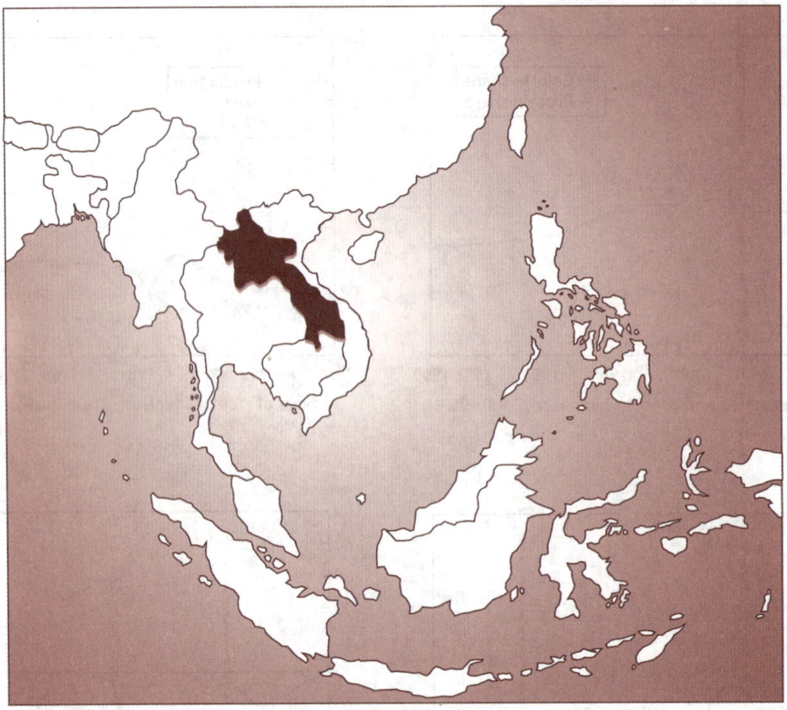

Lao PDR is an elongated, landlocked country between Thailand and Vietnam, also bordering Cambodia in the south and Myanmar and China in the north. The area is 237,000 km^2, 3.3% of which is arable. Most is mountainous and covered in rainforest. Logging of the forest is a major environmental problem. The Mekong River forms a fertile valley through most of the country, where agriculture, mainly rice farming, is carried out. Agriculture makes up half the country's GDP.

General information

- GNI per capita PPP$, 2000: 1,540
- Internal renewable water resources: 270 km^3
- Main food consumed: rice, meat, roots and tubers, maize, pulses
- Rice consumption, 1999: 171.4 kg milled rice per person per year

Production constraints

During the last five years, rice yield has steadily increased. A further increase in yield, however, may require the development of irrigation facilities. There are several constraints to sustainable rice production:

- Drought in rainfed areas
- Soil erosion and fertility losses in upland rice cultivation
- Periodic flooding because of swelling of the Mekong River
- Sandy soils with low fertility, especially of P and K
- Weeds, insects, and diseases
- Inadequate infrastructure
- Inadequate credit and input supply: fertilizer and other agrochemicals
- Small farm size
- Price fluctuations
- Popular preference for glutinous rice

Production season

	Planting	Harvesting
Wet season	May-Jul	Nov-Dec
Dry season	Dec-Jan	Apr-Jun

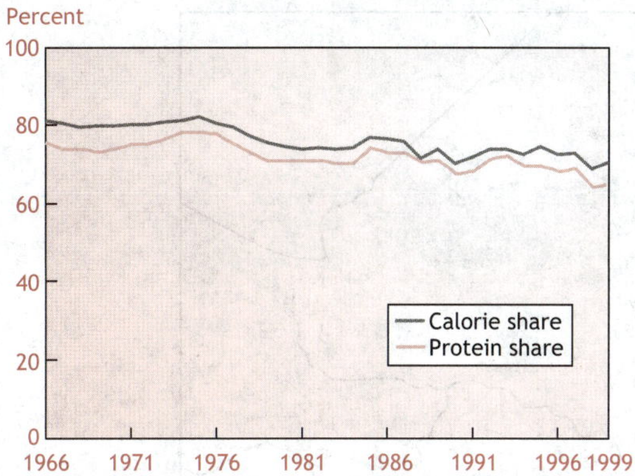

Share of calories and protein from rice, 1966-99

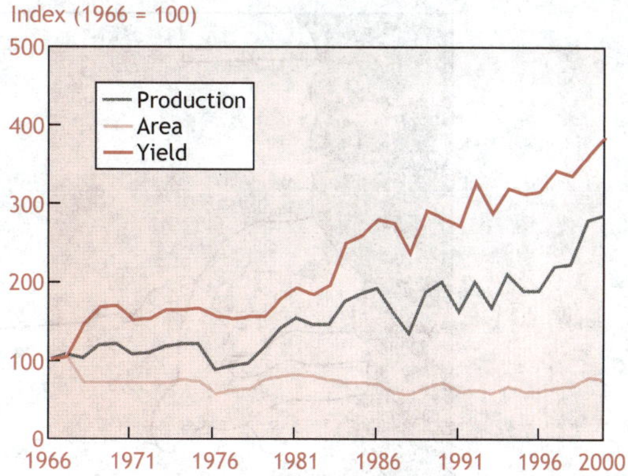

Indices of rice production, area, and yield, 1966-2000

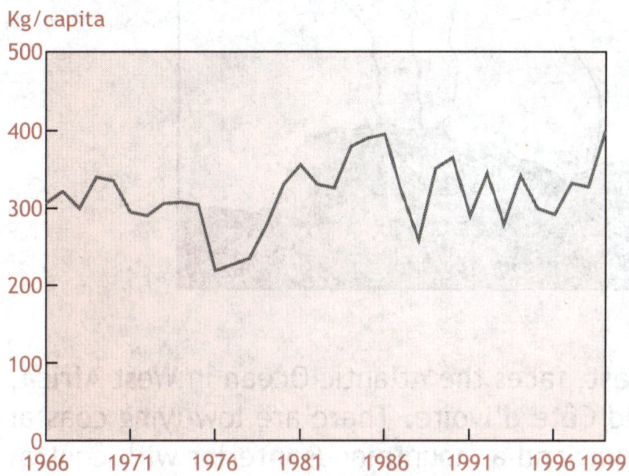

Per capita production (paddy terms), 1966-99

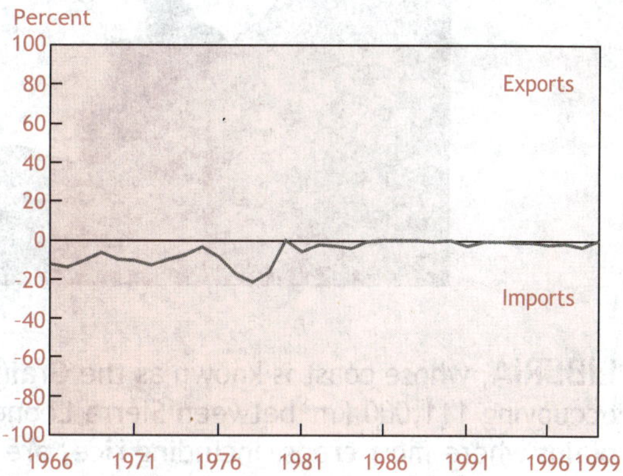

Net trade status, 1966-99

Basic statistics, Lao PDR

	1985	1990	1995	1998	1999	2000
Rice						
Area harvested (ha)	663,487	663,600	559,900	617,538	718,200	690,000
Yield (t/ha)	2.1	2.3	2.5	2.7	2.9	3.1
Production (t)	1,396,200	1,507,500	1,417,829	1,674,500	2,102,815	2,155,100
Rice imports (t)	6,800	4,240	15,939	40,585	4,707	na
Others						
Population, total ($\times10^3$)	3,594	4,152	4,773	5,163	5,297	na
Population, agriculture ($\times10^3$)	2,833	3,242	3,688	3,964	4,057	na
Agricultural area ($\times10^3$ ha)	1,670	1,650	1,650	1,678	na	na
Irrigated agricultural area ($\times10^3$ ha)	119	135	155	164	na	na
Total fertilizer consumption (t)	2,000	1,500	6,189	10,166	na	na
Tractors used in agric. (no.)	780	870	890	890	na	na

Source: FAOSTAT online database.

LIBERIA, whose coast is known as the Grain Coast, faces the Atlantic Ocean in West Africa, occupying 111,000 km^2 between Sierra Leone and Côte d'Ivoire. There are low-lying coastal plains where most crops, including rice, are grown, and a mountainous interior with central plateaus. Arable land is less than 5%, but agriculture contributes half of the country's GDP.

General information

- Internal renewable water resources: 200 km^3
- Incoming water flow: 32 km^3
- Main food consumed: rice, roots and tubers, oil and fat, fruits, sugar
- Rice consumption, 1999: 54.9 kg milled rice per person per year

Production constraints

There are several constraints to sustainable rice production in Liberia:

- Dominance of upland shifting cultivation
- Drought stress, low soil fertility, erosion of soil and soil fertility in upland areas
- Blast and weed competition
- Poor drainage, iron toxicity in undeveloped swamps
- Inadequate and irregular input supplies: seeds, fertilizer, and credit
- Lack of small farm equipment, especially for postharvest operations
- Lack of a well-defined rice policy
- Poor road networks and marketing systems
- Weak research and extension support
- Recently, deterioration of security situation

Production season

	Planting	Harvesting
Main season	Apr-Jul	Sep-Dec

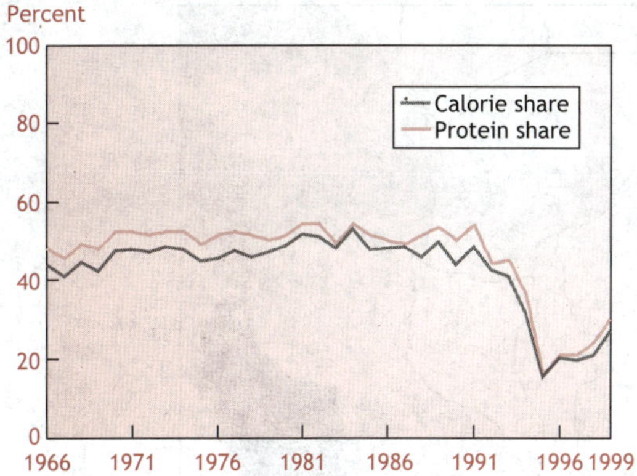

Share of calories and protein from rice, 1966-99

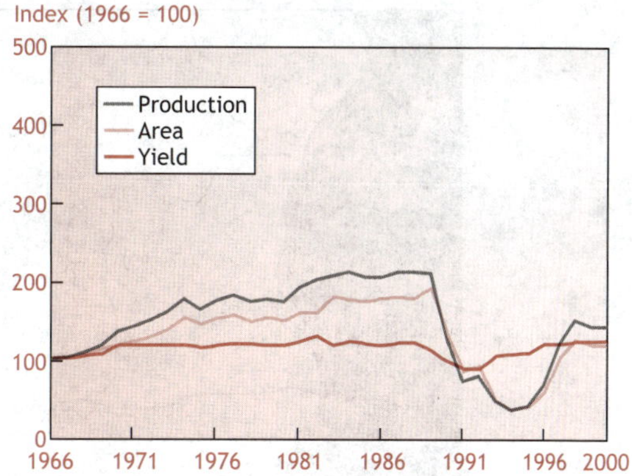

Indices of rice production, area, and yield, 1966-2000

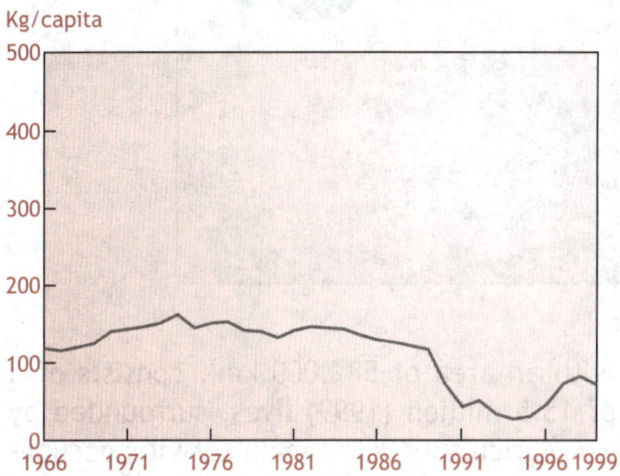

Per capita production (paddy terms), 1966-99

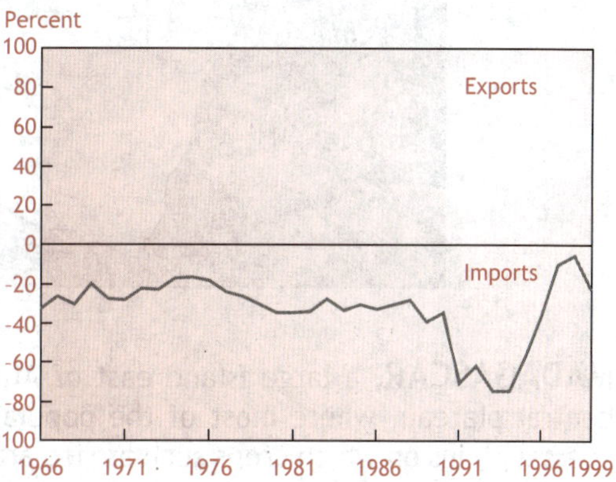

Net trade status, 1966-99

Basic statistics, Liberia

	1985	1990	1995	1998	1999	2000
Rice						
Area harvested (ha)	230,731	175,000	50,000	162,500	155,000	155,000
Yield (t/ha)	1.3	1.0	1.1	1.3	1.3	1.3
Production (t)	289,000	180,000	56,200	210,100	200,000	200,000
Rice imports (t)	85,318	65,000	50,000	9,100	40,700	na
Others						
Population, total ($\times 10^3$)	2,193	2,579	2,090	2,666	2,930	na
Population, agriculture ($\times 10^3$)	1,630	1,864	1,464	1,828	1,995	na
Agricultural area ($\times 10^3$ ha)	2,390	2,393	2,390	2,390	na	na
Total fertilizer consumption (t)	1,460	300	0	0	na	na
Tractors used in agric. (no.)	315	330	325	325	na	na

Source: FAOSTAT online database.

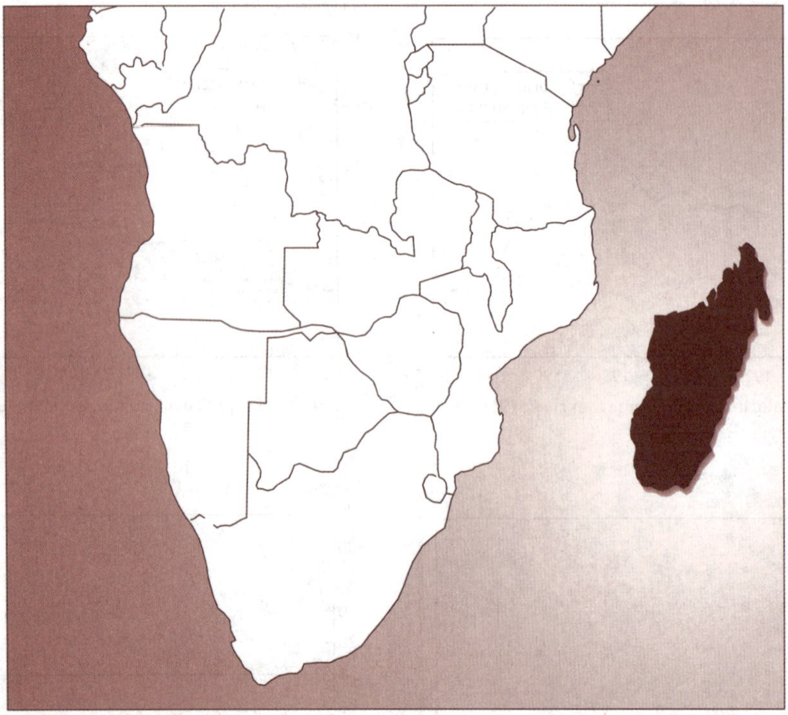

MADAGASCAR, a large island east of Africa, with an area of 587,000 km^2, consists of a central plateau, where most of the population of 15.5 million (1999) lives, surrounded by coastal plains on which crops such as rice are grown. Agriculture is the main activity, occupying 78% of the workforce and comprising 34% of the GDP. Arable land is 4.4% of the total. The plains are mainly rainforest on the wet and humid east side and grassland on the drier west side.

General information

- GNI per capita PPP$, 2000: 820
- Internal renewable water resources: 337 km^3
- Main food consumed: rice, roots and tubers, meat, oil and fat, maize
- Rice consumption, 1999: 91 kg milled rice per person per year

Production constraints

The following are the major constraints to sustainable rice production in Madagascar:

- Drought in upland areas and drought and flash flood in rainfed lowland areas because of irregular weather
- Low temperature in high plateaus, salinity in the western coast
- Disease (blast)
- Shortage of water in the dry season
- Irregular input supplies, especially fertilizers, improved seeds, and credit
- Poor maintenance of irrigation facilities
- Lack of clear understanding of rice ecosystems and a well-defined rice policy
- Poor road networks and marketing systems
- Weak research and extension support

Production season

	Planting	Harvesting
Main season, plateau	Oct-Nov	Apr-Jun
Vatomandry*, east coast	Oct-Nov	Jun-Jul
Hosy*, east coast	Apr-Jun	Oct-Nov
Asara*, west coast	Nov	Jan
Atriatry*, west coast	Feb	May
Jeby*, west coast	Jul	Oct

*Varieties of rice harvested

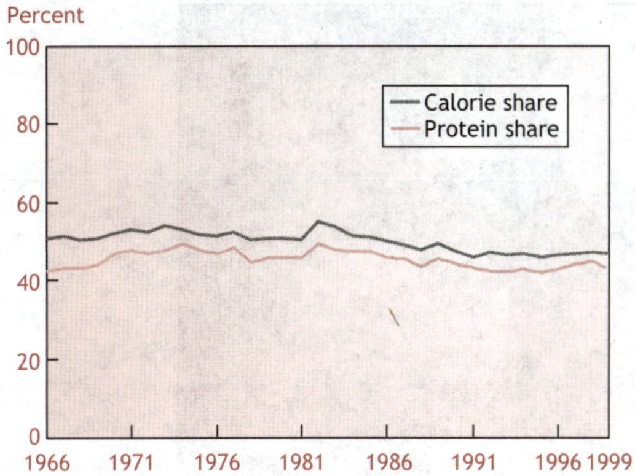

Share of calories and protein from rice, 1966-99

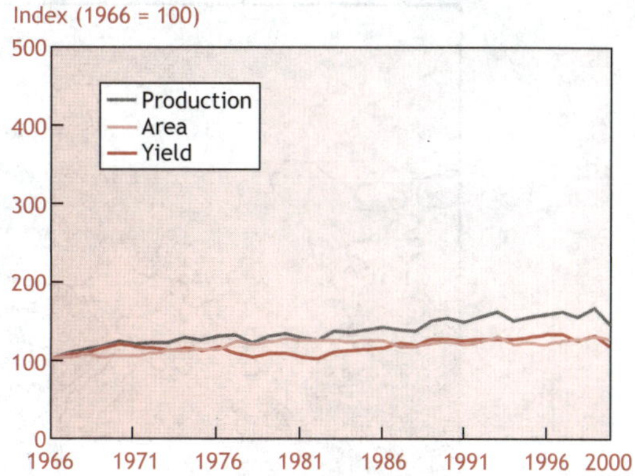

Indices of rice production, area, and yield, 1966-2000

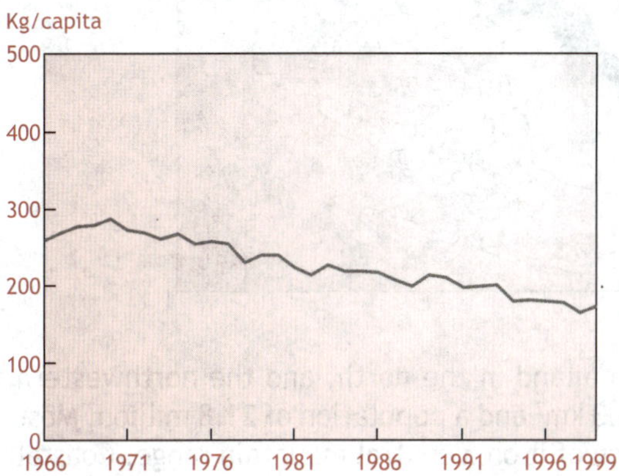

Per capita production (paddy terms), 1966-99

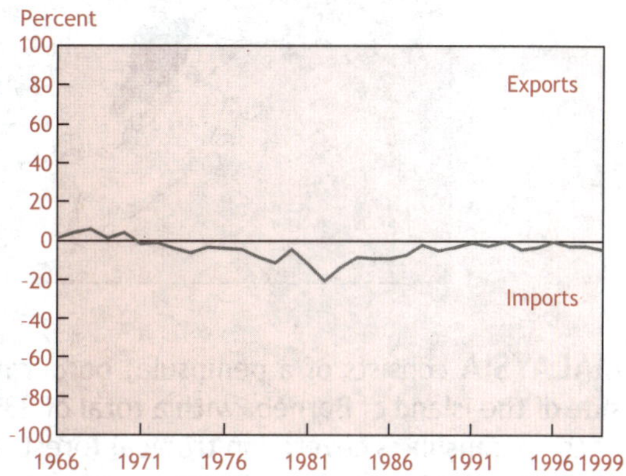

Net trade status, 1966-99

Basic statistics, Madagascar

	1985	1990	1995	1998	1999	2000
Rice						
Area harvested (ha)	1,183,520	1,165,000	1,150,000	1,203,000	1,227,000	1,206,940
Yield (t/ha)	1.8	2.1	2.1	2.0	2.1	1.9
Production (t)	2,177,680	2,420,000	2,450,000	2,447,000	2,637,000	2,300,000
Rice imports (t)	na	na	na	na	na	na
Others						
Population, total ($\times 10^3$)	10,123	11,632	13,744	15,057	15,497	na
Population, agriculture ($\times 10^3$)	8,093	9,098	10,493	11,313	11,578	na
Agricultural area ($\times 10^3$ ha)	27,040	27,102	27,105	27,108	na	na
Irrigated agricultural area ($\times 10^3$)	826	1,000	1,087	1,090	na	na
Total fertilizer consumption (t)	9,700	10,855	12,546	8,677	na	na
Tractors used in agric. (no.)	2,780	2,900	3,500	3,550	na	na

Source: FAOSTAT online database.

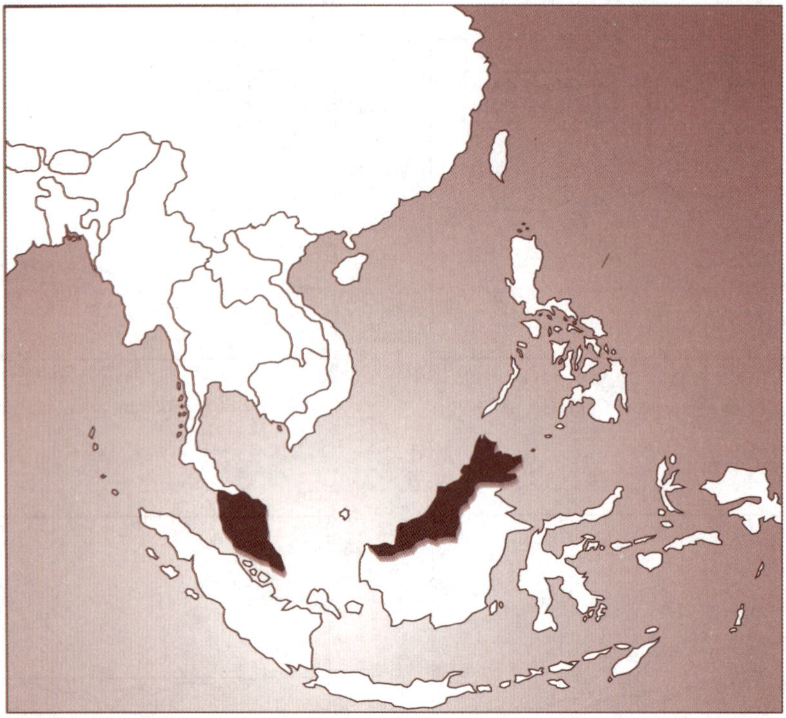

MALAYSIA consists of a peninsula, bordering Thailand in the north, and the northwestern side of the island of Borneo, with a total of 330,000 km^2 and a population of 21.8 million. Most of the peninsula is covered in tropical forest, much of it on a central mountain range. Coastal plains dominate the Borneo states and the interior is mountainous. About a quarter of the workforce is engaged in agriculture; 3% of the land is arable. Agriculture is 12% of the GDP.

General information

- GNI per capita PPP$, 2000: 8,330
- Internal renewable water resources: 456 km^3
- Main food consumed: rice, oil and fat, sugar and honey, wheat, meat
- Rice consumption, 1999: 88.4 kg milled rice per person per year

Production constraints

Yield has remained more or less unchanged during the last five years and is still low regardless of the large percentage of irrigated areas. There are several constraints to sustainable rice production in Malaysia:

- Periodic drought even under irrigation because of erratic rainfall
- Intrusion of seawater in low-lying rice fields
- Weeds: red rice because of the adoption of direct seeding; insects: brown planthopper, stem borer, and green leafhopper; and diseases: blast and rice tungro virus
- Labor shortage because of high labor wages in rural areas
- Unfavorable prices of rice compared with those of other food products

Production season

	Planting	Harvesting
Main season, Peninsular	Sep-Oct	Nov-Mar
Main season, Sabah	Jun-Aug	Jan-Mar
Main season, Sarawak	Oct-Nov	Mar-Apr

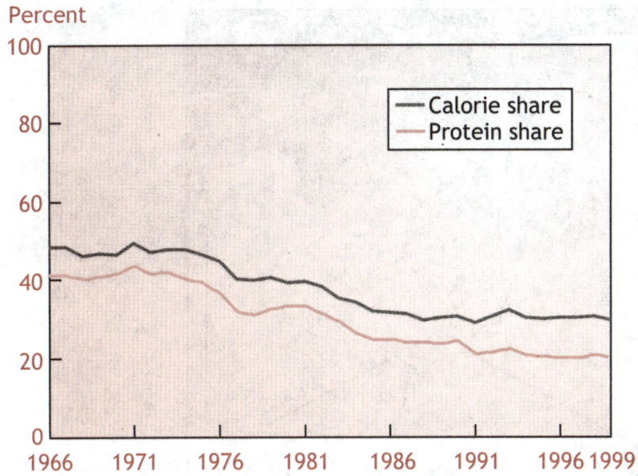

Share of calories and protein from rice, 1966-99

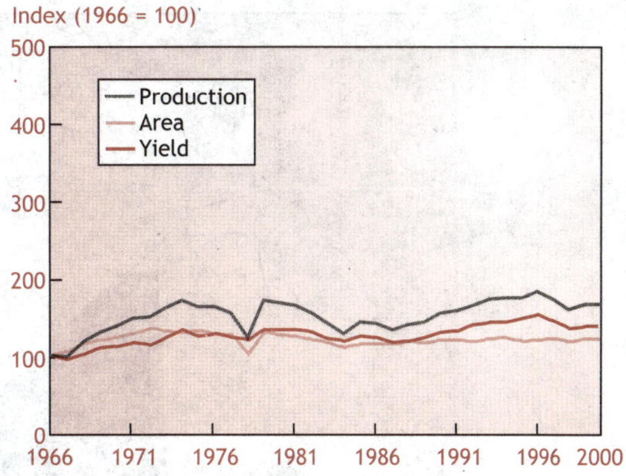

Indices of rice production, area, and yield, 1966-2000

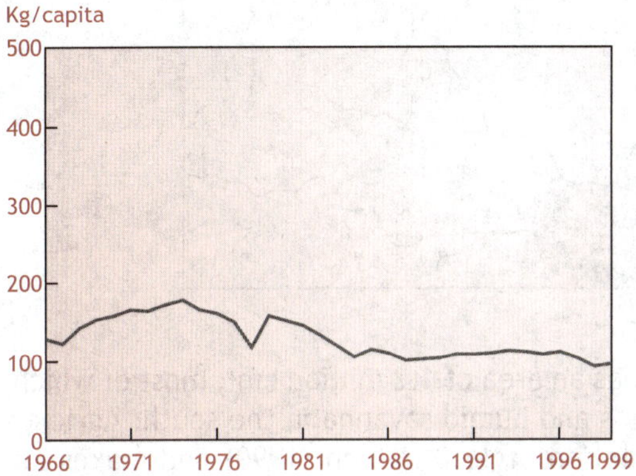

Per capita production (paddy terms), 1966-99

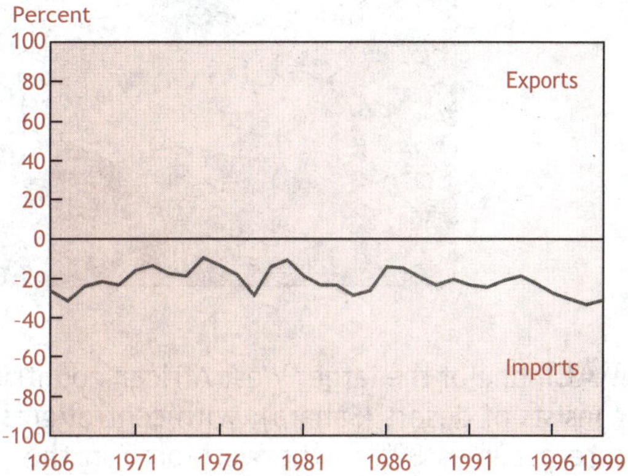

Net trade status, 1966-99

Basic statistics, Malaysia

	1985	1990	1995	1998	1999	2000
Rice						
Area harvested (ha)	654,974	680,647	672,787	674,404	692,389	692,389
Yield (t/ha)	2.7	2.8	3.2	2.9	2.9	2.9
Production (t)	1,745,367	1,884,984	2,127,271	1,944,240	2,036,641	2,036,641
Rice imports (t)	428,017	330,336	427,556	657,870	612,467	na
Rice exports (t)	2,002	111	2,430	2,088	117	na
Paddy exports (t)	0	0	24	na	na	na
Others						
Population, total ($\times10^3$)	15,677	17,845	20,108	21,410	21,830	na
Population, agriculture ($\times10^3$)	5,006	4,646	4,314	4,089	4,011	na
Agricultural area ($\times10^3$ ha)	5,798	7,176	7,885	7,890	na	na
Irrigated agricultural area ($\times10^3$ ha)	334	335	363	365	na	na
Total fertilizer consumption (t)	611,400	951,500	1,087,000	1,406,111	na	na
Tractors used in agric. (no.)	12,000	26,000	43,295	43,300	na	na

Source: FAOSTAT online database.

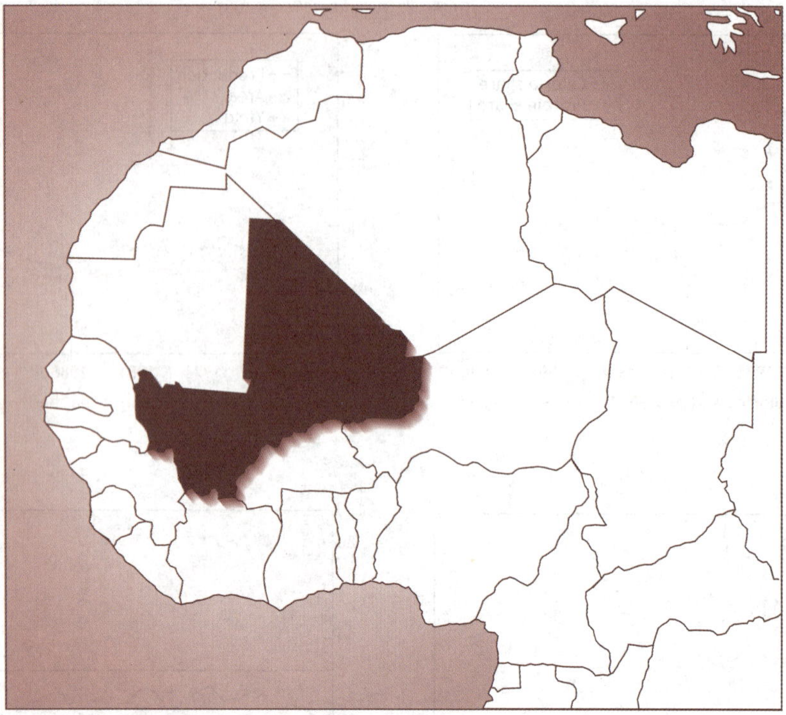

MALI, one of the larger West African countries, has an area of 1.2 million km^2, most of which consists of desert (Sahara), with central grasslands and humid savanna in the south. Agriculture occupies 80% of the workforce in the population of 11.0 million (1999) and makes up nearly half the GDP.

General information

- GNI per capita PPP$, 2000: 780
- Internal renewable water resources: 60 km^3
- Incoming water flow: 40 km^3
- Main food consumed: millet and sorghum, rice, maize, oil and fat, milk
- Rice consumption, 1999: 48.7 kg milled rice per person per year

Production constraints

The following are the major constraints to sustainable rice production in Mali:

- Heat-induced sterility during the off-season in irrigated areas
- Cold-induced sterility in December
- Recurrent drought and flooding in deepwater rice areas
- Shortage of water in the dry season
- Weed competition, especially wild rice (*Oryza barthi* and *O. longistaminata*)
- Poor maintenance of irrigation facilities
- Inadequate and irregular input supplies: seeds, fertilizer, and credit
- Lack of small farm equipment, especially for postharvest operations
- Competition for labor by other production activities
- Poor road networks and marketing systems
- Weak research and extension support

Production season

	Planting	Harvesting
Main season	May-Jul	Oct-Dec
Off-season	Jan-Mar	May-Jul
Deepwater rice	Jul-Aug	Dec-Jan

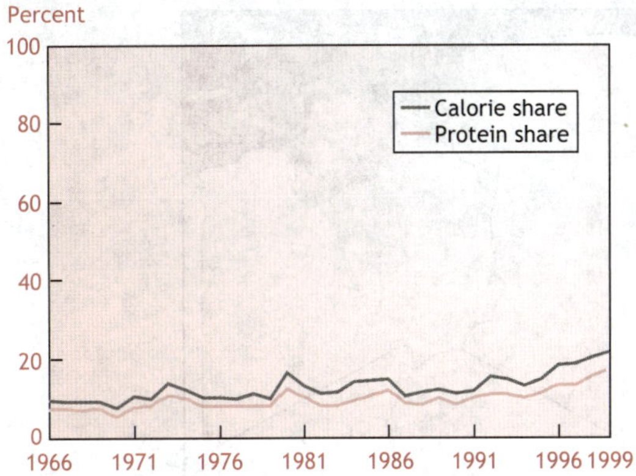

Share of calories and protein from rice, 1966-99

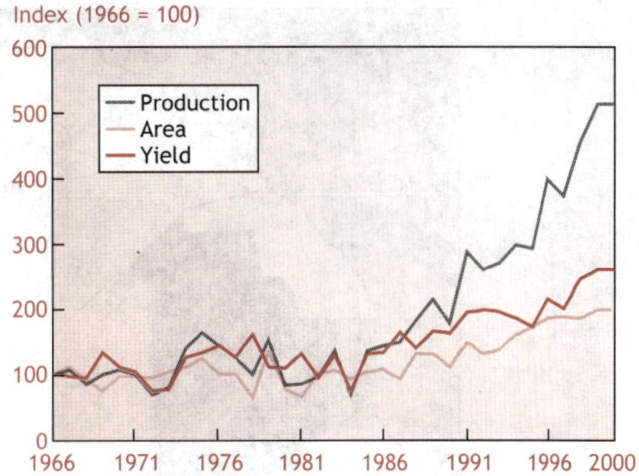

Indices of rice production, area, and yield, 1966-2000

The scale on this graph is not comparable with the scale on production, area, and yield graphs for most other countries.

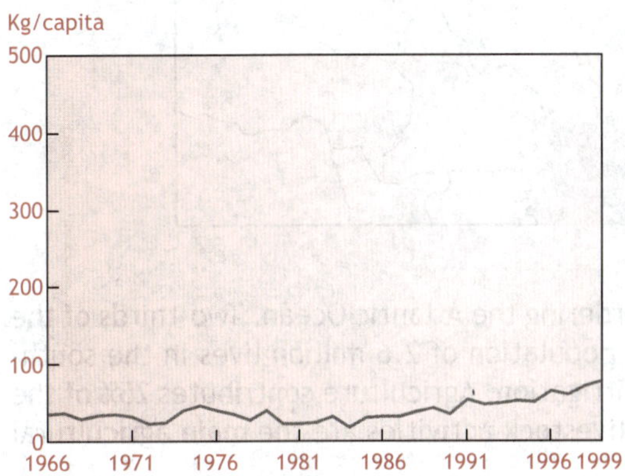

Per capita production (paddy terms), 1966-99

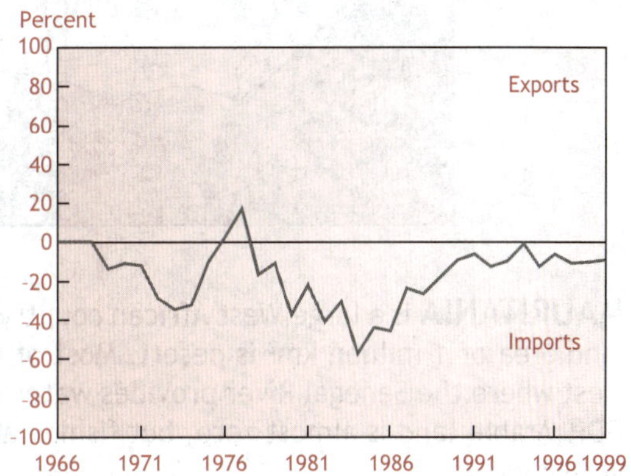

Net trade status, 1966-99

Basic statistics, Mali

	1985	1990	1995	1998	1999	2000
Rice						
Area harvested (ha)	184,833	196,631	302,669	330,000	350,000	350,000
Yield (t/ha)	1.2	1.4	1.5	2.2	2.3	2.3
Production (t)	213,841	282,366	462,702	717,860	809,555	809,555
Rice imports (t)	114,000	20,000.	45,700	55,300	55,000	na
Others						
Population, total (×10³)	7,915	8,842	9,944	10,694	10,960	na
Population, agriculture (×10³)	6,918	7,586	8,309	8,775	8,935	na
Agricultural area (×10³ ha)	32,073	32,093	33,419	34,650	na	na
Irrigated agricultural area (×10³ ha)	100	120	138	138	na	na
Total fertilizer consumption (t)	19,706	15,200	27,000	52,623	na	na
Tractors used in agric. (no.)	1,400	2,100	2,515	2,600	na	na

Source: FAOSTAT online database.

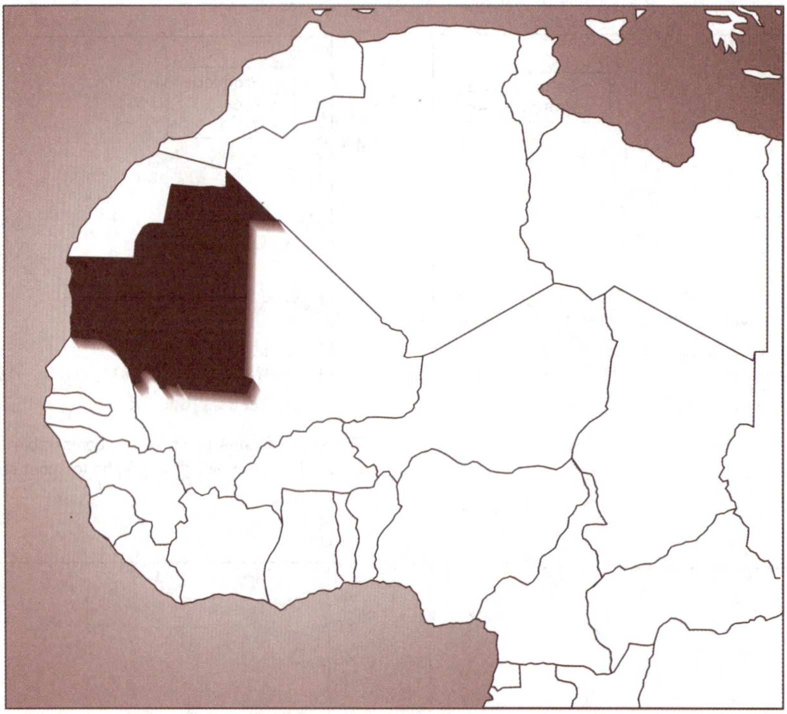

MAURITANIA is a large West African country bordering the Atlantic Ocean. Two-thirds of the land area of 1 million km^2 is desert. Most of the population of 2.6 million lives in the southwest where the Senegal River provides water for irrigation. Agriculture contributes 25% of the GDP. Arable land is almost zero, but fishing and livestock activities are the main agricultural activities.

General information

- GNI per capita PPP$, 2000: 1,630
- Internal renewable water resources: 0.4 km^3
- Incoming water flow: 11.4 km^3
- Main food consumed: wheat, rice, sugar and honey, fruits, roots and tubers
- Rice consumption, 1999: 62.4 kg milled rice per person per year

Production constraints

There are several constraints to sustainable rice production in Mauritania:

- Heat-induced sterility during the off-season in irrigated areas
- Saline soils
- Cold-induced sterility in December
- Poor maintenance of irrigation facilities
- Poor drainage in irrigated schemes
- High production costs
- Inadequate and irregular input supplies: seeds, fertilizer, and credit
- Lack of small farm equipment, especially for postharvest operations
- Lack of effective farmers' organizations and cooperatives
- Competition for labor by other production activities
- Lack of a well-defined rice policy
- Poor road networks and marketing systems
- Weak research and extension support

Production season

	Planting	Harvesting
Main season	Jun-Jul	Nov-Dec
Off-season	Feb-Mar	May-Jul

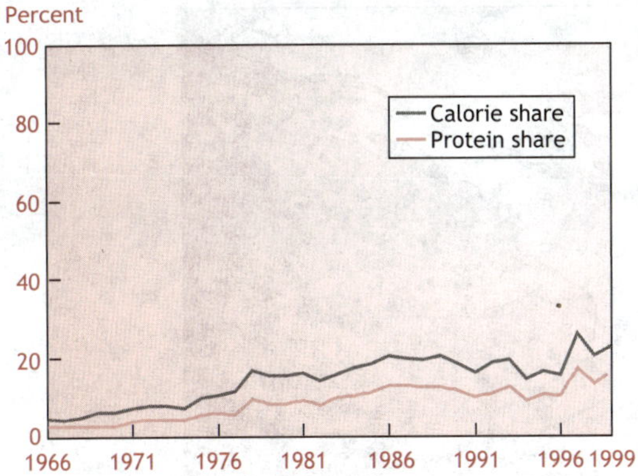

Share of calories and protein from rice, 1966-99

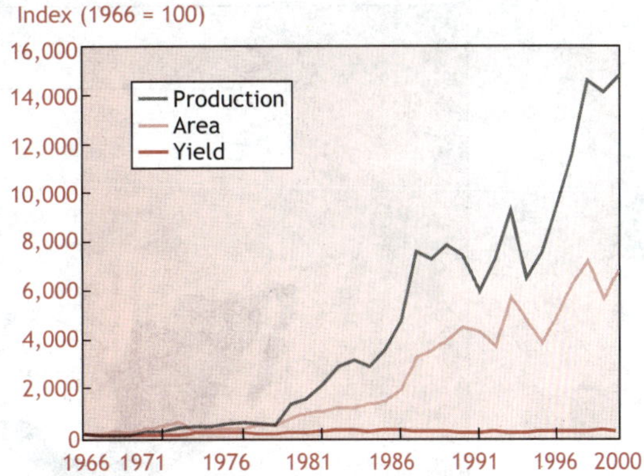

Indices of rice production, area, and yield, 1966-2000

The scale on this graph is not comparable with the scale on production, area, and yield graphs for most other countries.

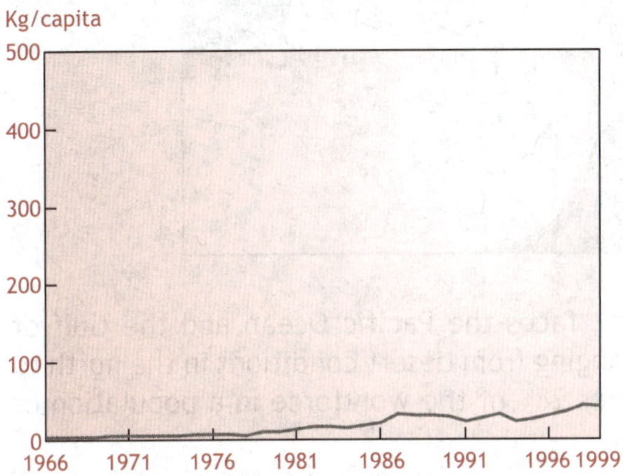

Per capita production (paddy terms), 1966-99

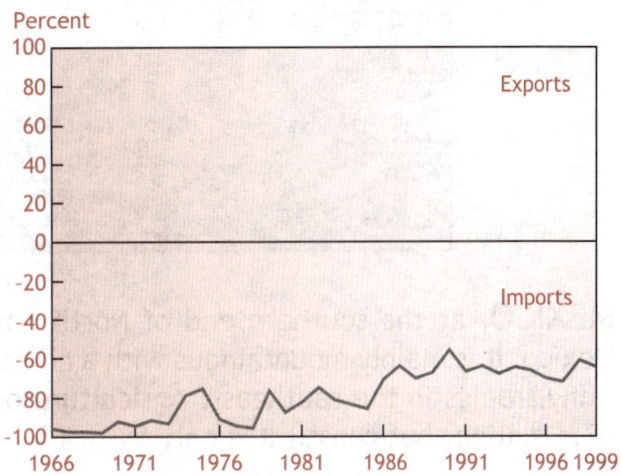

Net trade status, 1966-99

Basic statistics, Mauritania

	1985	1990	1995	1998	1999	2000
Rice						
Area harvested (ha)	5,000	15,551	13,418	25,100	19,600	24,000
Yield (t/ha)	5.0	3.3	3.9	4.1	5.0	4.3
Production (t)	25,000	51,796	52,813	101,900	98,500	103,400
Rice imports (t)	105,100	44,600	70,000	106,900	122,300	na
Others						
Population, total ($\times 10^3$)	1,766	2,026	2,329	2,529	2,598	na
Population, agriculture ($\times 10^3$)	1,119	1,118	1,259	1,349	1,379	na
Agricultural area ($\times 10^3$ ha)	39,555	39,656	39,760	39,750	na	na
Irrigated agricultural area ($\times 10^3$ ha)	49	49	49	49	na	na
Total fertilizer consumption (t)	2,010	3,900	4,000	2,100	na	na
Tractors used in agric. (no.)	320	335	330	380	na	na

Source: FAOSTAT online database.

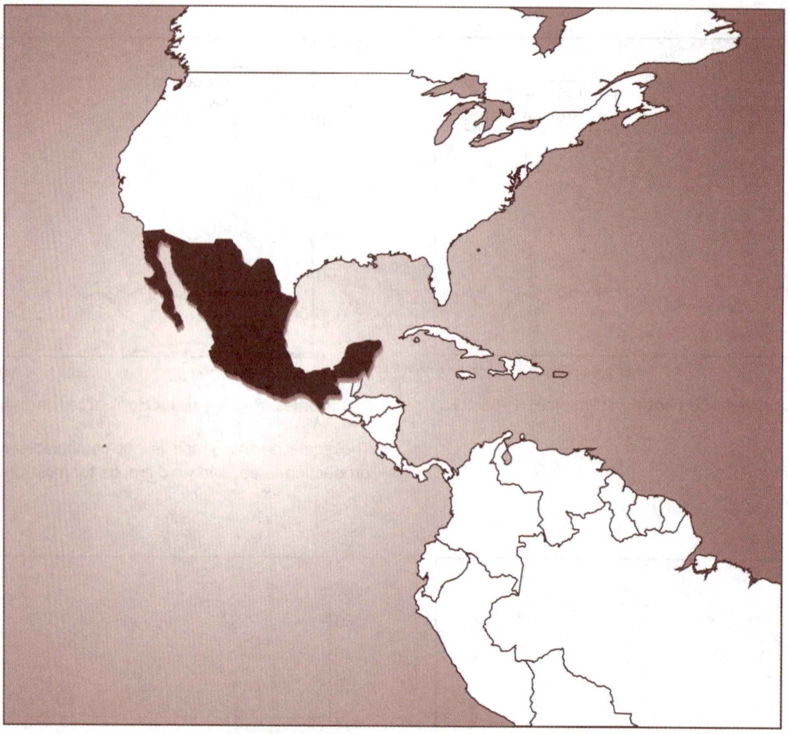

MEXICO, at the southern end of North America, faces the Pacific Ocean and the Gulf of Mexico. It is mainly mountainous with a climate ranging from desert conditions in the north to rainy tropics in the southeast. Agriculture occupies 28% of the workforce in a population of 97.4 million, but constitutes only 5% of the GDP.

General information

- GNI per capita PPP$, 2000: 8,790
- Internal renewable water resources: 357.4 km^3
- Main food consumed: maize, sugar and honey, oil and fat, wheat, meat
- Rice consumption, 1999: 5.8 kg milled rice per person per year

Production season

	Planting	Harvesting
Main season	Mar-Apr	Aug-Sep

Production constraints

The following are the major constraints to sustainable rice production in Mexico:

- Limited irrigation water supply; salinity in the northern and dry zone
- Blast
- Weed and red rice competition
- Drought stress in the upland system in the southeastern region
- High production costs

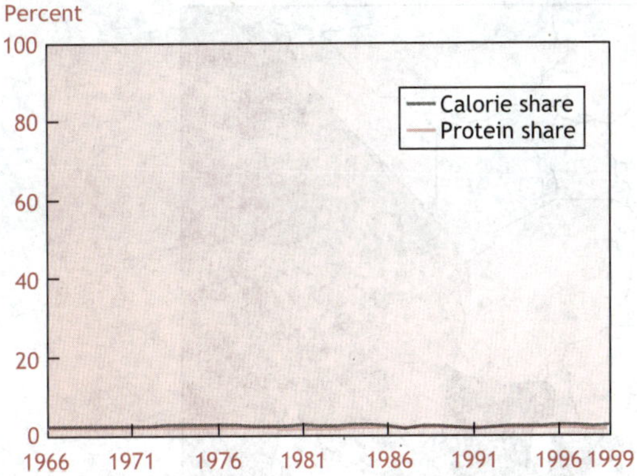

Share of calories and protein from rice, 1966-99

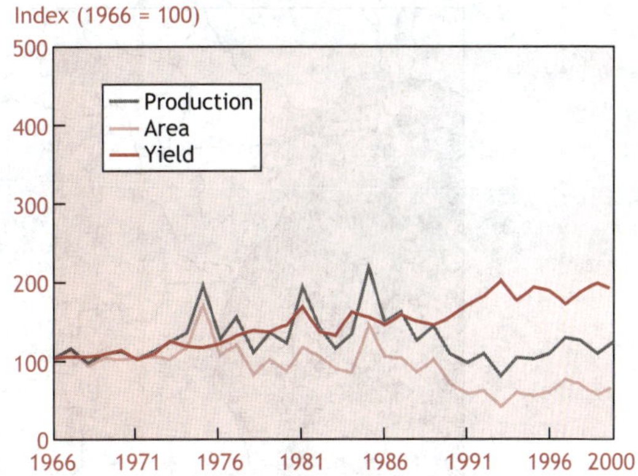

Indices of rice production, area, and yield, 1966-2000

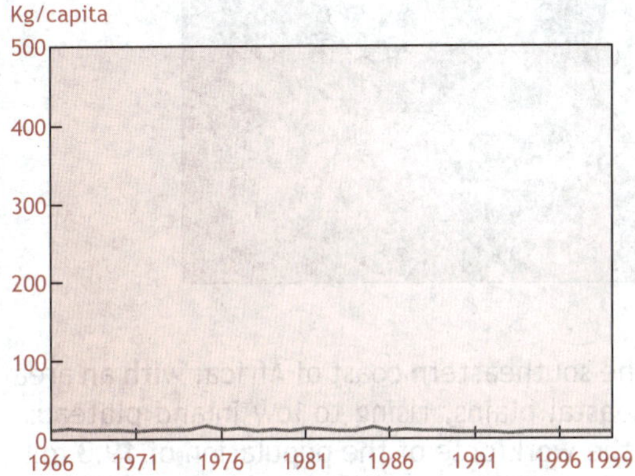

Per capita production (paddy terms), 1966-99

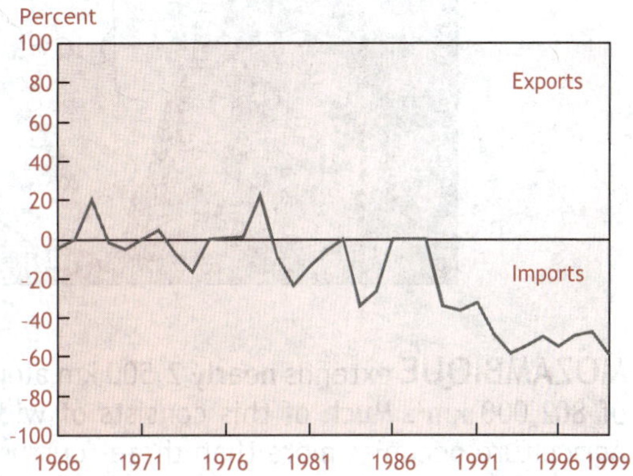

Net trade status, 1966-99

Basic statistics, Mexico

	1985	1990	1995	1998	1999	2000
Rice						
Area harvested (ha)	216,466	105,402	78,439	101,560	82,619	97,877
Yield (t/ha)	3.7	3.7	4.7	4.5	4.8	4.6
Production (t)	807,529	394,388	367,030	458,112	394,746	450,488
Rice imports (t)	199,532	150,670	246,432	291,775	405,023	na
Paddy imports (t)	0	18,114	264,941	380,363	513,577	na
Rice exports (t)	0	0	30	na	na	na
Paddy exports (t)	na	na	na	36	0	na
Others						
Population, total ($\times 10^3$)	75,465	83,226	91,145	95,831	97,365	na
Population, agriculture ($\times 10^3$)	25,680	24,795	24,276	23,864	23,709	na
Agricultural area ($\times 10^3$ ha)	100,000	103,400	107,200	107,200	na	na
Irrigated agricultural area ($\times 10^3$ ha)	5,285	5,600	6,400	6,500	na	na
Total fertilizer consumption (t)	1,714,100	1,798,600	1,286,000	1,705,700	na	na
Tractors used in agric. (no.)	157,000	170,000	172,000	172,000	na	na

Source: FAOSTAT online database.

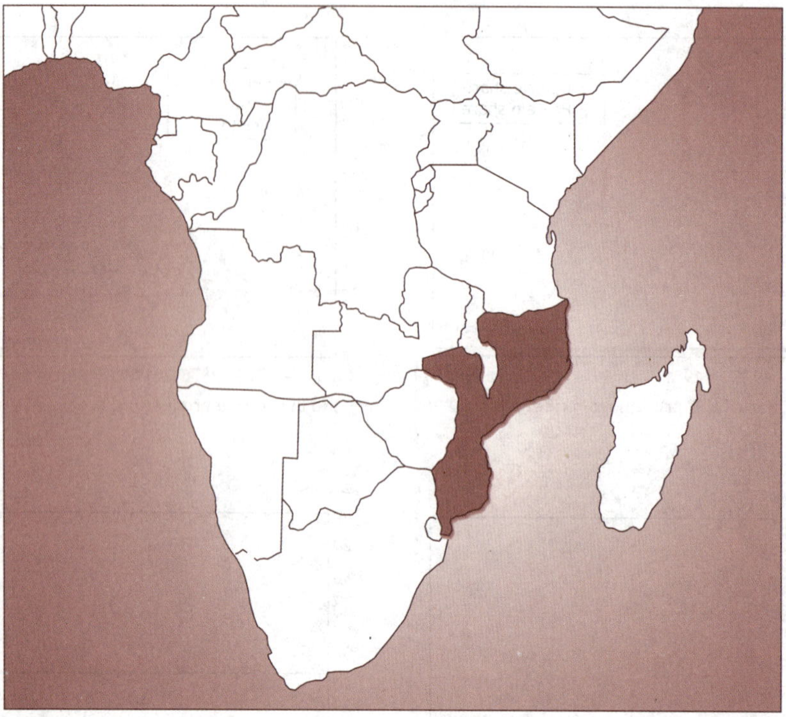

MOZAMBIQUE extends nearly 2,500 km along the southeastern coast of Africa, with an area of 802,000 km². Much of this consists of wide coastal plains, rising to low inland plateaus. Agriculture occupies more than three-fourths of the workforce of the population of 19.3 million (1999). Arable land is less than 5% of the total. Agriculture makes up 34% of the GDP.

General information

- GNI per capita PPP$, 2000: 800
- Internal renewable water resources: 100 km^3
- Incoming water flow: 116 km^3
- Main food consumed: roots and tubers, maize, oil and fat, millet and sorghum, rice
- Rice consumption, 1999: 7.5 kg milled rice per person per year

Production constraints

The major constraints to sustainable rice production in Mozambique are

- Drought and flash flood in rainfed lowland areas because of irregular weather
- Salinity in many irrigated areas
- Poor maintenance of irrigation facilities
- Lack of pure and good seeds
- Inadequate and irregular input supplies: seeds, fertilizer, and credit
- Lack of small farm equipment
- Lack of effective farmers' organizations and cooperatives
- Poor road networks and marketing systems
- Lack of a well-defined rice policy
- Weak research and extension support

Production season

	Planting	Harvesting
Main season	Nov-Jan	May-Jun

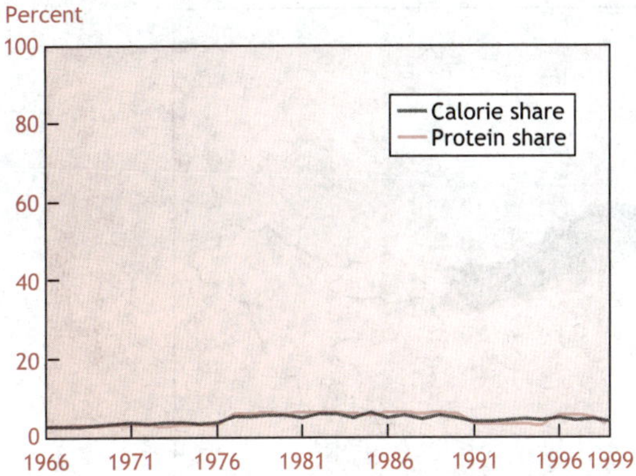

Share of calories and protein from rice, 1966-99

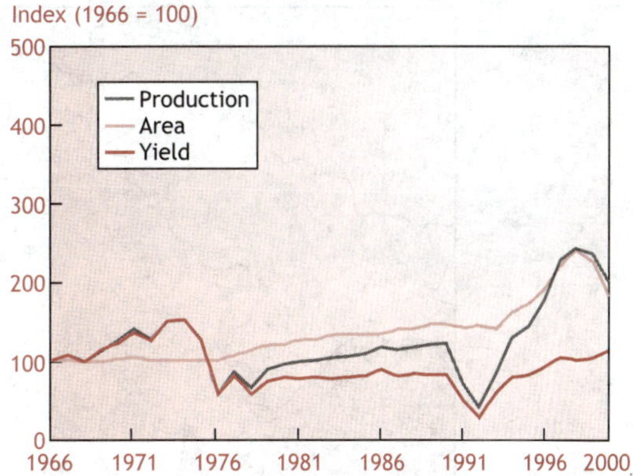

Indices of rice production, area, and yield, 1966-2000

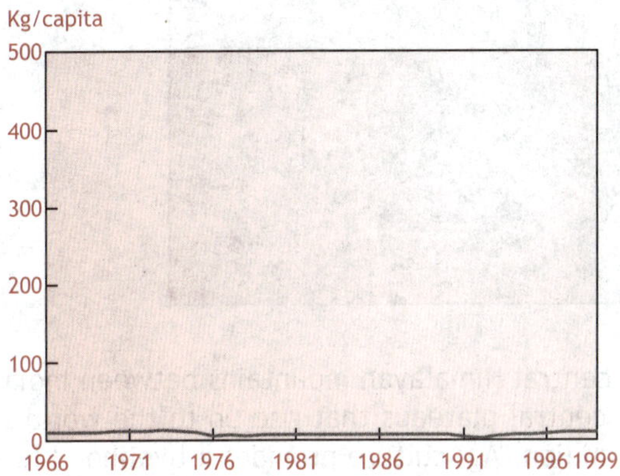

Per capita production (paddy terms), 1966-99

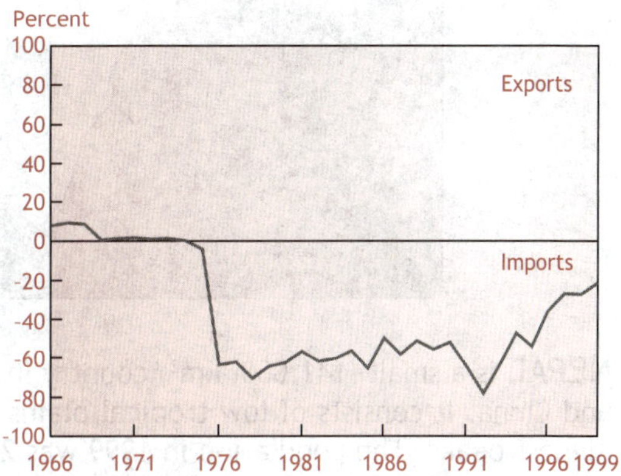

Net trade status, 1966-99

Basic statistics, Mozambique

	1985	1990	1995	1998	1999	2000
Rice						
Area harvested (ha)	100,000	110,160	129,605	181,000	170,000	136,000
Yield (t/ha)	0.9	0.9	0.9	1.1	1.1	1.2
Production (t)	86,000	96,355	112,982	191,000	186,000	157,937
Rice imports (t)	108,955	70,000	90,000	48,700	34,000	na
Others						
Population, total (×10^3)	13,535	14,198	17,388	18,880	19,286	na
Population, agriculture (×10^3)	10,762	11,160	13,447	14,459	14,722	na
Agricultural area (×10^3 ha)	47,150	47,300	47,350	47,350	na	na
Irrigated agricultural area (×10^3 ha)	93	105	107	107	na	na
Total fertilizer consumption (t)	3,803	2,600	7,800	5,035	na	na
Tractors used in agric. (no.)	5,750	5,750	5,750	5,750	na	na

Source: FAOSTAT online database.

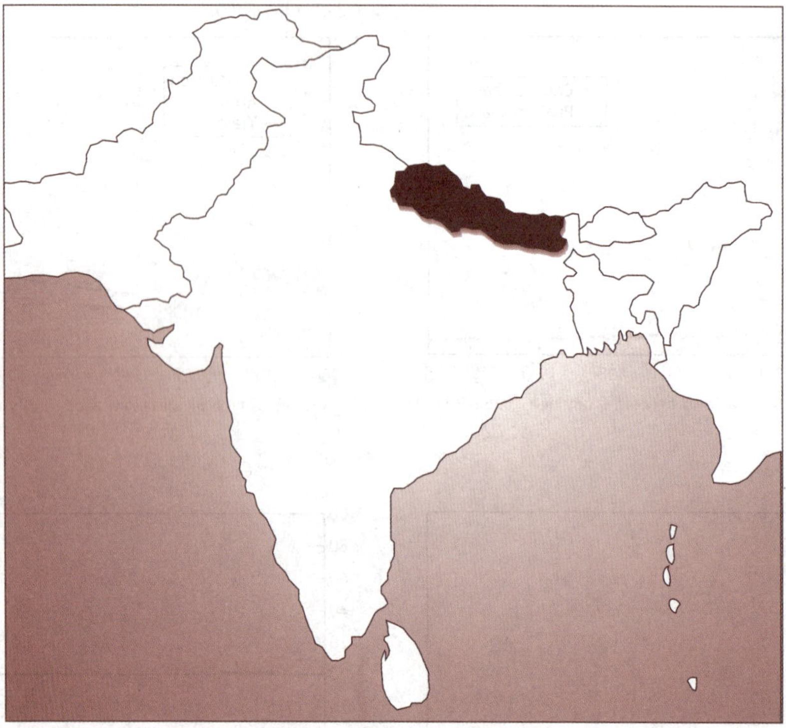

NEPAL is a small (141,000 km²) country in the central Himalayan mountains between India and China. It consists of low tropical plains and central plateaus that rise up to the world's highest peaks. The population in 1999 was 23.4 million. Agriculture provides a livelihood for 80% of the population and accounts for 41% of the GDP.

General information

- GNI per capita PPP$, 2000: 1,370
- Internal renewable water resources: 170 km³
- Main food consumed: rice, maize, wheat, oil and fat, millet and sorghum
- Rice consumption, 1999: 93.4 kg milled rice per person per year

Production constraints

Yields in 2000 remained low. There are several constraints to sustainable rice production in Nepal:

- Blast and bacterial blight
- Stem borers, hispa, leaf roller, and planthoppers
- Prolonged drought or prolonged submergence because of excess rain; unpredictable and erratic rainfall
- Adverse topography, declining soil fertility, and Zn deficiency
- Deterioration of irrigation facilities
- Cold injury, strong wind, and hail storms
- Ineffective and inadequate input/credit supplies
- Weak research and extension support
- Popular use of rice straw as feed for livestock. Farmers often sacrifice grain yields and grow tall varieties with high straw yields to meet feed requirements of their livestock

Production season

	Planting	Harvesting
Main season	Jun-Aug	Sep-Nov

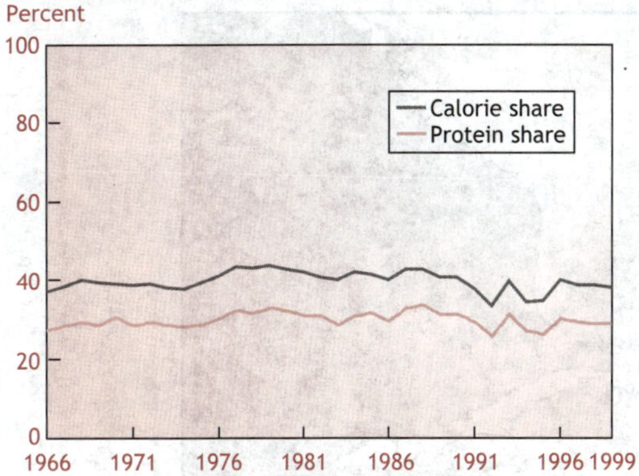

Share of calories and protein from rice, 1966-99

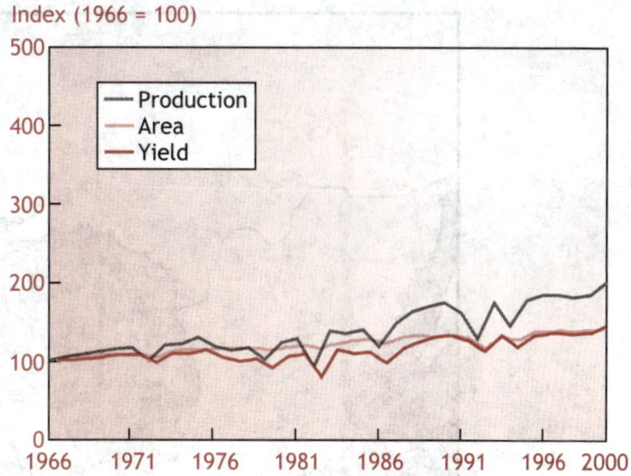

Indices of rice production, area, and yield, 1966-2000

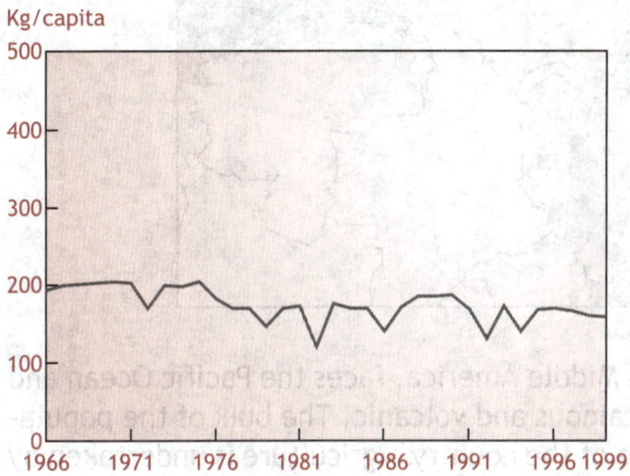

Per capita production (paddy terms), 1966-99

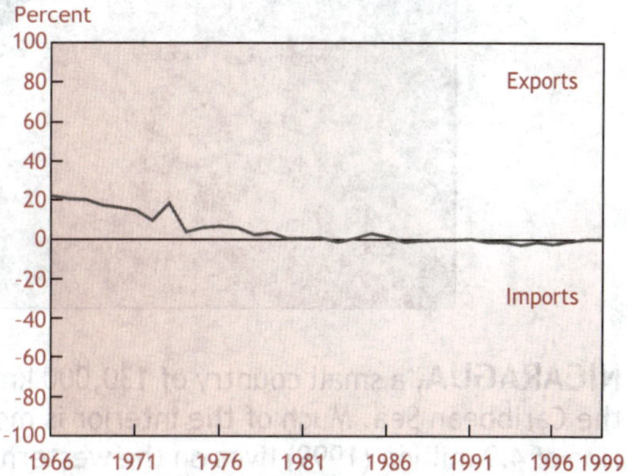

Net trade status, 1966-99

Basic statistics, Nepal

	1985	1990	1995	1998	1999	2000
Rice						
Area harvested (ha)	1,391,040	1,455,170	1,496,790	1,506,340	1,514,210	1,550,000
Yield (t/ha)	2.0	2.4	2.4	2.4	2.5	2.6
Production (t)	2,804,490	3,502,160	3,578,830	3,640,860	3,709,770	4,030,100
Rice imports (t)	7,412	11,594	40,000	11,100	11,100	na
Paddy imports (t)	3,619	5,500	0	na	na	na
Rice exports (t)	59,077	0	0	7,876	7,876	na
Paddy exports (t)	109	0	0	na	na	na
Others						
Population, total (×10^3)	16,503	18,772	21,272	22,847	23,385	na
Population, agriculture (×10^3)	15,458	17,580	19,862	21,294	21,782	na
Agricultural area (×10^3 ha)	4,313	4,150	4,725	4,725	na	na
Irrigated agricultural area (×10^3 ha)	760	950	1,134	1,135	na	na
Total fertilizer consumption (t)	43,409	72,682	93,699	121,300	na	na
Tractors used in agric. (no.)	183,373	182,228	172,596	155,000	na	na

Source: FAOSTAT online database.

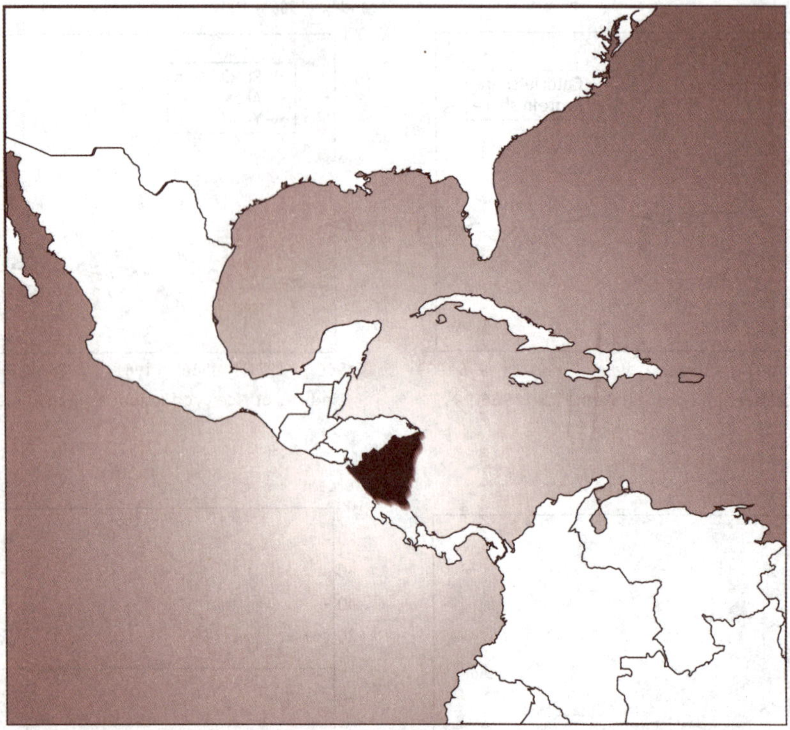

NICARAGUA, a small country of 130,000 km^2 in Middle America, faces the Pacific Ocean and the Caribbean Sea. Much of the interior is mountainous and volcanic. The bulk of the population of 4.9 million (1999) lives on the western side of the country. Agriculture is undertaken by 42% of the workforce and makes up 34% of the GDP.

General information

- GNI per capita PPP$, 2000: 2,080
- Internal renewable water resources: 175 km^3
- Main food consumed: maize, sugar and honey, rice, oil and fat, wheat
- Rice consumption, 1999: 31.2 kg milled rice per person per year

Production season

	Planting	Harvesting
Main season	Apr-May	Sep-Oct

Production constraints

The following are the major constraints to sustainable rice production in Nicaragua:

- Drought stress because of unreliable rainfall in upland areas
- Blast and weed competition
- Lack of input supply
- Weak research and extension, especially in on-farm water management

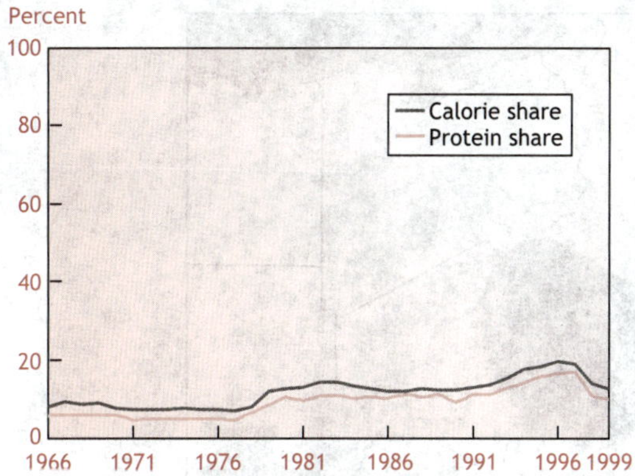

Share of calories and protein from rice, 1966-99

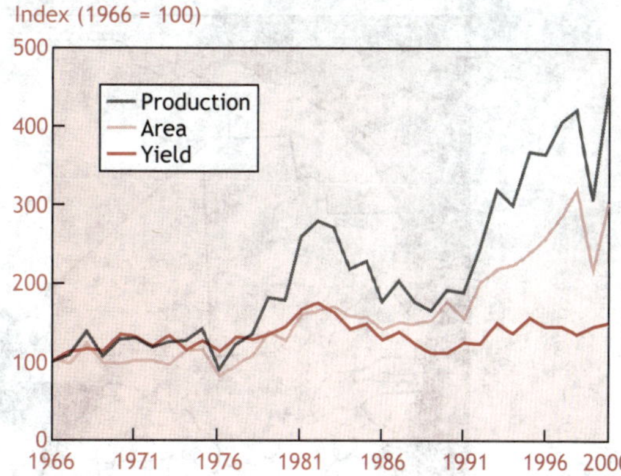

Indices of rice production, area, and yield, 1966-2000

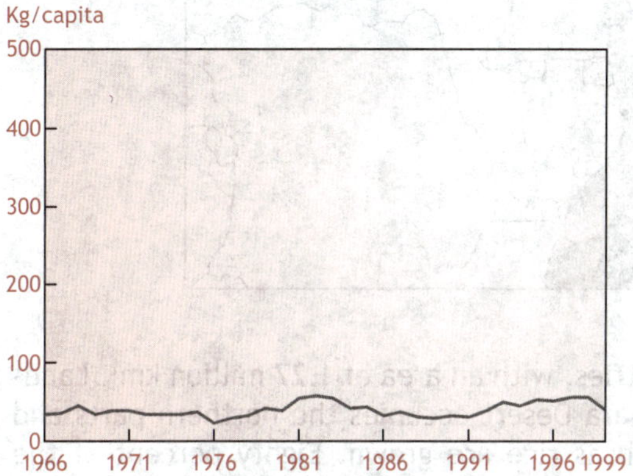

Per capita production (paddy terms), 1966-99

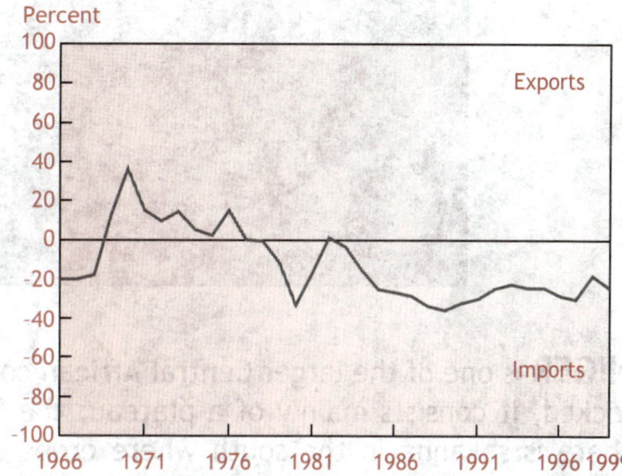

Net trade status, 1966-99

Basic statistics, Nicaragua

	1985	1990	1995	1998	1999	2000
Rice						
Area harvested (ha)	40,810	45,920	62,683	83,573	56,226	80,184
Yield (t/ha)	3.5	2.6	3.7	3.2	3.4	3.6
Production (t)	143,992	120,890	232,456	265,927	192,962	285,315
Rice imports (t)	33,008	38,730	52,893	40,657	42,334	na
Paddy imports (t)	0	0	1,156	13,762	3,322	na
Rice exports (t)	0	0	1,329	na	na	na
Paddy exports (t)	0	0	672	110	5	na
Others						
Population, total ($\times 10^3$)	3,404	3,827	4,426	4,807	4,938	na
Population, agriculture ($\times 10^3$)	1,228	1,194	1,162	1,139	1,130	na
Agricultural area ($\times 10^3$ ha)	6,587	7,029	7,561	7,561	na	na
Irrigated agricultural area ($\times 10^3$ ha)	83	85	88	88	na	na
Total fertilizer consumption (t)	62,985	40,010	31,000	52,600	na	na
Tractors used in agric. (no.)	2,430	2,600	2,700	2,700	na	na

Source: FAOSTAT online database.

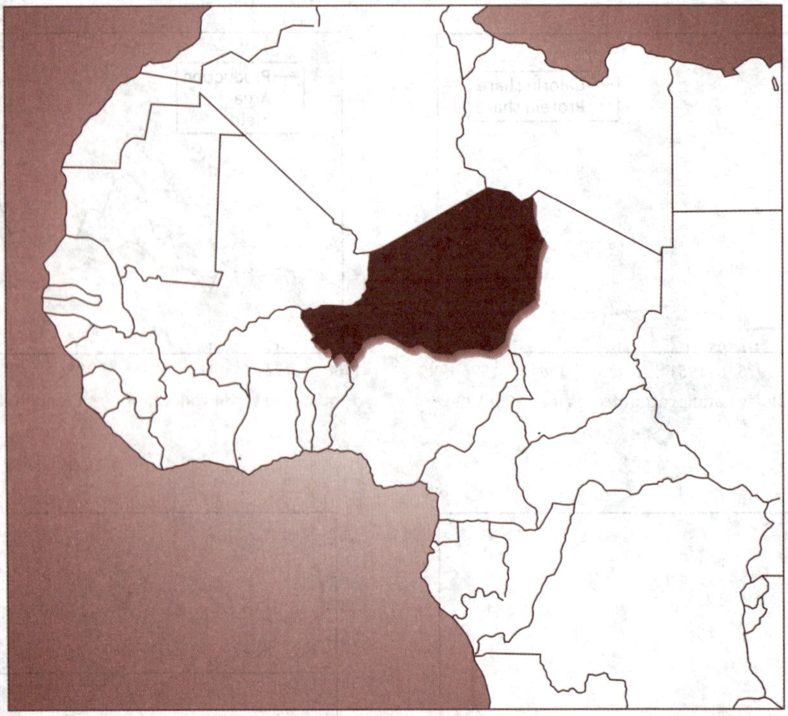

NIGER is one of the larger Central African countries, with an area of 1.27 million km^2. Landlocked, it consists mainly of a plateau; the Sahara Desert occupies the northern parts and there is savanna in the south where crops such as rice are grown. Eighty percent of the population of 10.4 million is rural, engaged in subsistence agriculture, which makes up 40% of the GDP.

General information

- GNI per capita PPP$, 2000: 740
- Internal renewable water resources: 3.5 km^3
- Incoming water flow: 29 km^3
- Main food consumed: millet and sorghum, pulses, rice, roots and tubers, oil and fat
- Rice consumption, 1999: 5.8 kg milled rice per person per year

Production constraints

The following are the major constraints to sustainable rice production:

- Heat-induced sterility during the off-season in irrigated areas
- Cold-induced sterility in December
- Shortage of water in the dry season
- Poor maintenance of irrigation facilities
- Inadequate and irregular input supplies: seeds, fertilizer, and credit
- Lack of small farm equipment, especially for postharvest operations
- Lack of effective farmers' organizations and cooperatives
- Competition for labor by other production activities
- Lack of a well-defined rice policy
- Poor road networks and marketing systems
- Weak research and extension support

Production season

	Planting	Harvesting
Deepwater rice	Jul-Aug	Dec-Jan
Other rice, main season	Jun-Jul	Oct-Dec

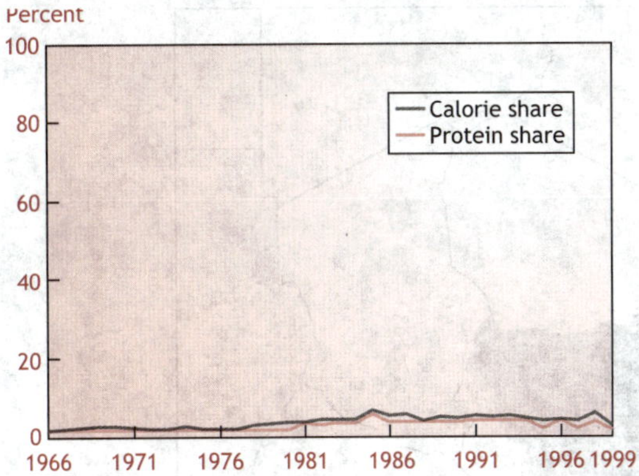

Share of calories and protein from rice, 1966-99

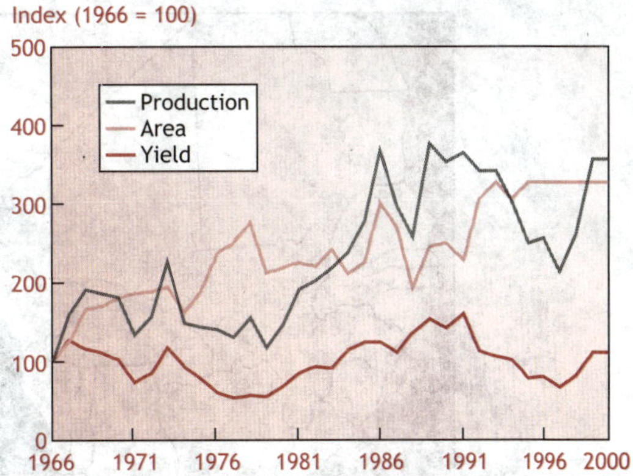

Indices of rice production, area, and yield, 1966-2000

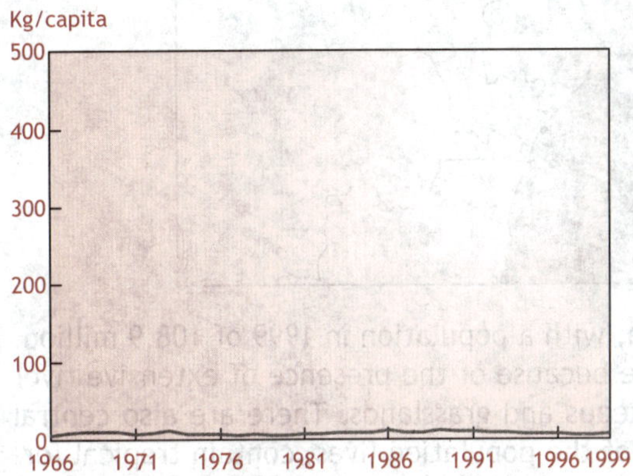

Per capita production (paddy terms), 1966-99

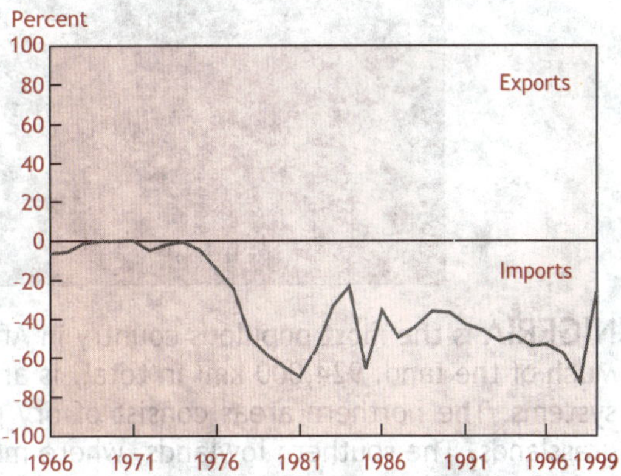

Net trade status, 1966-99

Basic statistics, Niger

	1985	1990	1995	1998	1999	2000
Rice						
Area harvested (ha)	20,600	22,989	30,000	30,000	30,000	30,000
Yield (t/ha)	2.8	3.1	1.7	1.8	2.4	2.4
Production (t)	56,700	72,260	51,000	53,500	73,000	73,000
Rice imports (t)	73,400	28,145	40,000	95,528	17,717	na
Paddy imports (t)	0	0	0	1,026	1,026	na
Rice exports (t)	0	0	0	266	266	na
Paddy exports (t)	0	0	0	34	34	na
Others						
Population, total ($\times10^3$)	6,608	7,731	9,150	10,078	10,400	na
Population, agriculture ($\times10^3$)	5,990	6,951	8,139	8,904	9,166	na
Agricultural area ($\times10^3$ ha)	12,570	14,047	16,500	17,000	na	na
Irrigated agricultural area ($\times10^3$ ha)	30	66	66	66	na	na
Total fertilizer consumption (t)	3,571	2,299	9,979	930	na	na
Tractors used in agric. (no.)	150	170	145	130	na	na

Source: FAOSTAT online database.

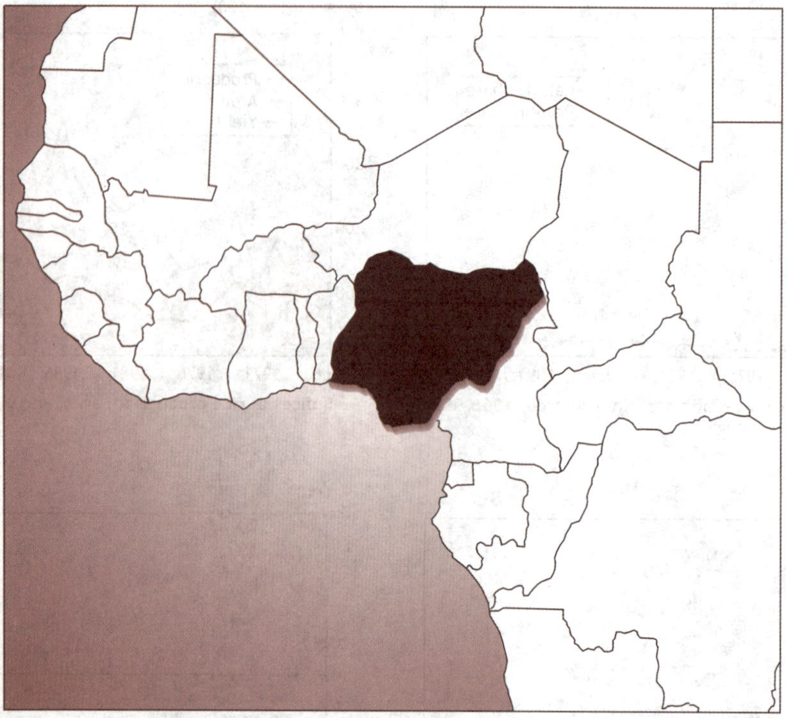

NIGERIA is the most populous country in Africa, with a population in 1999 of 108.9 million. Much of the land, 924,000 km² in total, is arable because of the presence of extensive river systems. The northern areas consist of dry plateaus and grasslands. There are also central grasslands. The southern lowlands, where most of the population lives, contain tropical forests and are the site of most agriculture. Agriculture occupies 54% of the workforce and constitutes 33% of the GDP.

General information

- GNI per capita PPP$, 2000: 800
- Internal renewable water resources: 221 km³
- Incoming water flow: 59 km³
- Main food consumed: millet and sorghum, roots and tubers, oil and fat, rice, maize
- Rice consumption, 1999: 23.3 kg milled rice per person per year

Production constraints

The following are the major constraints to sustainable rice production in Nigeria:

- Shifting cultivation in upland areas
- Drought in upland areas and drought and flash flood in rainfed lowland (or inland swamp) and tidal wetland (or mangrove) areas because of irregular weather
- Heat-induced sterility at the end of the off-season in irrigated areas in northern parts
- Cold-induced sterility in December
- Shortage of water in the dry season
- Poor maintenance of irrigation facilities
- Inadequate and irregular input supplies: seeds, fertilizer, and credit
- Lack of small farm equipment, especially for postharvest operations
- Poor maintenance of developed swamps
- Poor drainage, iron toxicity in undeveloped swamps
- Lack of effective farmers' organizations and cooperatives
- Lack of well-defined rice policy
- Poor road networks and marketing systems
- Weak research and extension support

Production season

	Planting	Harvesting
Main season, south	Apr-May	Aug-Oct
Main season, north	Jun-Jul	Nov-Dec
Off-season, south	Nov-Dec	Mar-Apr
Off-season, north	Jan-Feb	May-Jun

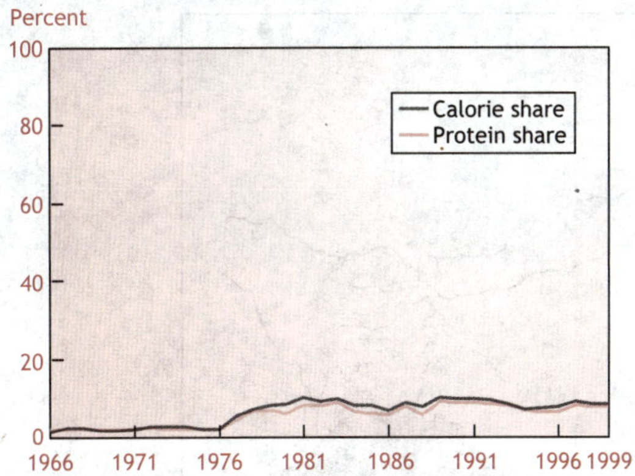

Share of calories and protein from rice, 1966-99

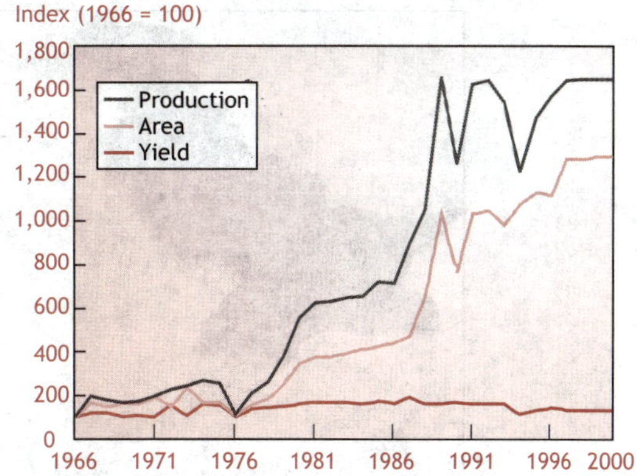

Indices of rice production, area, and yield, 1966-2000

The scale on this graph is not comparable with the scale on production, area, and yield graphs for most other countries.

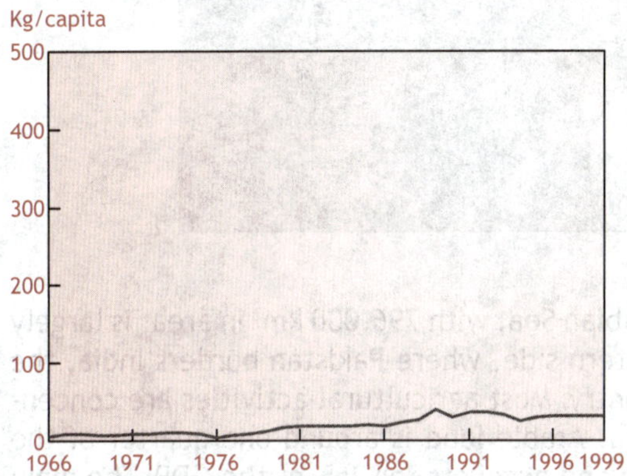

Per capita production (paddy terms), 1966-99

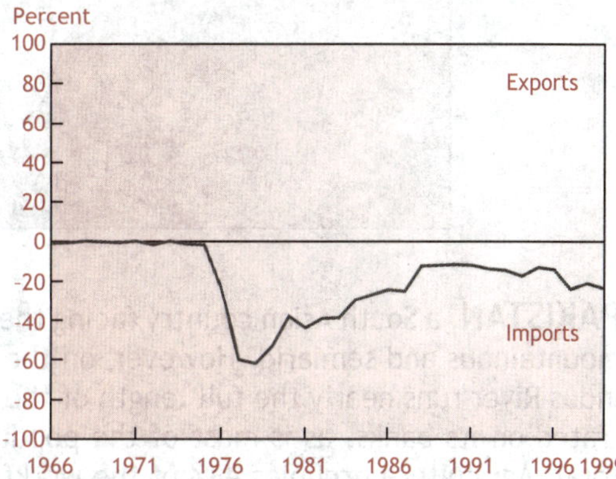

Net trade status, 1966-99

Basic statistics, Nigeria

	1985	1990	1995	1998	1999	2000
Rice						
Area harvested (ha)	670,000	1,208,000	1,796,000	2,044,000	2,061,000	2,061,000
Yield (t/ha)	2.1	2.1	1.6	1.6	1.6	1.6
Production (t)	1,430,000	2,500,000	2,920,000	3,275,000	3,277,000	3,277,000
Rice imports (t)	356,135	224,000	300,000	594,057	687,925	na
Others						
Population, total ($\times 10^3$)	75,805	87,031	98,952	106,409	108,945	na
Population, agriculture ($\times 10^3$)	36,745	37,431	37,646	37,433	37,309	na
Agricultural area ($\times 10^3$ ha)	71,035	72,074	72,830	69,938	na	na
Irrigated agricultural area ($\times 10^3$ ha)	200	230	235	233	na	na
Total fertilizer consumption (t)	292,000	400,340	183,000	188,300	na	na
Tractors used in agric. (no.)	18,000	23,000	28,500	30,000	na	na

Source: FAOSTAT online database.

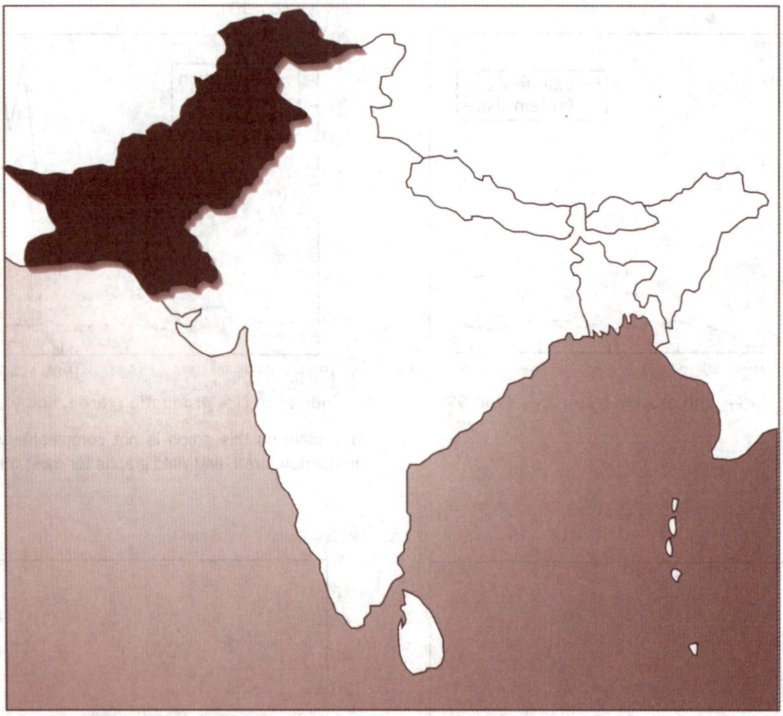

PAKISTAN, a South Asian country facing the Arabian Sea, with 796,000 km² in area, is largely mountainous and semiarid. However, on the eastern side, where Pakistan borders India, the Indus River runs nearly the full length of the country. Most agricultural activities are concentrated on its banks, as is most of the population. Arable land is around one-quarter of the total. Agriculture occupies 44% of the workforce and accounts for 25% of the GDP. The main crops are cotton and wheat.

General information

- GNI per capita PPP$, 2000: 1,860
- Internal renewable water resources: 88.1 km^3
- Incoming water flow: 410 km^3
- Main food consumed: wheat, meat, rice, maize, and fruits
- Rice consumption, 1999: 14.6 kg milled rice per person per year

Production constraints

Yield has increased steadily during the last five years, but in 2000 it was still low. The major constraints to sustainable rice production are

- Imbalance in the use of crop inputs
- The dominant use of varieties with high grain quality (basmati), especially in the Kalar tract
- Low plant populations
- Waterlogging and salinity
- Shortage of irrigation water
- Yield losses caused by insects, pests, diseases, and weeds
- Inadequate and inefficient postharvest operations
- Cold damage to the rice crop because of cool air temperature and cold irrigation water in northern mountain areas

Production season

	Planting	Harvesting
Main season	May-Jul	Oct-Nov

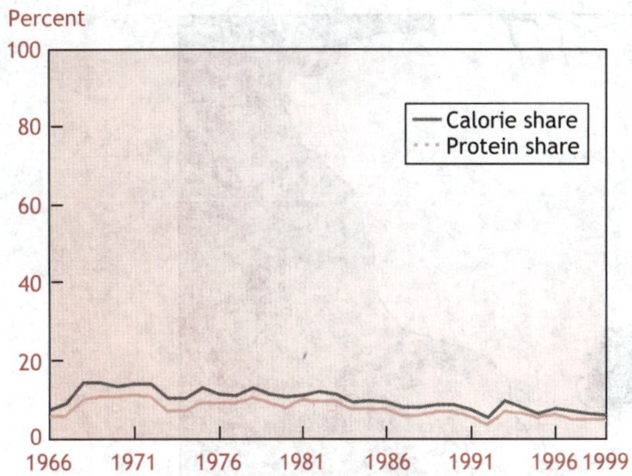

Share of calories and protein from rice, 1966-99

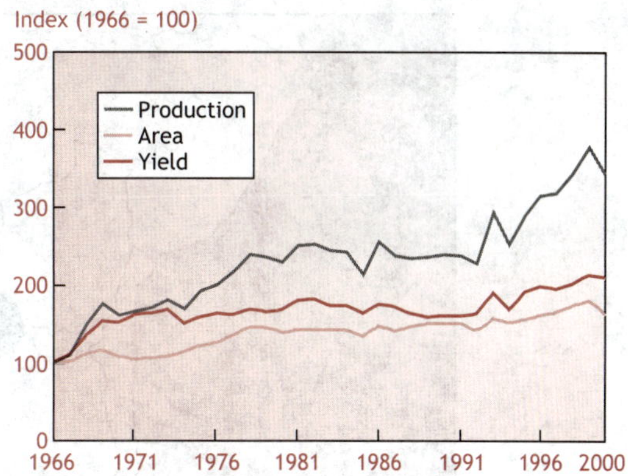

Indices of rice production, area, and yield, 1966-2000

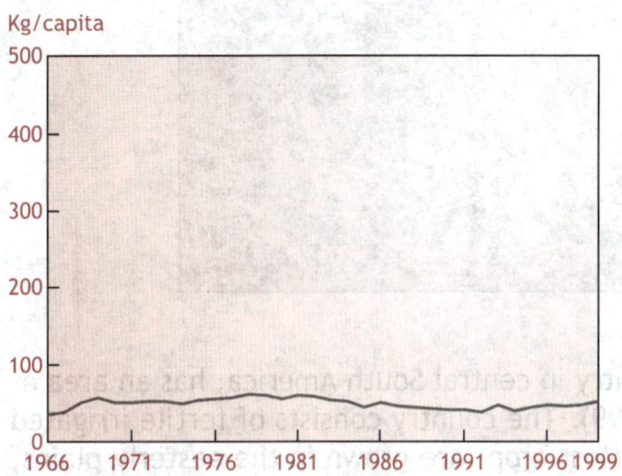

Per capita production (paddy terms), 1966-99

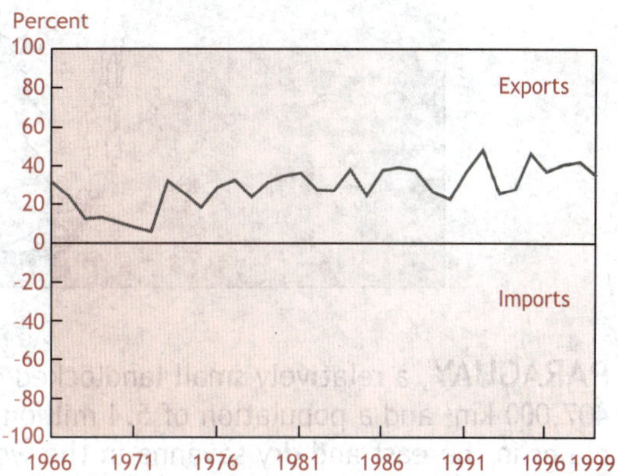

Net trade status, 1966-99

Basic statistics, Pakistan

	1985	1990	1995	1998	1999	2000
Rice						
Area harvested (ha)	1,863,200	2,112,700	2,161,800	2,423,600	2,515,400	2,312,200
Yield (t/ha)	2.3	2.3	2.8	2.9	3.1	3.0
Production (t)	4,378,400	4,891,200	5,949,750	7,010,700	7,733,400	7,000,000
Rice imports (t)	7	25	68	854	1,471	na
Paddy imports (t)	0	0	44	1	0	na
Rice exports (t)	718,686	743,889	1,852,267	1,971,601	1,791,193	na
Others						
Population, total (×10^3)	101,202	119,155	136,244	148,166	152,331	na
Population, agriculture (×10^3)	61,579	66,220	72,360	76,579	78,021	na
Agricultural area (×10^3 ha)	25,610	25,940	26,550	27,040	na	na
Irrigated agricultural area (×10^3 ha)	15,760	16,940	17,200	18,000	na	na
Total fertilizer consumption (t)	1,511,119	1,892,899	2,507,669	2,461,254	na	na
Tractors used in agric. (no.)	156,633	265,728	304,992	320,500	na	na

Source: FAOSTAT online database.

PARAGUAY, a relatively small landlocked country in central South America, has an area of 407,000 km^2 and a population of 5.4 million (1999). The country consists of fertile irrigated plains in the east and dry savanna in the west. Most crops are grown in the easterly plains, with soybeans being the major export crop. About half the workforce is engaged in mainly subsistence agriculture, which constitutes 28% of the GDP.

General information

- GNI per capita PPP$, 2000: 4,450
- Internal renewable water resources: 94 km^3
- Incoming water flow: 220 km^3
- Main food consumed: roots and tubers, meat, oil and fat, wheat, maize
- Rice consumption, 1999: 11 kg milled rice per person per year

Production season

	Planting	Harvesting
Main season	Sep-Dec	Jan-Mar

Production constraints

The following are the major constraints to sustainable rice production in Paraguay:

- Reduction in support from the public sector for rice research and extension
- Blast and straighthead
- Weed competition
- Inadequate milling facilities
- High production costs

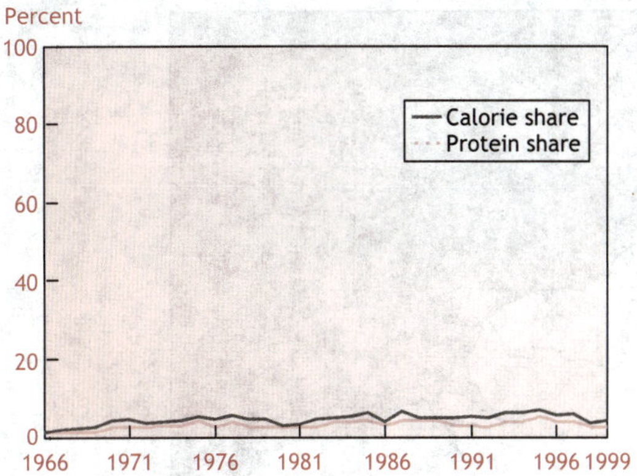

Share of calories and protein from rice, 1966-99

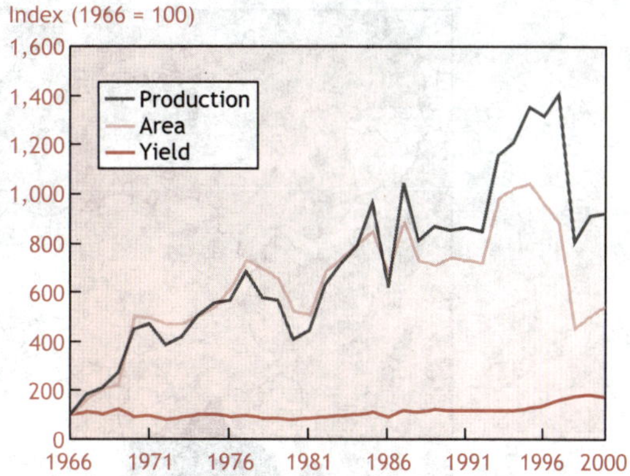

Indices of rice production, area, and yield, 1966-2000

The scale on this graph is not comparable with the scale on production, area, and yield graphs for most other countries.

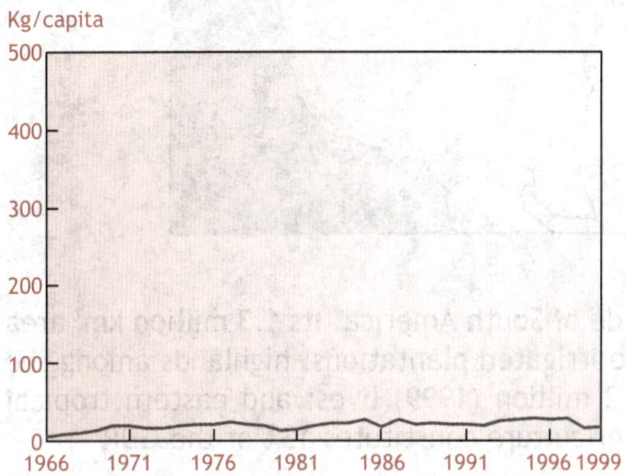

Per capita production (paddy terms), 1966-99

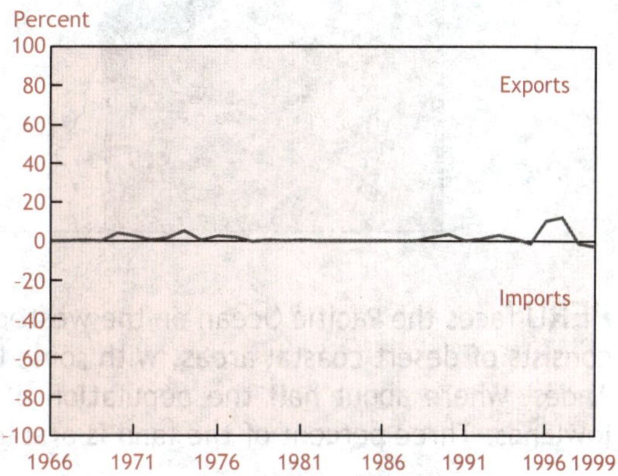

Net trade status, 1966-99

Basic statistics, Paraguay

	1985	1990	1995	1998	1999	2000
Rice						
Area harvested (ha)	39,000	34,000	48,000	20,860	23,000	25,000
Yield (t/ha)	2.5	2.5	2.8	3.9	4.0	3.7
Production (t)	97,207	85,701	136,260	80,921	92,000	93,000
Rice imports (t)	0	1	1,267	2,524	3,115	na
Paddy imports (t)	0	1	475	323	100	na
Rice exports (t)	0	1,800	0	1,765	1,295	na
Paddy exports (t)	0	0	0	720	1,270	na
Others						
Population, total ($\times10^3$)	3,609	4,219	4,828	5,222	5,358	na
Population, agriculture ($\times10^3$)	1,723	1,931	2,088	2,182	2,214	na
Agricultural area ($\times10^3$ ha)	19,999	23,299	23,985	23,985	na	na
Irrigated agricultural area ($\times10^3$ ha)	65	65	67	67	na	na
Total fertilizer consumption (t)	11,271	17,923	23,000	61,500	na	na
Tractors used in agric. (no.)	11,200	15,100	16,500	16,500	na	na

Source: FAOSTAT online database.

PERU faces the Pacific Ocean on the western side of South America. Its 1.3 million km² area consists of desert coastal areas, with some large irrigated plantations; highlands among the Andes, where about half the population of 25.2 million (1999) lives; and eastern tropical lowlands. Three percent of the land is arable. Agriculture constitutes 13% of the GDP.

General information

- GNI per capita PPP$, 2000: 4,660
- Internal renewable water resources: 400 km^3
- Main food consumed: rice, sugar and honey, wheat, roots and tubers, oil and fat
- Rice consumption, 1999: 50.8 kg milled rice per person per year

Production constraints

The following are the major constraints to sustainable rice production in Peru:

- Limited water supply and low temperature; salinity in coastal zones
- Competition from other high-value crops in coastal zones
- Yield potential of present varieties has reached a plateau
- In interior regions, lack of roads and transportation network
- Blast, rice hoja blanca virus, leaf spot, sheath blight
- Weed competition
- Acid soil, aluminum toxicity, and drought stress in the upland system in the Selva Baja

Production season

	Planting	Harvesting
Main season	Jan-Feb	May-Jun

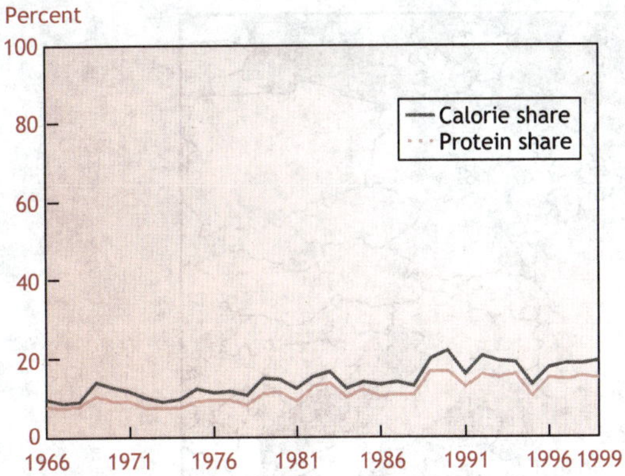

Share of calories and protein from rice, 1966-99

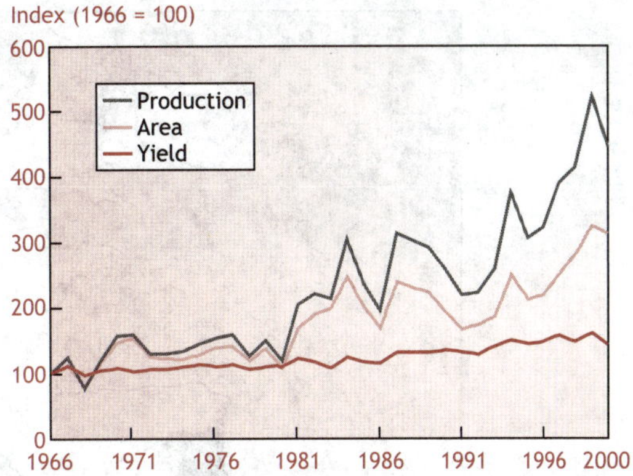

Indices of rice production, area, and yield, 1966-2000

The scale on this graph is not comparable with the scale on production, area, and yield graphs for most other countries.

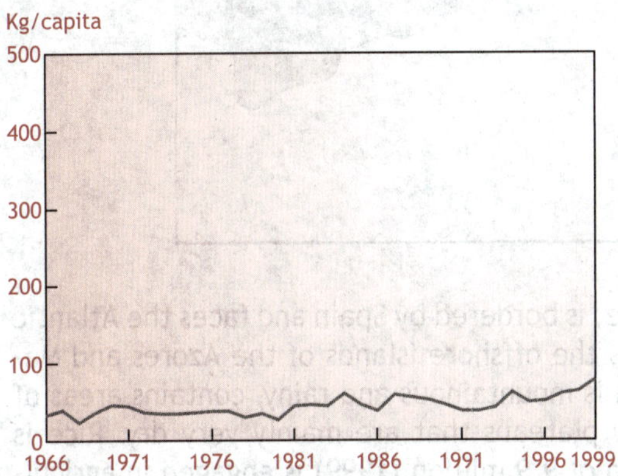

Per capita production (paddy terms), 1966-99

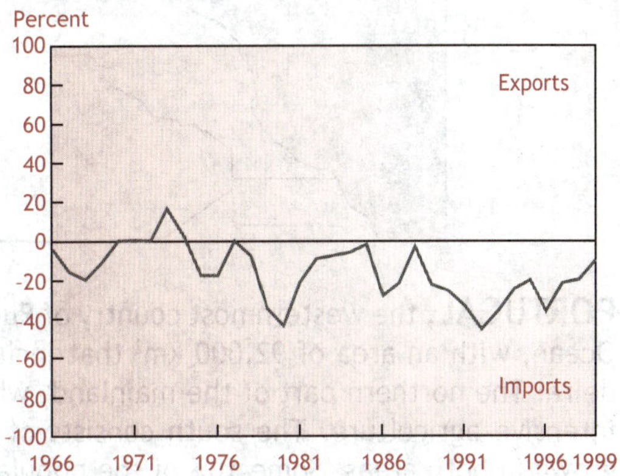

Net trade status, 1966-99

Basic statistics, Peru

	1985	1990	1995	1998	1999	2000
Rice						
Area harvested (ha)	192,711	184,758	203,196	269,080	311,565	300,000
Yield (t/ha)	4.6	5.2	5.6	5.8	6.3	5.5
Production (t)	878,279	966,101	1,141,550	1,548,780	1,955,030	1,664,700
Rice imports (t)	22,402	239,440	196,508	256,831	149,630	na
Paddy imports (t)	0	503	12,360	16,300	14	na
Rice exports (t)	10,200	11,000	0	43	2,241	na
Paddy exports (t)	0	0	0	2	5	n
Others						
Population, total (×10³)	19,492	21,569	23,532	24,797	25,230	na
Population, agriculture (×10³)	7,236	7,528	7,644	7,716	7,739	na
Agricultural area (×10³ ha)	30,856	31,040	31,174	31,270	na	na
Irrigated agricultural area (×10³ ha)	1,170	1,190	1,193	1,195	na	na
Total fertilizer consumption (t)	74,143	125,156	163,200	190,700	na	na
Tractors used in agric. (no.)	12,000	12,700	13,191	13,191	na	na

Source: FAOSTAT online database.

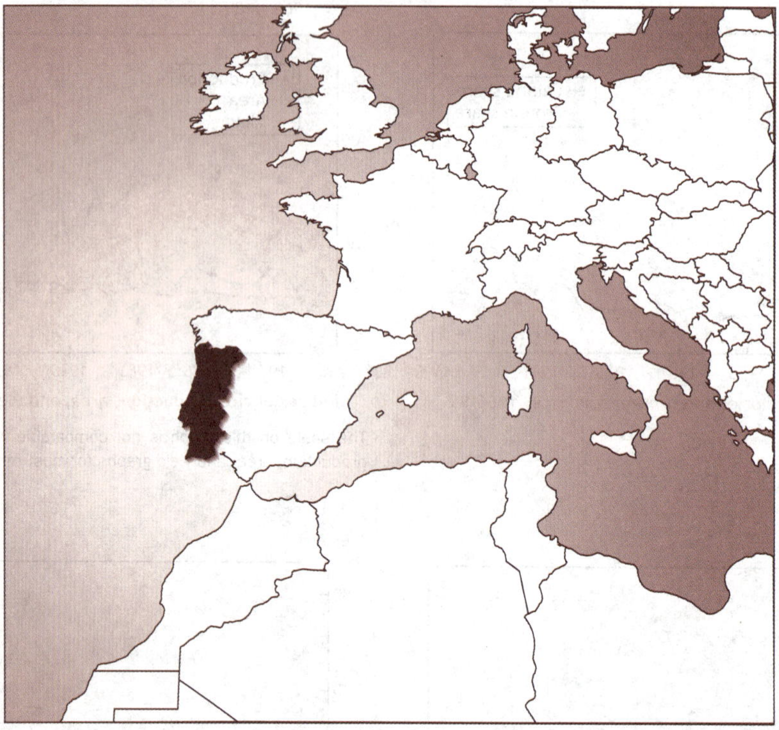

PORTUGAL, the westernmost country of Europe, is bordered by Spain and faces the Atlantic Ocean, with an area of 92,000 km^2 that includes the offshore islands of the Azores and Madeira. The northern part of the mainland, which is mountainous and rainy, contains areas of intensive agriculture. The south consists of low plateaus that are mainly very dry. Rice is grown in both areas. Some 10% of the population of 9.9 million (1999) is engaged in agriculture. Arable land is 24% of the total, but agriculture forms only 4% of the GDP.

General information

- GNI per capita PPP$, 2000: 16,990
- Internal renewable water resources: 34 km^3
- Incoming water flow: 31.6 km^3
- Main food consumed: wheat, oil and fat, meat, sugar and honey, alcohol beverages
- Rice consumption, 1999: 16.2 kg milled rice per person per year

Production season

	Planting	Harvesting
Main season	Apr-May	Sep-Oct

Production constraints

Rice yield has steadily increased during the last decade. The following are the major constraints to sustainable rice production in Portugal:

- Low temperature
- Stagnation of the yield potential of rice varieties
- High production costs
- Blast
- Limited water supply

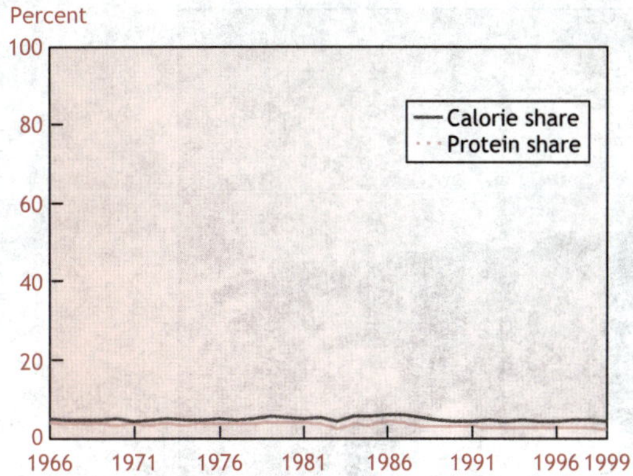

Share of calories and protein from rice, 1966-99

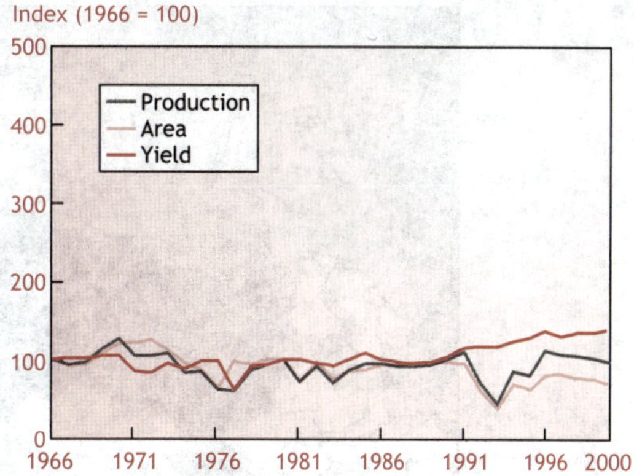

Indices of rice production, area, and yield, 1966-2000

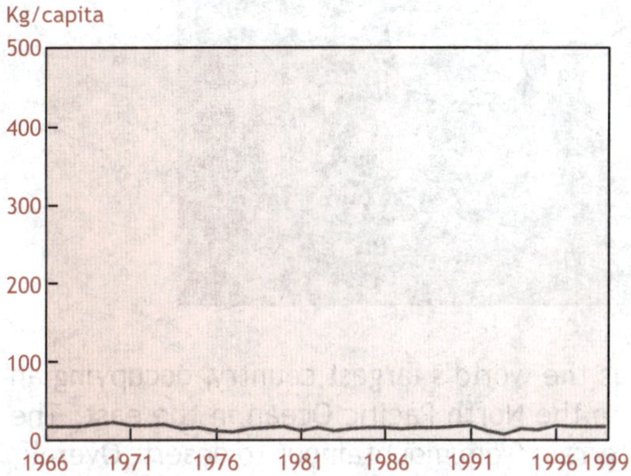

Per capita production (paddy terms), 1966-99

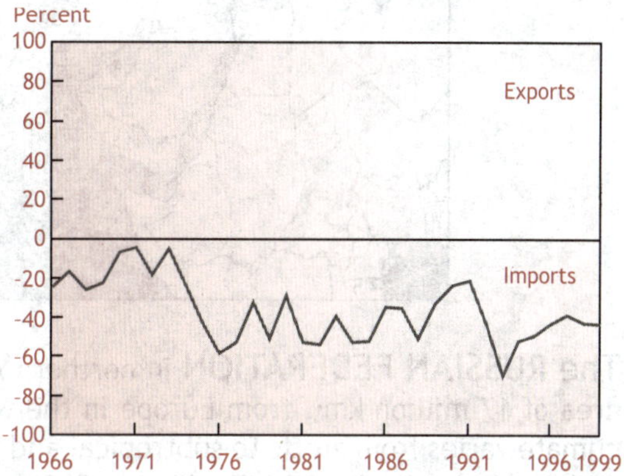

Net trade status, 1966-99

Basic statistics, Portugal

	1985	1990	1995	1998	1999	2000
Rice						
Area harvested (ha)	30,302	33,824	21,726	27,020	26,147	24,000
Yield (t/ha)	4.8	4.6	5.7	6.0	6.0	6.2
Production (t)	146,525	156,072	124,554	161,774	157,252	149,000
Rice imports (t)	107,790	76,733	103,228	108,365	98,558	na
Paddy imports (t)	25	5,977	46,305	50,447	38,849	na
Rice exports (t)	21	32,746	10,350	14,558	9,140	na
Paddy exports (t)	6	0	1,055	11	4	na
Others						
Population, total (×10^3)	9,904	9,869	9,856	9,869	9,873	na
Population, agriculture (×10^3)	2,366	1,962	1,662	1,508	1,460	na
Agricultural area (×10^3 ha)	3,994	4,011	3,924	3,572	na	na
Irrigated agricultural area (×10^3 ha)	630	631	632	632	na	na
Total fertilizer consumption (t)	240,800	278,400	244,000	248,000	na	na
Tractors used in agric. (no.)	111,400	132,000	150,087	155,000	na	na

Source: FAOSTAT online database.

The **RUSSIAN FEDERATION** in northern Asia is the world's largest country, occupying an area of 17 million km², from Europe in the west to the North Pacific Ocean in the east. The climate varies from arctic to subtropical and the terrain from mountainous to desert. Overall, 15% of the population of 147 million (1999) is engaged in agriculture, which forms 8.4% of the GDP.

General information

- GNI per capita PPP$, 2000: 8,010
- Internal renewable water resources: 4,030 km^3
- Incoming water flow: 227 km^3
- Rice consumption, 1999: 5.5 kg milled rice per person per year

Production season

	Planting	Harvesting
Main season	May-Jun	Aug-Sep

Production constraints

Rice yields have declined during the last five years. The following are the major problems:

- Low temperature and short growing season
- High production costs
- Lack of research and extension support
- Salinity in rice soil

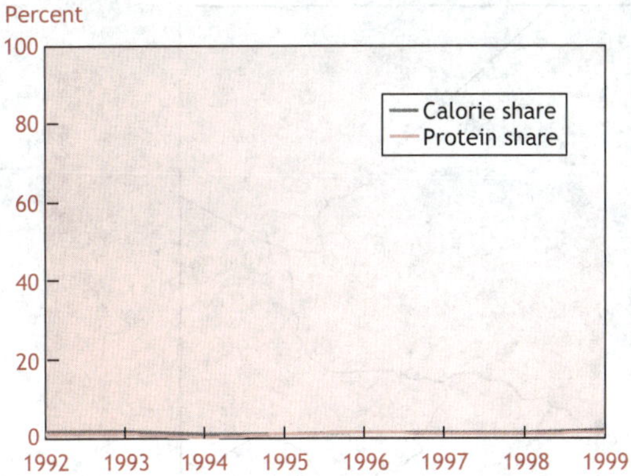

Share of calories and protein from rice, 1992-99

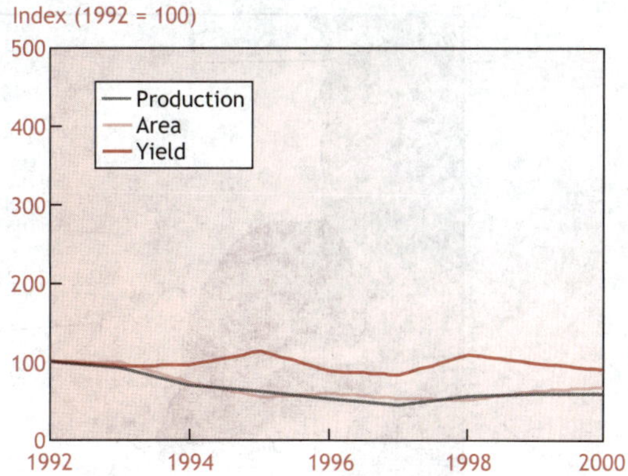

Indices of rice production, area, and yield, 1966-2000
Indices of rice production, area, and yield, 1992-99

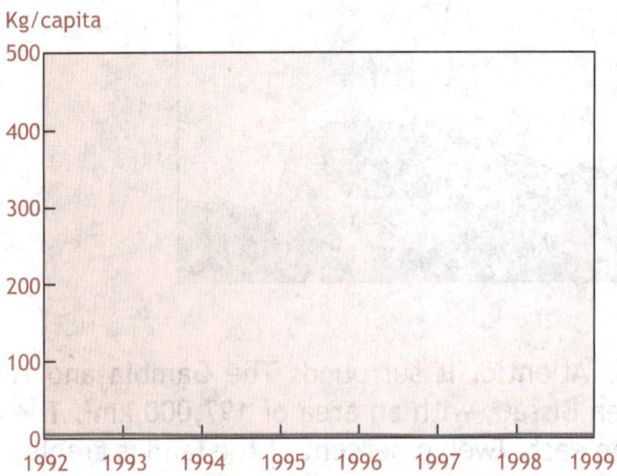

Per capita production (paddy terms), 1992-99

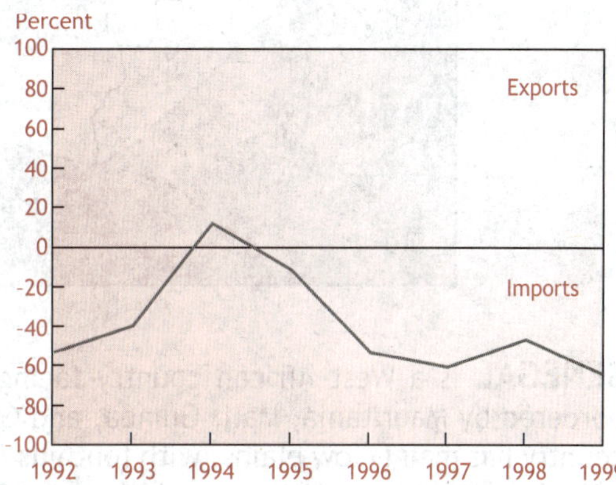

Net trade status, 1992-99

Basic statistics, Russian Federation

	1985	1990	1995	1998	1999	2000
Rice						
Area harvested (ha)	na	na	144,070	135,900	162,300	175,800
Yield (t/ha)	na	na	3.2	3.0	2.7	2.5
Production (t)	na	na	461,900	412,750	444,000	440,000
Rice imports (t)	na	na	145,667	264,995	558,048	na
Paddy imports (t)	na	na	0	3,358	10,879	na
Rice exports (t)	na	na	90,727	11,035	4,764	na
Paddy exports (t)	na	na	8,393	2,974	na	
Others						
Population, total (×10³)	na	na	148,097	147,434	147,196	na
Population, agriculture (×10³)	na	na	17,755	16,303	15,844	na
Agricultural area (×10³ ha)	na	na	216,400	217,155	na	na
Irrigated agricultural area (×10³ ha)	na	na	5,362	4,663	na	na
Total fertilizer consumption (t)	na	na	1,750,000	1,087,000	na	na
Tractors used in agric. (no.)	na	na	1,052,105	856,708	na	na

Source: FAOSTAT online database.

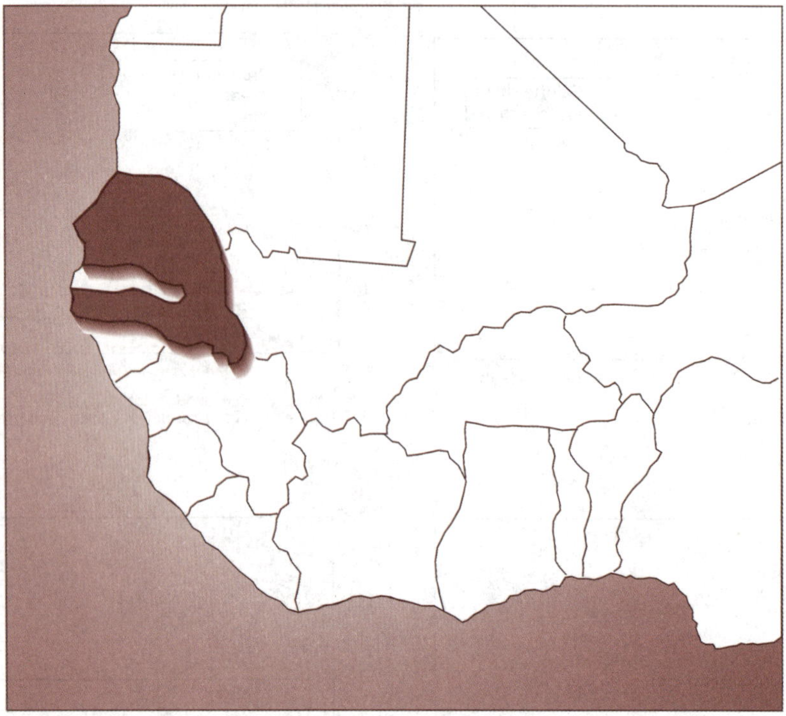

SENEGAL is a West African country facing the Atlantic. It surrounds The Gambia and is bordered by Mauritania, Mali, Guinea, and Guinea Bissau, with an area of 197,000 km². The country has mainly low plains, with foothills in the east. Twelve percent of the land is arable; 60% of the workforce in the population of 9.2 million (1999) is engaged in agriculture, which makes up 19% of the GDP.

General information

- GNI per capita PPP$, 2000: 1,480
- Internal renewable water resources: 26.2 km³
- Incoming water flow: 13 km³
- Main food consumed: rice, millet and sorghum, oil and fat, wheat, sugar and honey
- Rice consumption, 1999: 71.2 kg milled rice per person per year

Production constraints

There are several constraints to sustainable rice production in Senegal:

- Heat-induced sterility at the end of the off-season in irrigated areas
- Cold-induced sterility in December
- Shortage of water in the dry season
- Poor maintenance of irrigation facilities
- Cheap, low-quality rice imports
- Inadequate and irregular input supplies: seeds, fertilizer, and credit
- Lack of small farm equipment, especially for postharvest operations
- Competition for labor by other production activities
- Lack of a well-implemented rice policy
- Weak research and extension support
- Acid sulfate soils in the south (Casamance)
- Periodic drought, affecting rainfed lowlands and uplands

Production season

	Planting	Harvesting
Main season	Jun-Jul	Oct-Dec
Off-season	Feb-Mar	Jun-Jul

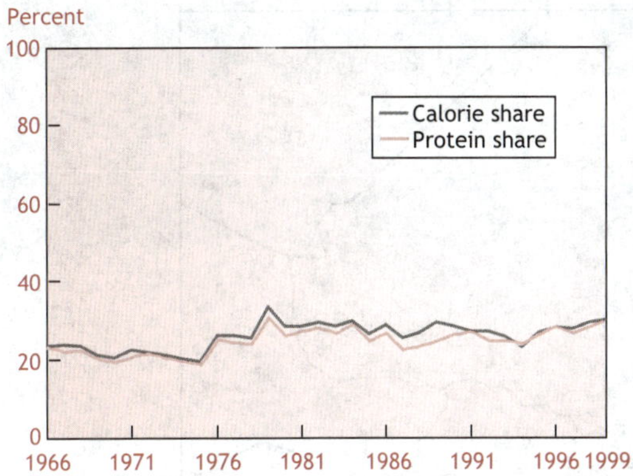

Share of calories and protein from rice, 1966-99

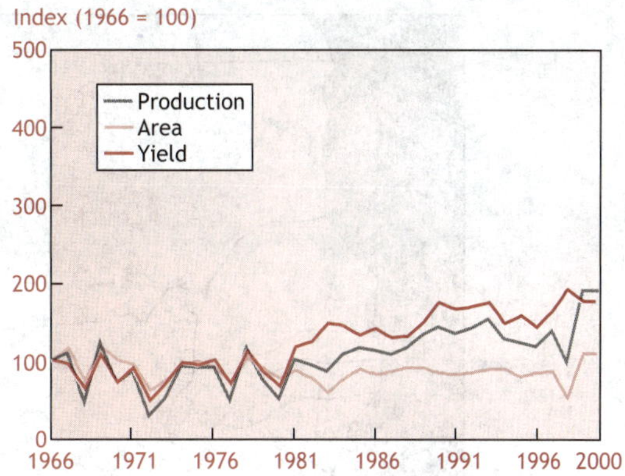

Indices of rice production, area, and yield, 1966-2000

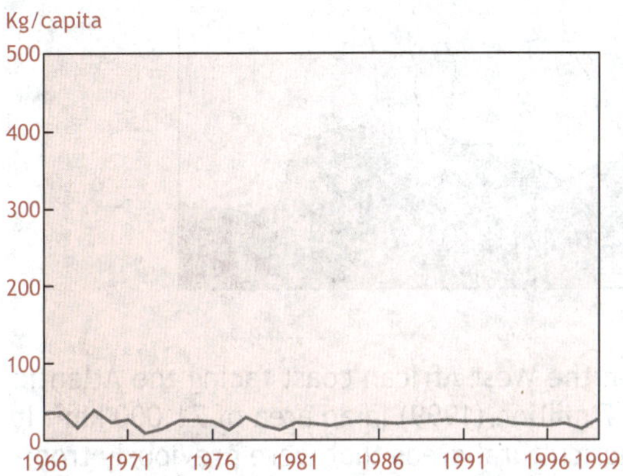

Per capita production (paddy terms), 1966-99

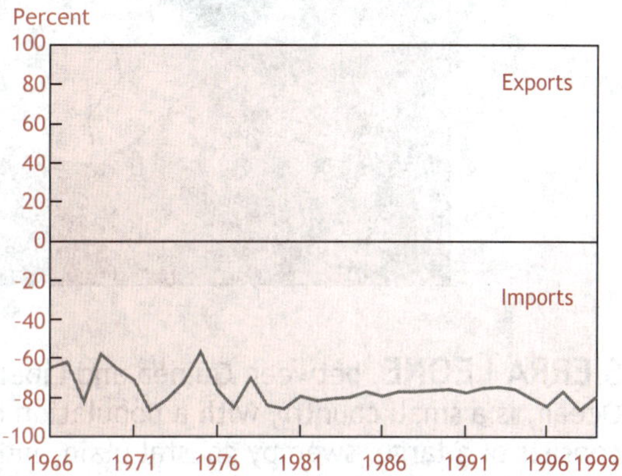

Net trade status, 1966-99

Basic statistics, Senegal

	1985	1990	1995	1998	1999	2000
Rice						
Area harvested (ha)	78,153	72,957	68,966	45,405	95,884	95,884
Yield (t/ha)	1.9	2.5	2.2	2.7	2.5	2.5
Production (t)	147,005	181,119	155,152	123,519	239,786	239,786
Rice imports (t)	342,372	391,513	441,205	557,066	625,160	na
Paddy imports (t)	0	20,942	258	na	na	na
Rice exports (t)	0	1	1	800	7	na
Paddy exports (t)	0	1	0	na	na	na
Others						
Population, total (×10³)	6,375	7,327	8,330	9,003	9,240	na
Population, agriculture (×10³)	5,019	5,621	6,271	6,695	6,841	na
Agricultural area (×10³ ha)	8,050	8,094	7,965	7,916	na	na
Irrigated agricultural area (×10³ ha)	90	94	71	71	na	na
Total fertilizer consumption (t)	20,500	12,004	16,200	26,800	na	na
Tractors used in agric. (no.)	460	490	550	550	na	na

Source: FAOSTAT online database.

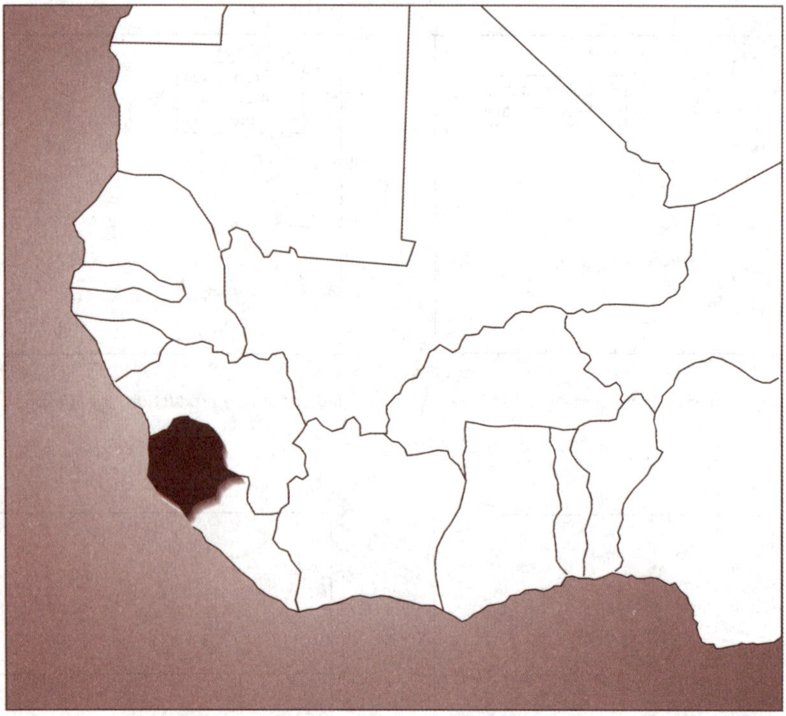

SIERRA LEONE, between Guinea and Liberia on the West African coast facing the Atlantic Ocean, is a small country, with a population of 4.7 million (1999) in an area of 71,000 km^2. It consists of a large, swampy coastal plain, inner agricultural areas that were previously tropical forest, and an eastern plateau. About 12% of the land is arable and just over half the workforce is engaged in agriculture, mainly at the subsistence level.

General information

- GNI per capita PPP$, 2000: 480
- Internal renewable water resources: 160 km^3
- Main food consumed: rice, oil and fat, roots and tubers, pulses, wheat
- Rice consumption, 1999: 99.2 kg milled rice per person per year

Production constraints

- Insecurity
- Poor road network
- Rice policy environment

Production season

	Planting	Harvesting
Main season	Apr-Jul	Sep-Jan

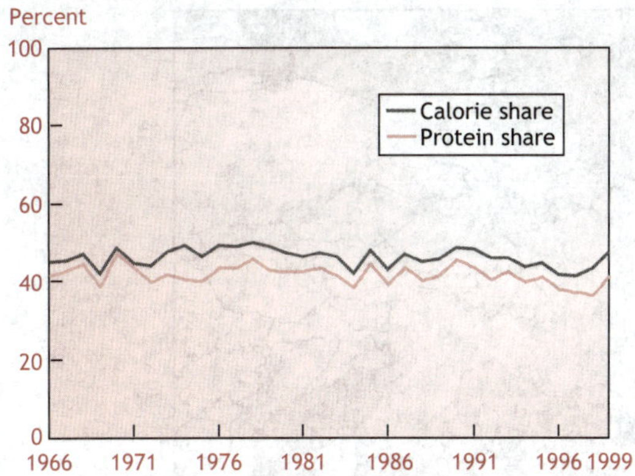

Share of calories and protein from rice, 1966-99

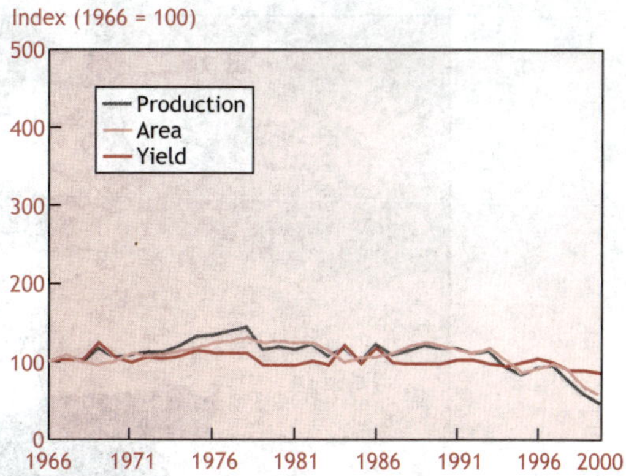

Indices of rice production, area, and yield, 1966-2000

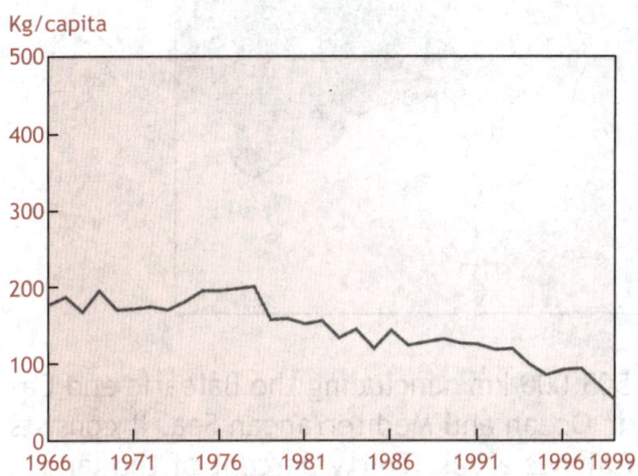

Per capita production (paddy terms), 1966-99

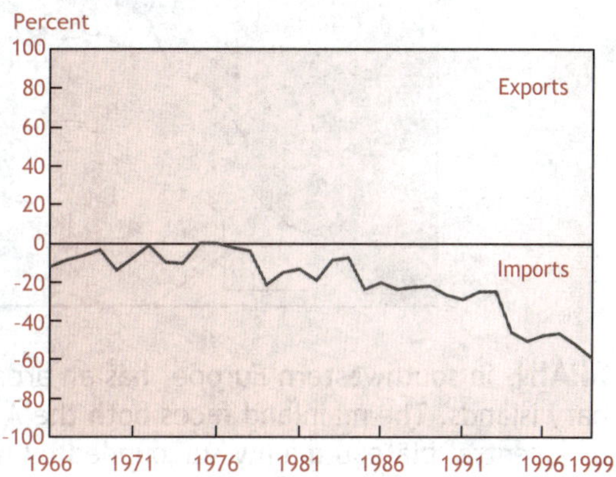

Net trade status, 1966-99

Basic rice statistics, Sierra Leone

	1985	1990	1995	1998	1999	2000
Rice						
Area harvested (ha)	339,865	392,600	274,500	284,770	213,054	183,214
Yield (t/ha)	1.3	1.3	1.3	1.2	1.2	1.1
Production (t)	430,000	503,700	355,500	328,310	247,235	199,134
Rice imports (t)	92,193	123,700	243,200	243,200	243,200	na
Others						
Population, total ($\times 10^3$)	3,583	3,994	4,188	4,568	4,717	na
Population, agriculture ($\times 10^3$)	2,458	2,694	2,716	2,889	2,958	na
Agricultural area ($\times 10^3$ ha)	2,732	2,744	2,741	2,740	na	na
Irrigated agricultural area ($\times 10^3$ ha)	28	28	29	29	na	na
Total fertilizer consumption (t)	3,600	1,300	3,000	3,000	na	na
Tractors used in agric. (no.)	470	200	100	81	na	na

Source: FAOSTAT online database.

SPAIN, in southwestern Europe, has an area of 505,000 km², including the Balearic and Canary islands. The mainland faces both the Atlantic Ocean and Mediterranean Sea. It consists of a central plateau mainly surrounded by mountainous areas. Thirty percent of the land is arable, but agriculture now forms only 3.2% of the GDP and occupies 8% of the workforce. There are extensive rice-growing areas in the southwest and northeast.

General information

- GNI per capita PPP$, 2000: 19,260
- Internal renewable water resources: 110.3 km³
- Incoming water flow: 1 km³
- Main food consumed: meat, oil and fat, wheat, sugar and honey, milk
- Rice consumption, 1999: 7.6 kg milled rice per person per year

Production season

	Planting	Harvesting
Main season	Apr-May	Sep-Oct

Production constraints

Rice production and yield have increased substantially during the last five years. There are several constraints to sustainable rice production:

- Low temperature
- Stagnation of the yield potential of rice varieties
- High production costs
- Blast
- Limited water supply
- Salinity in rice soils
- Red rice

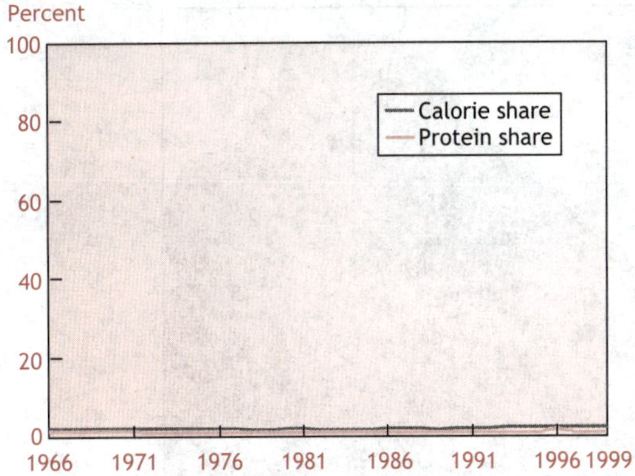

Share of calories and protein from rice, 1966-99

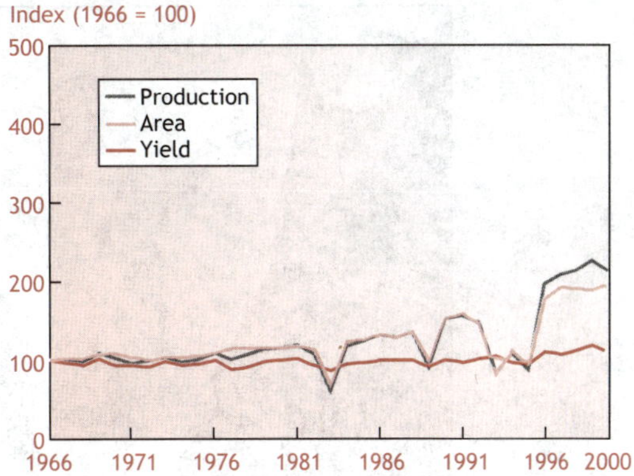

Indices of rice production, area, and yield, 1966-2000

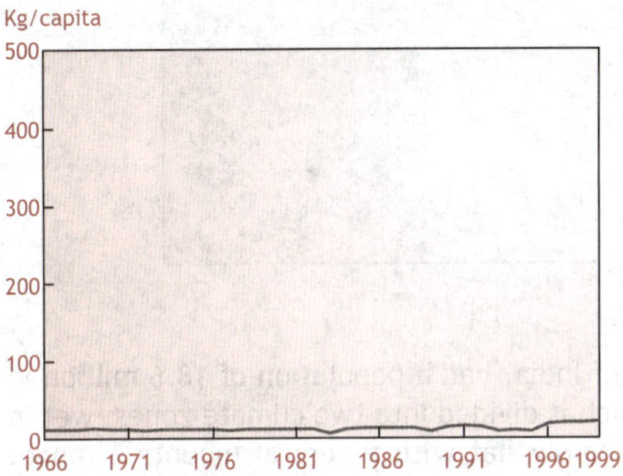

Per capita production (paddy terms), 1966-99

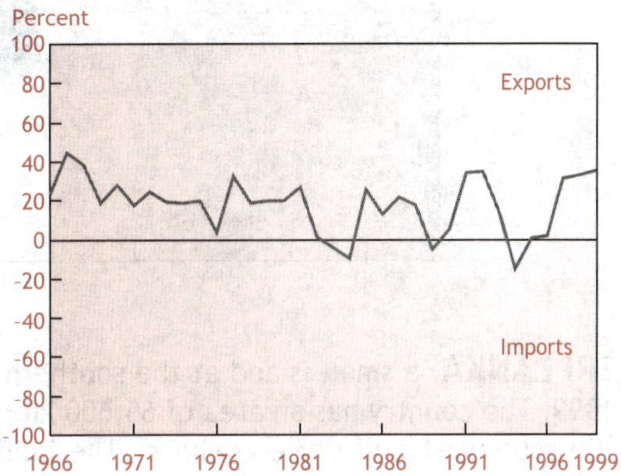

Net trade status, 1966-99

Basic statistics, Spain

	1985	1990	1995	1998	1999	2000
Rice						
Area harvested (ha)	74,581	90,387	54,500	112,700	112,100	115,200
Yield (t/ha)	6.2	6.3	6.0	7.1	7.5	6.9
Production (t)	462,253	571,330	329,500	796,300	845,100	797,800
Rice imports (t)	26,382	152,806	179,317	82,426	85,836	na
Paddy imports (t)	10,540	81,840	151,955	24,346	21,912	na
Rice exports (t)	111,915	178,060	179,645	284,716	314,634	na
Paddy exports (t)	13	15,649	11,968	21,334	17,230	na
Others						
Population, total ($\times10^3$)	38,474	39,303	39,568	39,628	39,634	na
Population, agriculture ($\times10^3$)	5,736	4,762	3,778	3,280	3,128	na
Agricultural area ($\times10^3$ ha)	30,712	30,472	29,719	30,080	na	na
Irrigated agricultural area ($\times10^3$ ha)	3,217	3,402	3,527	3,640	na	na
Total fertilizer consumption (t)	1,734,144	1,975,901	1,868,700	2,106,800	na	na
Tractors used in agric. (no.)	633,210	740,830	805,593	841,932	na	na

Source: FAOSTAT online database.

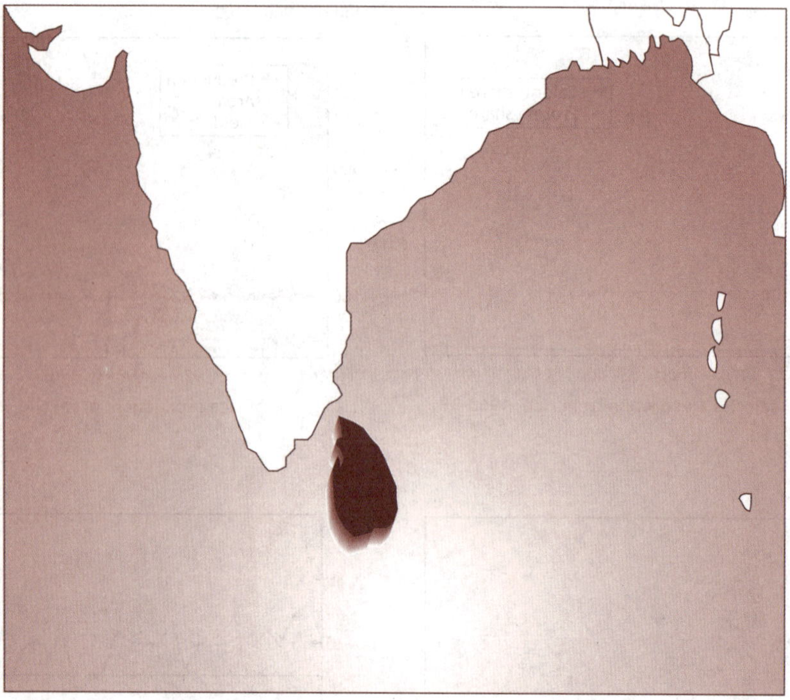

SRI LANKA, a small island at the southern tip of India, had a population of 18.6 million in 1999. The country has an area of 65,600 km², which is divided into two climate zones, wet in the southwest and dry elsewhere. The land is mainly flat with a central mountain range. Arable land is 14% of the total. Agriculture accounts for 38% of the workforce and 21% of the GDP. Rice is the second largest crop after tea.

General information

- GNI per capita PPP$, 2000: 3,460
- Internal renewable water resources: 43.2 km^3
- Main food consumed: rice, wheat, nuts, sugar and honey, oil and fat
- Rice consumption, 1999: 99.2 kg milled rice per person per year

Production season

	Planting	Harvesting
Major (Maha)	Oct-Nov	Feb-Mar
Minor (Yala)	Apr-May	Aug-Sep

Production constraints

Both the harvested area and yield have remained more or less unchanged during the last 10 years, indicating the need to create a more dynamic rice research and development program. The major constraints to sustainable rice production in Sri Lanka are

- Small holdings, limited ownership of land, and high costs of inputs, especially labor
- Lack of credit and inputs (fertilizer and seed)
- Inadequacy of trained personnel for decentralization of agricultural extension
- Inconsistent government policies on rice imports
- Low soil fertility
- Degradation of irrigation facilities

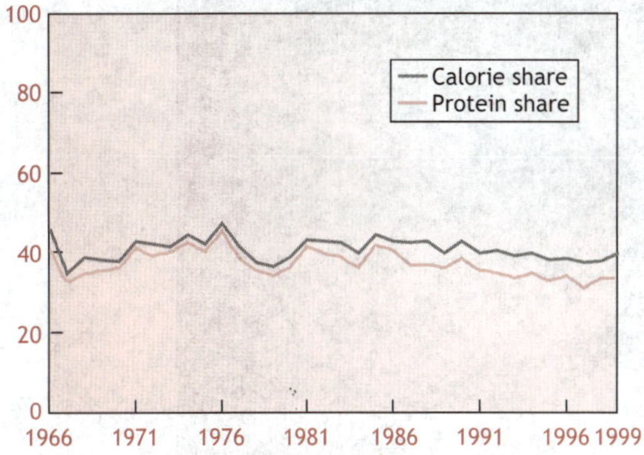

Share of calories and protein from rice, 1966-99

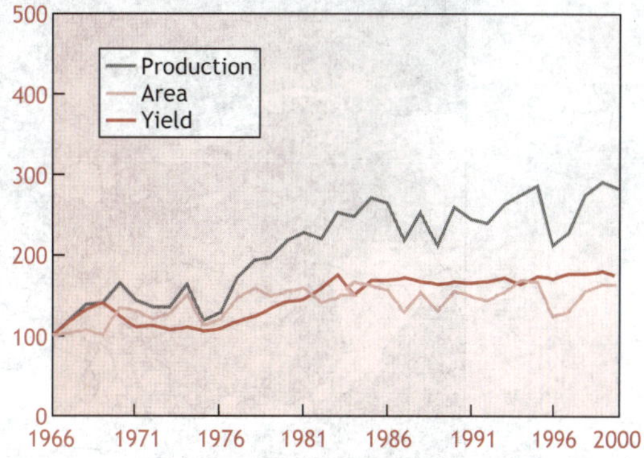

Indices of rice production, area, and yield, 1966-2000

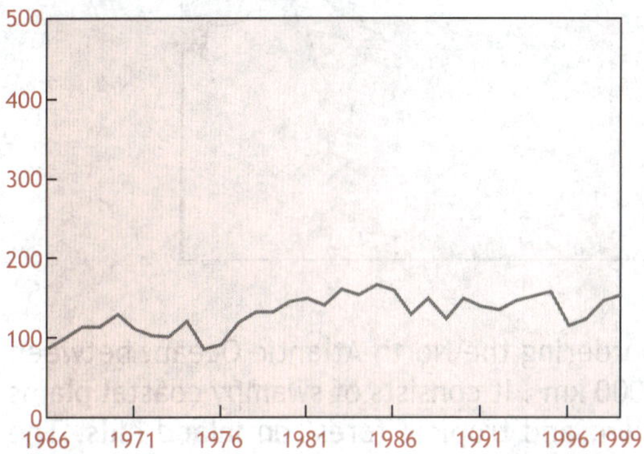

Per capita production (paddy terms), 1966-99

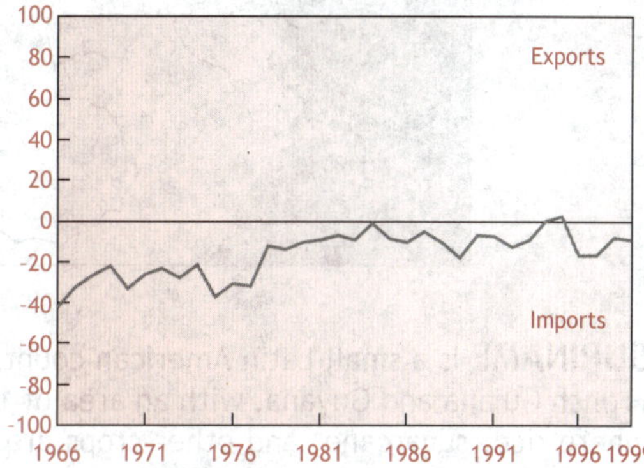

Net trade status, 1966-99

Basic statistics, Sri Lanka

	1985	1990	1995	1998	1999	2000
Rice						
Area harvested (ha)	864,677	828,246	889,589	829,098	870,880	870,880
Yield (t/ha)	3.1	3.1	3.2	3.2	3.3	3.1
Production (t)	2,661,211	2,538,000	2,809,890	2,692,340	2,857,110	2,767,000
Rice imports (t)	176,863	131,771	9,106	167,547	206,137	na
Paddy imports (t)	29	0	21	60	2,761	na
Rice exports (t)	5	246	43,832	2,739	1,404	na
Paddy exports (t)	2	29	9	9	9	na
Others						
Population, total (×10^3)	16,046	17,045	17,920	18,455	18,639	na
Population, agriculture (×10^3)	8,176	8,402	8,563	8,656	8,688	na
Agricultural area (×10^3 ha)	2,315	2,339	2,326	2,329	na	na
Irrigated agricultural area (×10^3 ha)	583	520	570	651	na	na
Total fertilizer consumption (t)	195,483	171,143	199,861	233,128	na	na
Tractors used in agric. (no.)	8,500	6,500	7,417	7,356	na	na

Source: FAOSTAT online database.

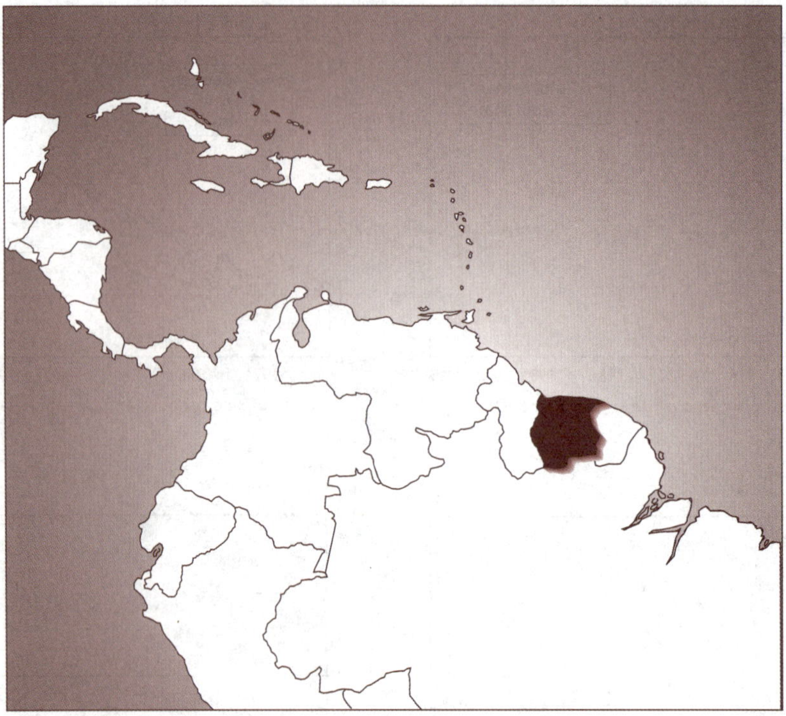

SURINAME is a small Latin American country bordering the North Atlantic Ocean, between French Guiana and Guyana, with an area of 163,000 km^2. It consists of swampy coastal plains where rice, sugarcane, and other crops are grown, and tropical forest on inland hills. The population in 1999 was 415,000. There is little arable land; bauxite mining dominates the economy. However, agriculture accounted for 13% of the GDP in 1998; it occupied 21% of the workforce in 1996.

General information

- GNI per capita PPP$, 2000: 3,480
- Internal renewable water resources: 200 km^3
- Main food consumed: rice, wheat, sugar and honey, oil and fat, meat
- Rice consumption, 1999: 77.9 kg year per person

Production season

	Planting	Harvesting
Main season	Jan-Feb	Apr-May

Production constraints

There are several constraints to sustainable rice production in Suriname:

- Unfavorable government policy, especially on input supply, milling, price control, and taxation of exports
- Blast
- Weed and red rice competition
- Degradation of irrigation infrastructure
- Yield potential of current varieties has reached a plateau

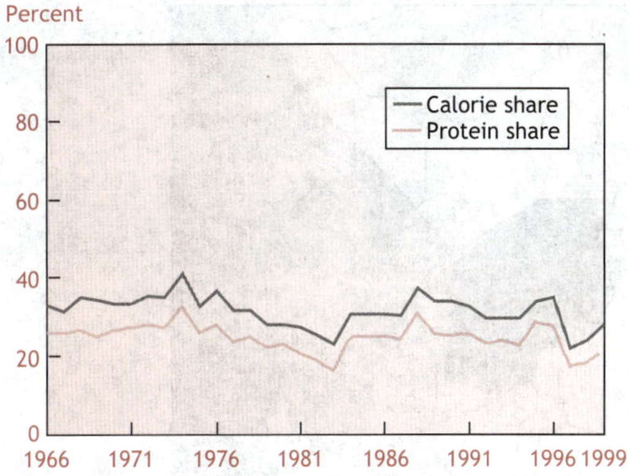

Share of calories and protein from rice, 1966-99

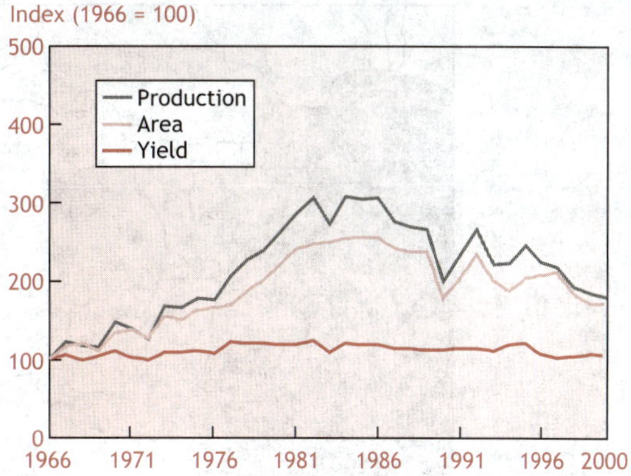

Indices of rice production, area, and yield, 1966-2000

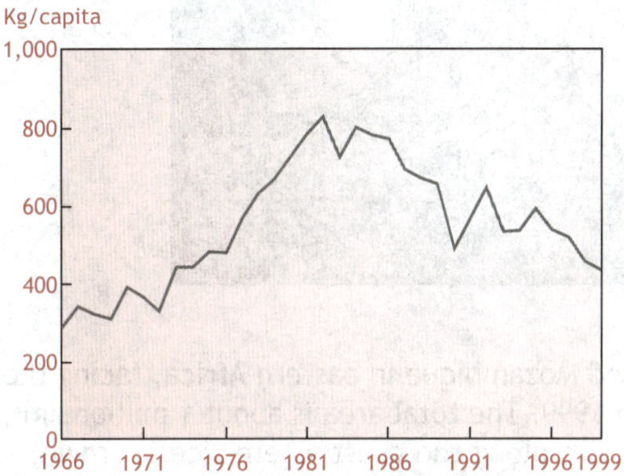

Per capita production (paddy terms), 1966-99

The scale on this graph is not comparable with the scale on per capita production graphs for most other countries.

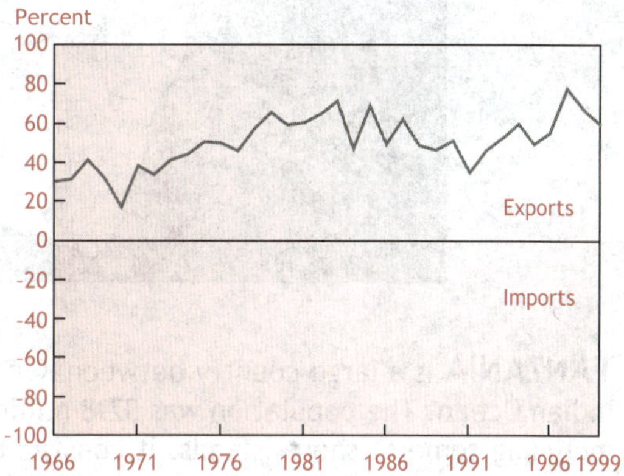

Net trade status, 1966-99

Basic statistics, Suriname

	1985	1990	1995	1998	1999	2000
Rice						
Area harvested (ha)	74,890	52,005	60,000	54,000	50,000	50,000
Yield (t/ha)	4.0	3.8	4.0	3.5	3.6	3.5
Production (t)	299,185	196,010	242,000	188,400	180,300	175,000
Rice imports (t)	0	0	0	80	110	na
Rice exports (t)	137,735	66,650	79,329	84,100	71,300	na
Others						
Population, total ($\times 10^3$)	384	402	409	414	415	na
Population, agriculture ($\times 10^3$)	86	85	82	80	80	na
Agricultural area ($\times 10^3$ ha)	81	88	89	88	na	na
Irrigated agricultural area ($\times 10^3$ ha)	45	47	49	49	na	na
Total fertilizer consumption (t)	11,658	1,000	4,300	5,500	na	na
Tractors used in agric. (no.)	1,202	1,290	1,330	1,330	na	na

Source: FAOSTAT online database.

TANZANIA is a large country between Kenya and Mozambique in eastern Africa, facing the Indian Ocean. The population was 32.8 million in 1999. The total area is about 1 million km^2, including some offshore islands. It consists of plains along the coast where rice is grown, a large central plateau separating the rift valleys, and highlands in the north and south. Three percent of the land is arable. The economy is heavily dependent on agriculture, which accounts for 48% of the GDP, provides 85% of exports, and occupies 90% of the workforce.

General information

- GNI per capita PPP$, 2000: 520
- Internal renewable water resources: 80 km^3
- Incoming water flow: 9 km^3
- Main food consumed: maize, roots and tubers, rice, pulses, oil and fat
- Rice consumption, 1999: 14.3 kg milled rice per person per year

Production constraints

The major constraints to sustainable rice production in Tanzania are

- Drought in upland areas and drought and flash flood in rainfed lowland (or inland swamp) areas because of irregular weather
- Infestation of red rice in the irrigated schemes
- Inadequate and irregular input supplies: seeds, fertilizer, and credit
- Lack of small farm equipment, especially for postharvest operations
- Lack of effective farmers' organizations and cooperatives
- Poor maintenance of irrigation facilities
- Lack of a well-defined rice policy
- Poor road networks and marketing systems
- Labor shortage because of competition from other crops
- Weak research and extension support

Production season

	Planting	Harvesting
Main season	Dec-Feb	May-Jul
Off-season	Jun-Jul	Nov-Dec

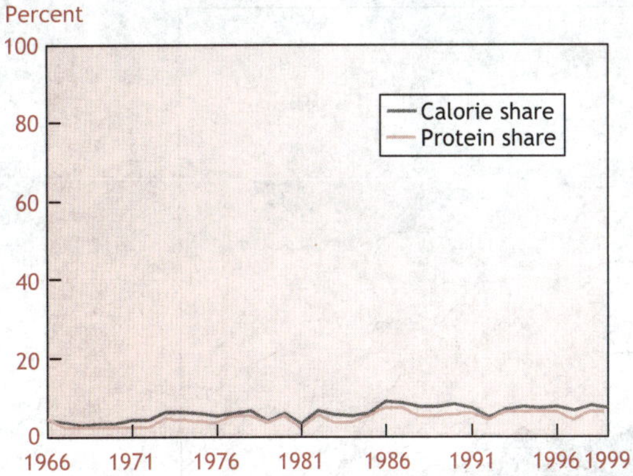

Share of calories and protein from rice, 1966-99

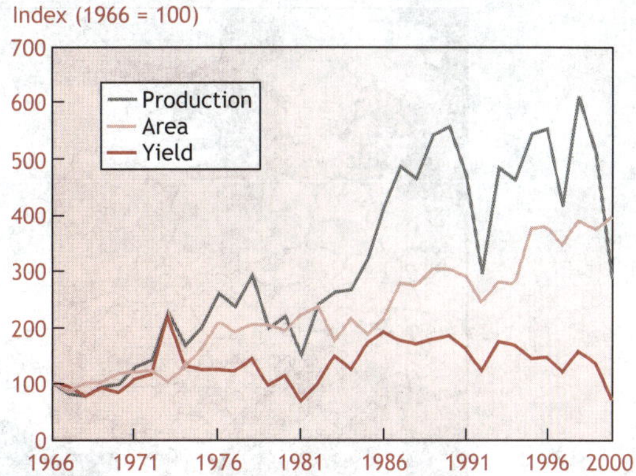

Indices of rice production, area, and yield, 1966-2000

The scale on this graph is not comparable with the scale on production, area, and yield graphs for most other countries.

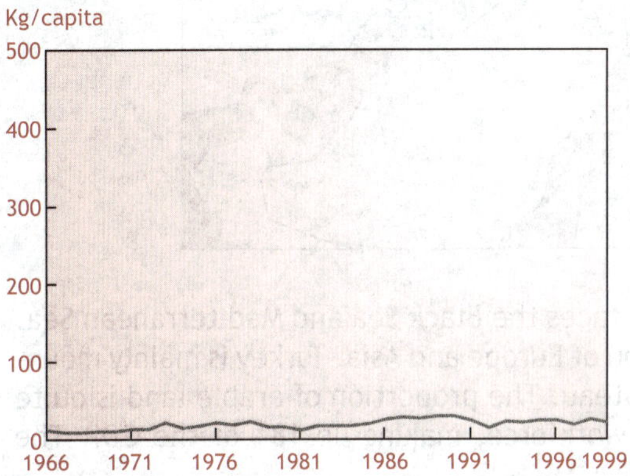

Per capita production (paddy terms), 1966-99

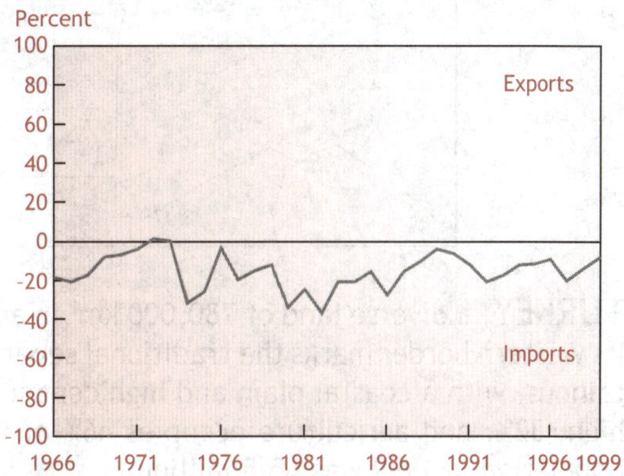

Net trade status, 1966-99

Basic statistics, Tanzania

	1985	1990	1995	1998	1999	2000
Rice						
Area harvested (ha)	236,540	384,500	477,900	492,306	473,909	503,533
Yield (t/ha)	1.8	1.9	1.5	1.6	1.4	0.8
Production (t)	427,692	740,000	722,700	810,800	676,000	378,562
Rice imports (t)	53,192	34,000	65,000	108,500	46,700	na
Rice exports (t)	0	0	0	14,504	5,000	na
Paddy exports (t)	na	na	na	160	0	na
Others						
Population, total ($\times10^3$)	21,775	25,470	29,925	32,102	32,793	na
Population, agriculture ($\times10^3$)	18,127	21,012	24,031	25,366	25,773	na
Agricultural area ($\times10^3$ ha)	39,250	39,400	39,648	39,650	na	na
Irrigated agricultural area ($\times10^3$ ha)	127	144	150	155	na	na
Total fertilizer consumption (t)	38,927	51,249	27,000	27,826	na	na
Tractors used in agric. (no.)	8,000	6,800	7,525	7,600	na	na

Source: FAOSTAT online database.

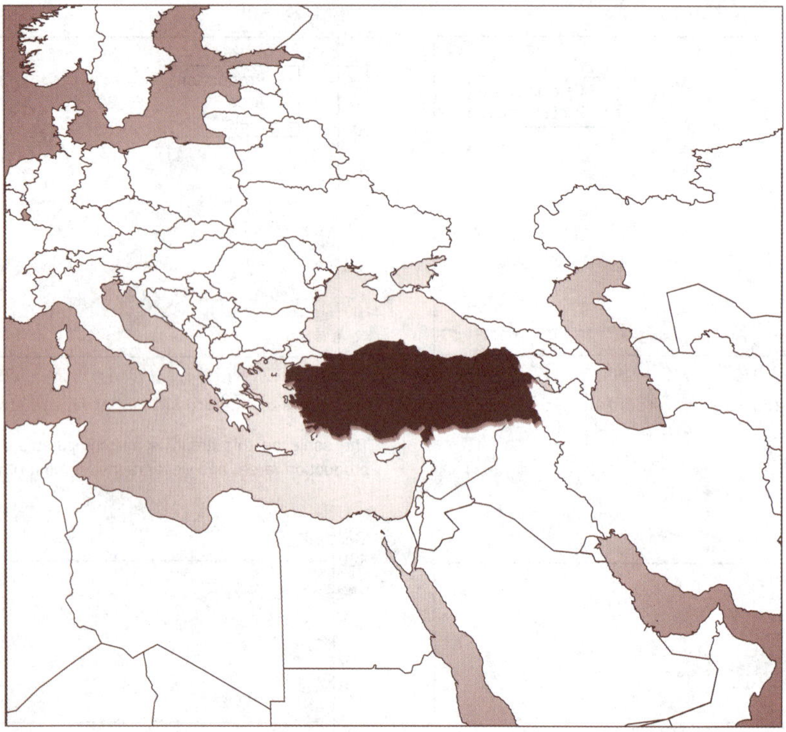

TURKEY, a diverse land of 780,000 km² in area, faces the Black Sea and Mediterranean Sea. Its western border marks the traditional separation of Europe and Asia. Turkey is mainly mountainous, with a coastal plain and high central plateau. The proportion of arable land is quite high, 32%, and agriculture occupies 46% of the workforce, making up 18% of the GDP. The population in 1999 was 65.5 million.

General information

- GNI per capita PPP$, 2000: 7,030
- Internal renewable water resources: 186.1 km^3
- Incoming water flow: 7 km^3
- Main food consumed: wheat, oil and fat, sugar and honey, milk, fruits
- Rice consumption, 1999: 7.2 kg milled rice per person per year

Production season

	Planting	Harvesting
Main season	May-Jun	Sep-Oct

Production constraints

Yield has increased steadily during the last five years. However, in 2000, yields were still low compared with those in other Mediterranean countries. The following are the major problems:

- Low temperature during early stages of rice growth
- Limited supply of irrigation water
- High production costs
- Blast

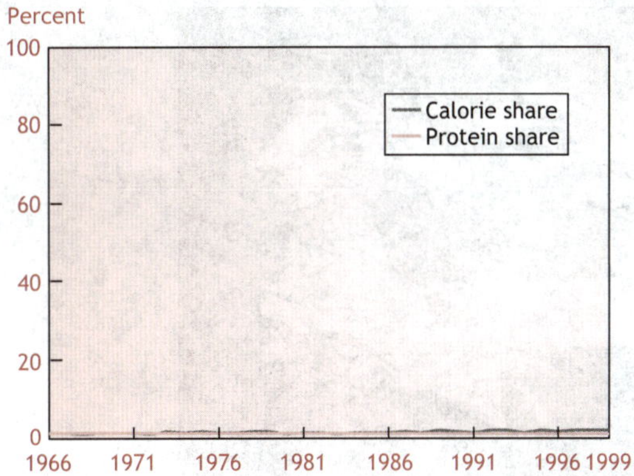

Share of calories and protein from rice, 1966-99

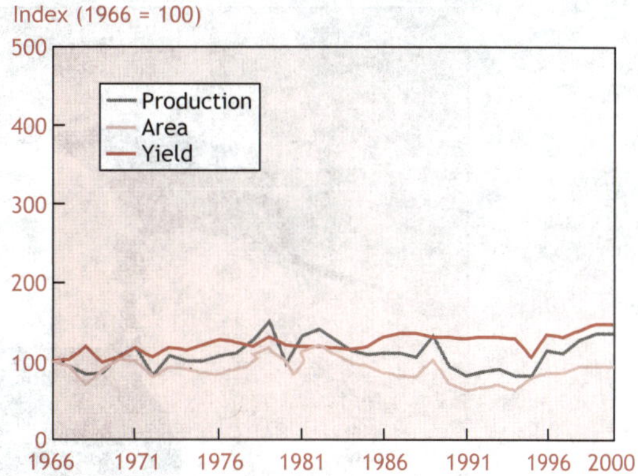

Indices of rice production, area, and yield, 1966-2000

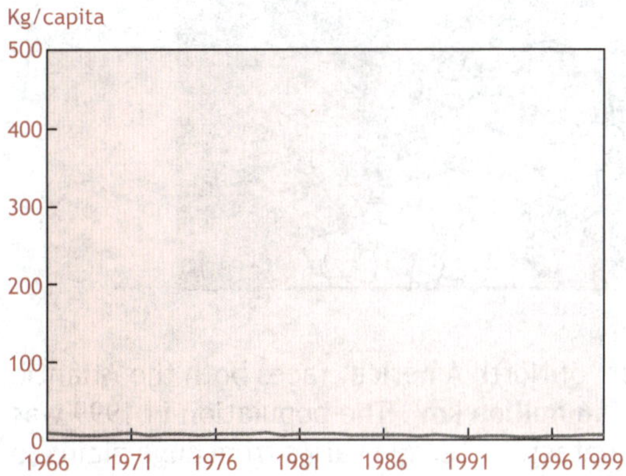

Per capita production (paddy terms), 1966-99

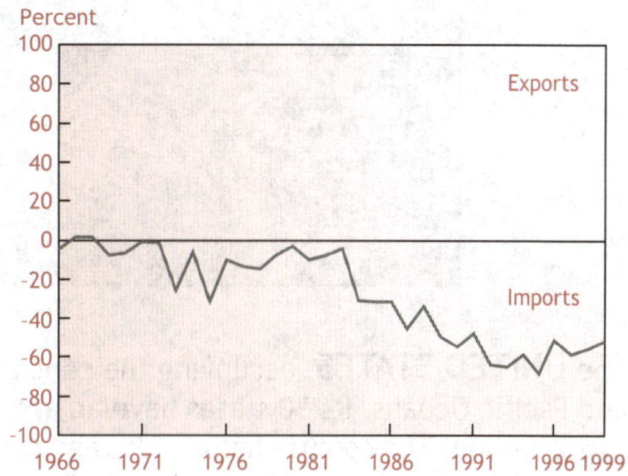

Net trade status, 1966-99

Basic statistics, Turkey

	1985	1990	1995	1998	1999	2000
Rice						
Area harvested (ha)	59,780	46,348	50,000	60,000	60,000	60,000
Yield (t/ha)	4.5	5.0	4.0	5.3	5.6	5.6
Production (t)	270,000	230,000	200,000	317,000	338,000	338,000
Rice imports (t)	85,379	190,717	300,112	275,268	246,935	na
Paddy imports (t)	809	21,252	22,101	42,186	164,917	na
Rice exports (t)	50	1,170	949	593	1,468	na
Paddy exports (t)	2	45	116	169	248	na
Others						
Population, total ($\times 10^3$)	50,345	56,098	61,276	64,479	65,546	na
Population, agriculture ($\times 10^3$)	21,368	20,950	20,763	20,610	20,547	na
Agricultural area ($\times 10^3$ ha)	38,130	39,677	39,493	39,346	na	na
Irrigated agricultural area ($\times 10^3$ ha)	3,200	3,800	4,186	4,200	na	na
Total fertilizer consumption (t)	1,426,948	1,887,520	1,700,376	2,180,700	na	na
Tractors used in agric. (no.)	582,291	689,650	776,863	875,000	na	na

Source: FAOSTAT online database.

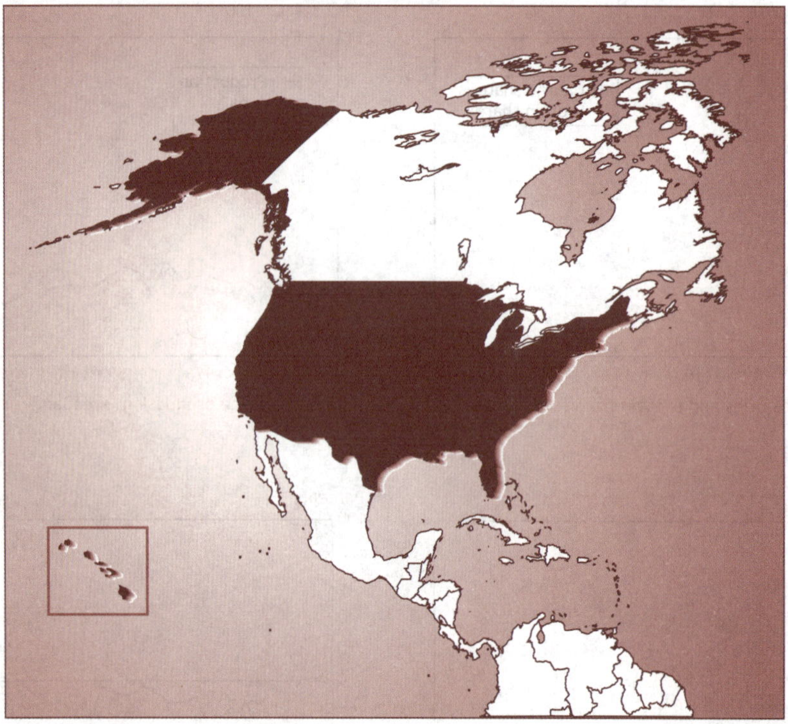

The **UNITED STATES**, occupying the central part of North America, faces both the Atlantic and Pacific Oceans. Its 50 states have an area of 9.6 million km^2. The population in 1999 was 276.2 million. The country has a wide range of climates; the terrain varies from huge plains to river valleys and mountain ranges. Arable land is 19% of the total. Agriculture is a minor sector, only 2% of the GDP and employing 2.8% of the workforce.

General information

- GNI per capita PPP$, 2000: 34,100
- Internal renewable water resources: 2,478 km^3
- Main food consumed: oil and fat, sugar and honey, wheat, meat, milk
- Rice consumption, 1999: 9 kg milled rice per person per year

Production season

	Planting	Harvesting
Main season, Gulf	Apr-Jun	Aug-Oct
Main season, California	Apr-Jun	Sep-Nov

Production constraints

Rice yield has increased steadily during the last five years. The following are the major constraints to sustainable rice production in the U.S.:

- High production costs
- Low temperature during the early and reproductive stages
- Blast, stem rot, rice water weevils
- Weed and red rice competition
- Yield potential of current varieties has reached a plateau
- Increasing pressure from legislation because of government concern for the environment, especially for the burning of rice straw

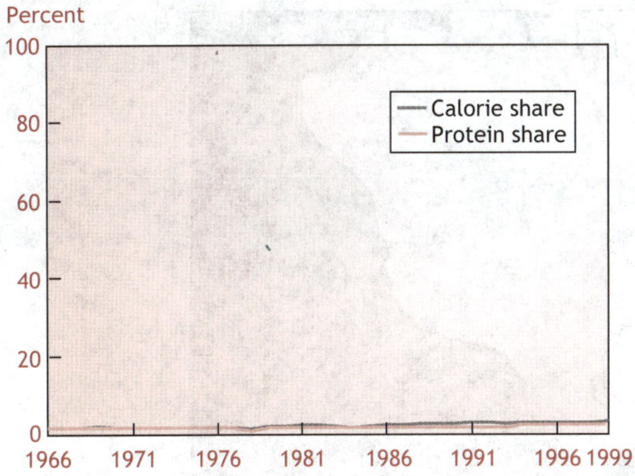

Share of calories and protein from rice, 1966-99

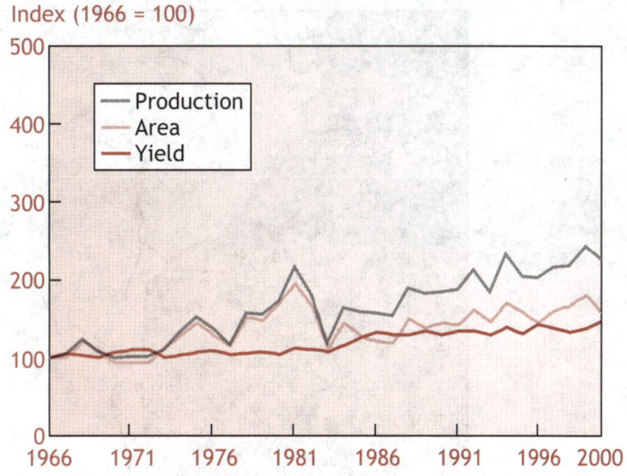

Indices of rice production, area, and yield, 1966-2000

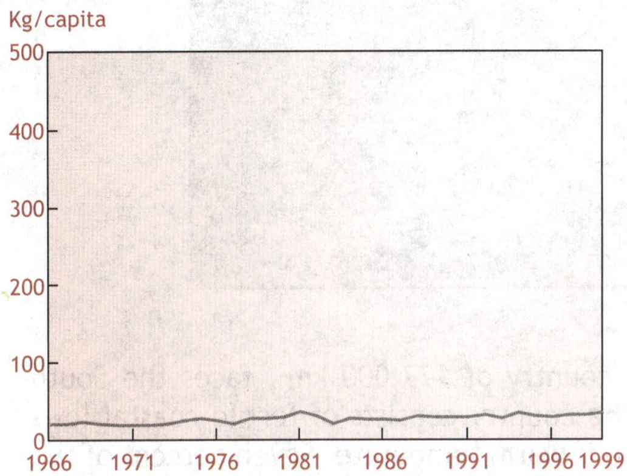

Per capita production (paddy terms), 1966-99

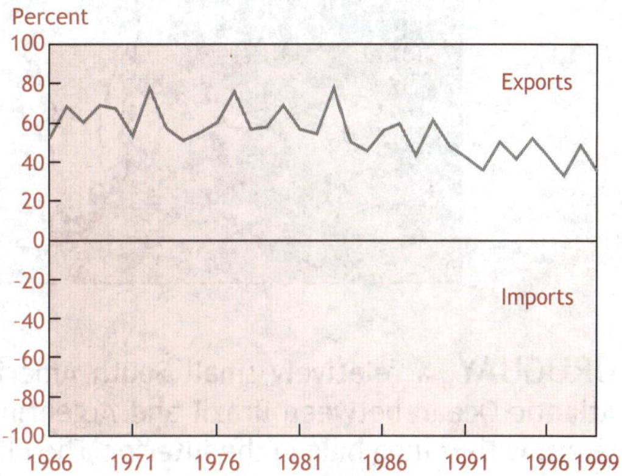

Net trade status, 1966-99

Basic statistics, United States

	1985	1990	1995	1998	1999	2000
Rice						
Area harvested (ha)	1,008,500	1,142,400	1,251,700	1,318,000	1,421,300	1,232,000
Yield (t/ha)	6.1	6.2	6.3	6.3	6.6	7.0
Production (t)	6,122,000	7,080,000	7,887,000	8,366,000	9,345,000	8,669,000
Rice imports (t)	58,897	148,270	224,338	278,592	353,644	na
Paddy imports (t)	271	728	0	12	0	na
Rice exports (t)	1,939,975	2,473,948	3,083,609	3,112,693	2,668,066	na
Paddy exports (t)	75,867	194,594	553,520	1,744,630	779,959	na
Others						
Population, total (×10^3)	241,855	254,076	267,020	274,028	276,218	na
Population, agriculture (×10^3)	8,225	7,668	6,913	6,470	6,325	na
Agricultural area (×10^3 ha)	431,399	426,948	418,250	418,250	na	na
Irrigated agricultural area (×10^3 ha)	19,831	20,900	21,400	21,400	na	na
Total fertilizer consumption (t)	17,830,541	18,586,936	20,037,976	19,773,802	na	na
Tractors used in agric. (no.)	4,670,000	4,800,000	4,800,000	4,800,000	na	na

Source: FAOSTAT online database.

URUGUAY, a relatively small South American country of 177,000 km², faces the South Atlantic Ocean between Brazil and Argentina. The country consists of fertile coastal lowlands, with rolling hills in the interior. The climate is warm temperate. Seven percent of the land is arable. Of its 3.3 million population, 14% is engaged in agriculture, which constitutes 12% of the GDP.

General information

- GNI per capita PPP$, 2000: 8,880
- Internal renewable water resources: 59 km³
- Main food consumed: wheat, meat, sugar and honey, oil and fat, milk
- Rice consumption, 1999: 12.1 kg milled rice per person per year

Production season

	Planting	Harvesting
Main season	Oct-Dec	Mar-May

Production constraints

Both rice production and yield have increased steadily during the last five years. However, there are several constraints to sustainable rice production in Uruguay:

- Low price in the world market
- Dependence on external markets for marketing of rice products
- Straighthead, especially on sandy soils
- Low temperature during the early and/or reproductive phases
- Yield potential of rice varieties has reached a plateau

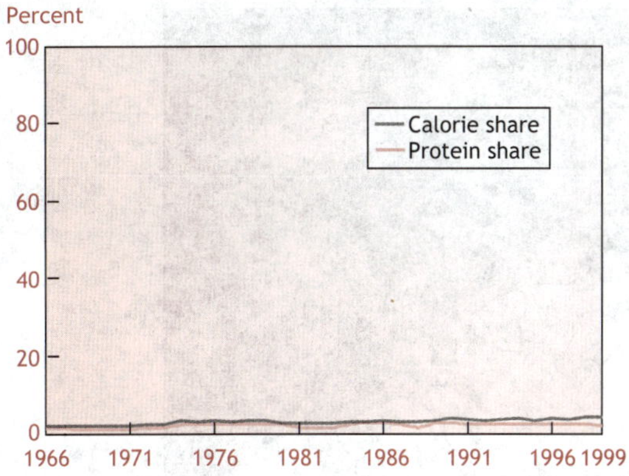

Share of calories and protein from rice, 1966-99

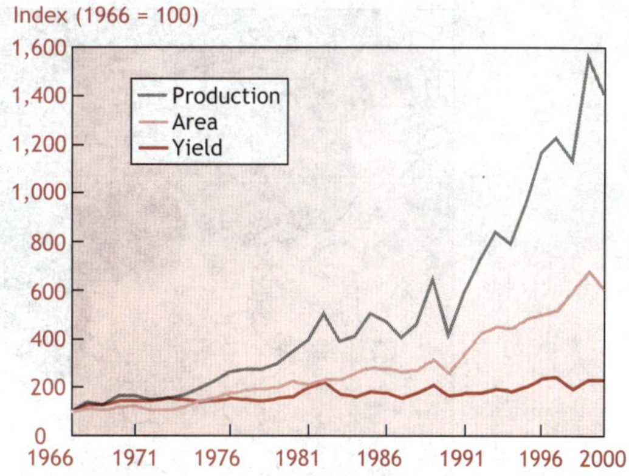

Indices of rice production, area, and yield, 1966-2000

The scale on this graph is not comparable with the scale on production, area, and yield graphs for most other countries.

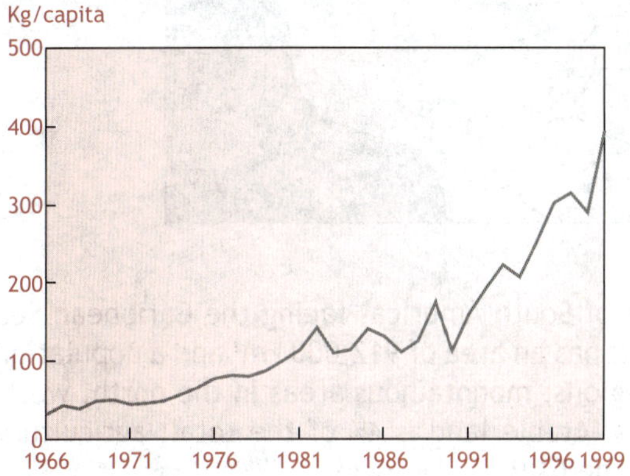

Per capita production (paddy terms), 1966-99

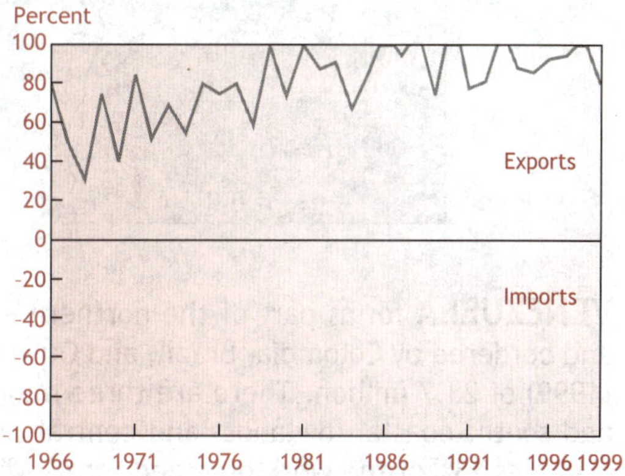

Net trade status, 1966-99

Basic statistics, Uruguay

	1985	1990	1995	1998	1999	2000
Rice						
Area harvested (ha)	84,929	78,091	146,300	180,200	206,000	185,000
Yield (t/ha)	5.0	4.4	5.5	5.3	6.3	6.4
Production (t)	420,700	347,300	806,100	949,800	1,301,900	1,174,800
Rice imports (t)	0	6,428	63	318	2,040	na
Paddy imports (t)	0	1,427	58	460	74	na
Rice exports (t)	241,609	290,426	462,471	659,192	699,044	na
Paddy exports (t)	180	4,652	67,570	138,896	118,719	na
Others						
Population, total ($\times10^3$)	3,009	3,106	3,218	3,289	3,313	na
Population, agriculture ($\times10^3$)	408	383	374	369	367	na
Agricultural area ($\times10^3$ ha)	14,891	14,825	14,825	14,827	na	na
Irrigated agricultural area ($\times10^3$ ha)	97	125	160	180	na	na
Total fertilizer consumption (t)	60,200	71,878	66,000	133,329	na	na
Tractors used in agric. (no.)	34,600	32,804	33,000	33,000	na	na

Source: FAOSTAT online database.

VENEZUELA forms part of the northern edge of South America, facing the Caribbean Sea and bordered by Colombia, Brazil, and Guyana. It has an area of 912,000 km² and a population (1999) of 23.7 million. There are three main regions: mountainous areas in the north, west, and south; coastal lowlands; and central plains. Arable land is 4% of the total; agriculture occupies 13% of the workforce and forms 4% of the GDP.

General information

- GNI per capita PPP$, 2000: 5,740
- Internal renewable water resources: 200 km³
- Main food consumed: maize, oil and fat, sugar and honey, wheat, fruits
- Rice consumption, 1999: 15.6 kg milled rice per person per year

Production season

	Planting	Harvesting
Wet season	Jul-Aug	Nov-Dec
Dry season	Nov-Dec	May-Jun

Production constraints

The following are the major constraints to sustainable rice production in Venezuela:

- Rice hoja blanca virus, blast, brown spot, and whitebacked planthopper
- Weed and red rice competition
- Iron toxicity, poor internal and external drainage because of heavy soils in Portuguesa and Guárico
- Yield decline in intensive irrigated areas
- Yield potential of present irrigated varieties has reached a plateau
- Poor land preparation

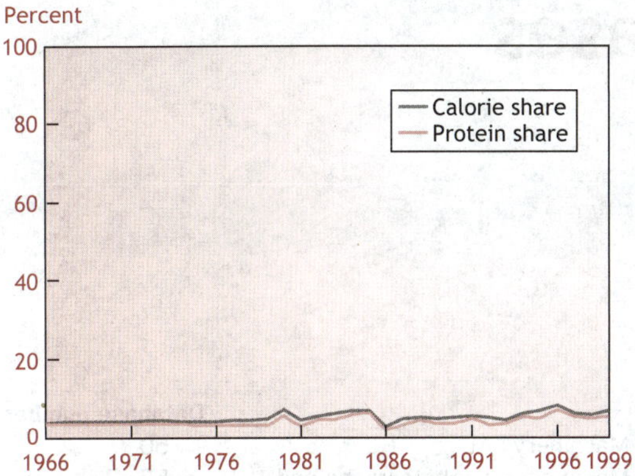

Share of calories and protein from rice, 1966-99

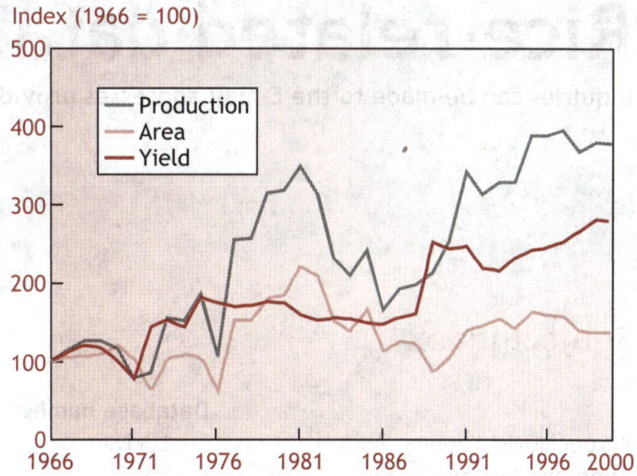

Indices of rice production, area, and yield, 1966-2000

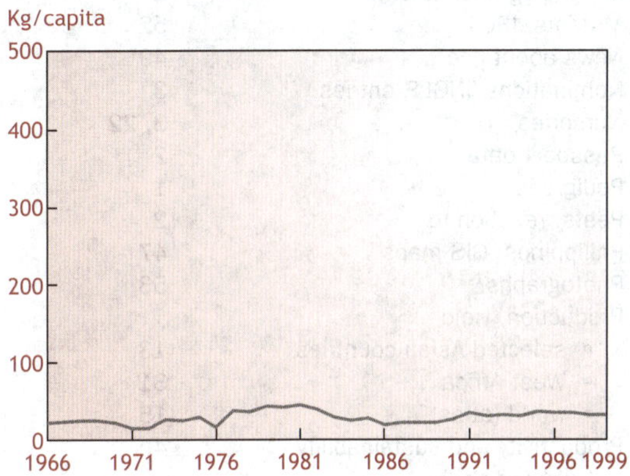

Per capita production (paddy terms), 1966-99

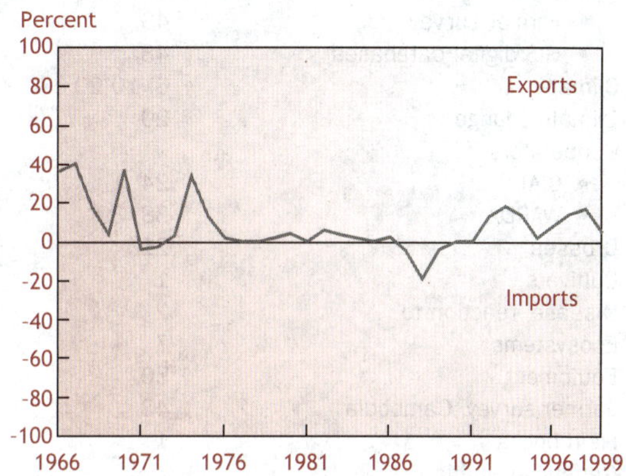

Net trade status, 1966-99

Basic statistics, Venezuela

	1985	1990	1995	1998	1999	2000
Rice						
Area harvested (ha)	180,769	114,755	177,430	151,899	148,883	150,000
Yield (t/ha)	2.6	4.3	4.3	4.7	5.0	4.9
Production (t)	471,722	495,000	756,950	716,090	740,000	737,000
Rice imports (t)	755	352	68,077	141	127	na
Paddy imports (t)	0	519	104,429	26	8	na
Rice exports (t)	28	0	76,927	81,235	22,958	na
Paddy exports (t)	0	0	335	7,508	1	na
Others						
Population, total ($\times 10^3$)	17,138	19,502	21,844	23,242	23,706	na
Population, agriculture ($\times 10^3$)	2,631	2,763	2,541	2,402	2,355	na
Agricultural area ($\times 10^3$ ha)	21,740	22,145	21,790	21,730	na	na
Irrigated agricultural area ($\times 10^3$ ha)	470	480	536	540	na	na
Total fertilizer consumption (t)	477,639	427,000	298,000	242,800	na	na
Tractors used in agric. (no.)	43,500	48,000	49,000	49,000	na	na

Source: FAOSTAT online database.

Rice-related databases

Inquiries can be made to the E-mail addresses provided

Subject index

1. Name: **INTERNATIONAL RICE INFORMATION SYSTEM (IRIS)**
Location: IRRI Biometrics and Bioinformatics Unit
Description: Integrated information on over 1,000,000 rice accessions, breeding lines, and cultivars from around the world. Stores information on origin, development, nomenclature, characterization, and evaluation. Manages information for breeding projects, produces field books, and allows pedigree analysis, data mining, and management and integration of local data.
Contact: C.G. McLaren (g.mclaren@cgiar.org)

2. Name: **INTERNATIONAL RICE GENEBANK COLLECTION INFORMATION SYSTEM (IRGCIS)**
Location: IRRI Genetic Resources Center
Description: Information system that supports genebank daily operations in acquisition of germplasm, multiplication, conservation, characterization, rejuvenation, and distribution to end-users. It also provides users with on-line information on the passport data, morpho-agronomic characterization, response to biotic and abiotic stresses, grain quality, and isozyme characteristics of the registered accessions in the genebank.
Contacts: A. Alcantara (a.alcantara@cgiar.org) and E. Guevarra (e.guevarra@cgiar.org)

3. Name: **INTERNATIONAL NETWORK FOR GENETIC EVALUATION OF RICE INFORMATION SYSTEM (INGERIS)**
Location: IRRI Genetic Resources Center (INGER)
Description: Information system that supports day-to-day operations of INGER, from processing of incoming seeds to seed multiplication, seed storage and inventory, nursery composition and germplasm distribution, and nursery data processing. It also facilitates retrieval of information on the multilocation nursery trials conducted by NARES through INGER.
Contacts: V. Lopez (v.lopez@cgiar.org) and E. Javier (e.javier@cgiar.org)

4. Name: **SOIL FERTILITY, PRODUCTIVITY, AND SUSTAINABILITY IN IRRIGATED RICE**
Location: IRRI Crop, Soil, and Water Sciences Division
Description: Stores fertilizer field-plot trial data. Used in production of "Progress Reports of Different INSURF Sub-Networks" and "Progress Reports on Experiments and Trials."
Contact: R.R. Pamplona (r.r.pamplona@cgiar.org)

5. Name: **RICESTAT DATABASE**
Location: IRRI Social Sciences Division
Description: Stores rice-related statistics such as production, consumption, trade, prices, and other socioeconomic information related to rice.
Contact: J. Narciso (j.narciso@.cgiar.org)

6. Name: **CLIMATOLOGY DATABASE**
Location: IRRI Crop, Soil, and Water Sciences Division
Description: Contains daily weather data from 10 Philippine stations + 5 rice-weather stations outside the Philippines.
Contacts: J. Sheehy (j.sheehy@cgiar.org) and G. Centeno (g.centeno@cgiar.org)

7. Name: **RICE ECOSYSTEMS**
Location: IRRI Crop, Soil, and Water Sciences Division
Description: Mega-level information on the five major rice ecosystems; soil and agroclimatic parameters
Contact: N. Fabellar (n.fabellar@cgiar.org)

8. Name: **RICE BIBLIOGRAPHY**
Location: IRRI Library and Documentation Services
Description: The world's technical rice literature (published and unpublished) produced since 1970 in more than 80 languages (English 65%, Japanese 20%, Chinese 6%, others 9%)
Contact: Carmelita Austria (c.austria@cgiar.org)

9. Name: **IRRI LIBRARY CATALOG**
Location: IRRI Library and Documentation Services
Description: Bibliographic records of all books, serials, conference proceedings, maps, and reports acquired
Contact: Allan Vallarta (a.vallarta@cgiar.org)

10. Name: **DIGITAL MAP OF AGROECOLOGICAL ZONES (AEZ) OF SOUTH AND SOUTHEAST ASIA**
Location: IRRI GIS Laboratory
Description: Contains AEZ information for South and Southeast Asia. Description and AEZ boundaries, length of growing period, climate description, and country labels.
Contact: N. Fabellar (n.fabellar@cgiar.org)

11. Name: **GIS DATABASE FOR RICE CULTURE**
Location: IRRI GIS Laboratory
Description: Digital point data (lat./long.) for rice culture by type for South and Southeast Asia (digitized from the rice culture maps prepared by R.E. Huke in 1982)
Contacts: S.P. Kam (s.kam@cgiar.org) and N. Fabellar (n.fabellar@cgiar.org)

12. Name: **GIS DATABASE FOR RICE AREA, BY TYPE OF CULTURE (South, Southeast, and East Asia)**
Location: IRRI GIS Laboratory
Description: Electronic database for rice culture by type prepared by R.E. Huke in 1996
Contacts: S.P. Kam (s.kam@cgiar.org), N. Fabellar (n.fabellar@cgiar.org), and A. Rala (a.rala@cgiar.org)

13. Name: **GIS DIGITAL DATABASES FOR ASIAN COUNTRIES**
Location: IRRI GIS Laboratory
Description: Contains information on administrative boundaries and thematic maps. Enables mapping of rice production, yield, and area harvested in Bangladesh, Bhutan, Cambodia, China, India, Lao PDR, Myanmar, Nepal, Pakistan, Philippines, Sri Lanka, Thailand, and Vietnam.
Contacts: S.P. Kam (s.kam@cgiar.org) and A. Rala (a.rala@cgiar.org)

14. Name: **IRRI IN-COUNTRY ACTIVITY DATABASE**
Location: IRRI, International Programs Management Office
Description: List of activities for each major country where IRRI works, showing staff involved and collaborating institutes and scientists
Contact: A. Quilloy (a.quilloy@cgiar.org)

15. Name: **WORLD ATLAS OF RICE PRODUCTION STATISTICS**
Location: IRRI GIS Laboratory
Description: Enables mapping at the global scale of rice production. Data from FAO statistical database.
Contacts: S.P. Kam (s.kam@cgiar.org) and A. Rala (a.rala@cgiar.org)

16. Name: **INTERNATIONAL RICE GENEBANK HERBARIUM INFORMATION SYSTEM (IRGHerbIS)**
Location: IRRI Genetic Resources Center
Description: Information on herbarium specimens of rice germplasm conserved in the IRRI genebank
Contact: A.P. Alcantara (a.alcantara@cgiar.org) and M.S. Almazan (s.almazan@cgiar.org)

17. Name: **DATABASE ON RICE METHANE**
Location: IRRI Crop, Soil, and Water Sciences Division
Description: Contains information on methane emissions, ebullition, and dissolved methane as affected by rice variety, spacing, crop establishment, water regime, and mineral/organic amendments.
Contact: Rhoda Lantin (r.lantin@cgiar.org)

18. Name: **SYSTEMWIDE INFORMATION NETWORK FOR GENETIC RESOURCES (SINGER)**
Location: International Plant Genetic Resources Institute, Rome, Italy
Description: Information on all the genetic resources held by the international research (CGIAR) centers
Contacts: singer@cgiar.org and Samy Gaiji (s.gaiji@cgiar.org)

19. Name: **FUNCTIONAL GENOMIC DATABASES**
Location: IRRI Entomology and Plant Pathology Division
Description: Rice functional genomic databases
Contact: Hei Leung (h.leung@cgiar.org)

20. Name: **RICE GLOBAL CLIMATE CHANGE DATABASE**
Location: IRRI Biometrics Unit
Description: Data bank for experiments conducted on CO_2, insects, methane, rice blast, and UV-B; also includes rainfall ionic content; other daily weather variables to be added.
Contact: C.G. McLaren (g.mclaren@cgiar.org)

21. Name: **RICE CROSSES**
Location: CIAT Rice Program
Description: Information on crosses made by CIAT's Rice Program from 1958 (under a Rockefeller project at that time) and by FLAR (since its creation in 1995) to 2000. Provides type of cross, year, purpose, parents, and targeted ecosystem.
Contacts: J. Barona, Rice Project (j.barona@cgiar.org), and L.R. Sanint, Executive Director, FLAR (l.sanint@cgiar.org), Fax 57-2-4450094, Cali, Colombia

22. Name: **NURSERIES**
Location: CIAT Rice Program
Description: Information from national breeders in the Latin American and Caribbean section of INGER (see also INGER-Database No. 3) on evaluation of INGER material. Main data are on reaction to disease, yield, days to flowering, and grain quality.
Contacts: J. Barona, Rice Project (j.barona@cgiar.org), and L.R. Sanint, Executive Director, FLAR (l.sanint@cgiar.org). Fax 57-2-4450094, Cali, Colombia

23. Name: **RICE VARIETIES**
Location: CIAT Rice Program
Description: Information on all rice varieties released in Latin America and the Caribbean. Provides year of release, parents, releasing entity, and country
Contacts: J. Barona, Rice Project (j.barona@cgiar.org), and L.R. Sanint, Executive Director, FLAR (l.sanint@cgiar.org). Fax 57-2-4450094, Cali, Colombia.

24. Name: **CIAT COOPERATORS**
Location: CIAT Rice Program
Description: Data on Latin American and Caribbean scientists collaborating in INGER—name, institution, education, address, phone, etc.
Contacts: J. Barona, Rice Project (j.barona@cgiar.org), and L.R. Sanint, Executive Director, FLAR (l.sanint@cgiar.org). Fax 57-2-4450094, Cali, Colombia

25. Name: **NEW RICE FOR AFRICA (NERICA) FOOD SECURITY PROGRAM DATABASE**
Location: WARDA Rainfed Rice Program
Description: Includes information on NERICA seed distribution and accompanying technologies in Sub-Saharan Africa.
Contact: Monty P. Jones (m.jones@cgiar.org)

26. Name: **INTERNATIONAL NETWORK FOR GENETIC EVALUATION OF RICE IN AFRICA (INGER-AFRICA) MULTILOCATION TRIAL DATABASE**
Location: INGER-Africa, WARDA
Description: Stores information on varietal evaluation and testing in 23 countries in Sub-Saharan Africa.
Contact: Robert Guei (r.guei@cgiar.org)

27. Name: **WEST AFRICAN INLAND VALLEY INFORMATION SYSTEM (WAIVIS)**
Location: WARDA Systems Analysis and GIS Unit
Description: Contains information on multiscale inland valley characterization studies conducted in 10 West African countries within the framework of the Inland Valley Consortium (IVC).
Contact: Mahaman Moussa (m.moussa@cgiar.org)

28. Name: **WARDA REGIONAL PARTICIPATORY RICE IMPROVEMENT AND GENDER/USER ANALYSIS (PRIGA) DATABASE**
Location: WARDA Upland Breeding Unit
Description: Includes information on researcher-led and extension-led participatory varietal selection work in WARDA's 17 member countries.
Contact: Monty Jones (m.jones@cgiar.org)

29. Name: **LONG-TERM SOIL FERTILITY TRIALS, SENEGAL**
Location: WARDA Sahel Station
Description: Includes information on rice performance and soil quality evolution at two locations in the Senegal River valley under an irrigated double-cropped system (6 fertilizer treatments) since 1991.
Contacts: Stephan Haefele (s.haefele@cgiar.org) and Toon Defoer (t.defoer@cgiar.org)

30. Name: **WEATHER DATABASE**
Location: WARDA Systems Analysis and GIS Unit
Description: Contains daily weather data from rice-growing areas in West Africa.
Contact: Mahaman Moussa (m.moussa@cgiar.org)

31. Name: **IRRIGATED RICE AGRONOMIC SURVEYS DATABASE**
Location: WARDA Sahel Station
Description: Includes information on crop management, yields, profitability, soil quality, nutrient uptake, and fertilizer recoveries from surveys conducted in irrigation schemes in four Sahelian countries: Burkina Faso, Mali, Senegal, and Mauritania.
Contact: Stephan Haefele (s.haefele@cgiar.org)

32. Name: **RICE GRAIN QUALITY**
Location: WARDA Grain Quality Laboratory
Description Contains results of grain quality analyses obtained at IITA (Ibadan) and WARDA (Bouaké) grain quality labs, including grain physiochemical, milling, and cooking characteristics as well as information on trial, year, location, source, ecology, etc. Not every record has the full range of measurements.
Contact: Ibrahim Teslim (i.teslim@cgiar.org)

33. Name: **WEST AFRICAN RICE STATISTICS DATABANK (WARSDB)**
Location: WARDA Economics Unit
Description: Data series from 1961 to 1998/1999 or 2000, depending on the indicator. The FAOSTAT database was used as the main data source to set up the WARSDB.
Contact: Ali Touré A. (a.toure@cgiar.org)

34. Name: **IRRIGATION SCHEMES IN CÔTE D'IVOIRE**
Location: WARDA Economics Unit
Description: Contains information on the major characteristics of irrigation schemes in Côte d'Ivoire.
Contact: Ali Touré A. (a.toure@cgiar.org)

35. Name: **RICETECH**
Location: WARDA Technology Transfer Unit
Description: Contains technologies available at WARDA and within the West and Central Africa subregion related to rice (in progress).
Contact: Moustapha Gaye (m.gaye@cgiar.org)

36. Name: **WEST AFRICA RICE BIBLIOGRAPHIC DATABASE (WARBI)**
Location: WARDA Library
Description: Contains annotated bibliographic references for the materials available in the WARDA library.
Contact: Florent Diouf (f.diouf@cgiar.org)

37. Name: **PERIO**
Location: WARDA Library
Description: Periodicals database, bibliographic references, and related information for all serials.
Contact: Florent Diouf (f.diouf@cgiar.org)

38. Name: **WARDA MAILING LIST**
Location: WARDA Library
Description: Addresses of institutions and individuals, including WARDA former trainees.
Contact: Florent Diouf (f.diouf@cgiar.org)

39. Name: **RSCWA**
Location: WARDA Library
Description: Addresses, biodata, research profiles, and ongoing research programs of rice scientists working at WARDA or in its member countries.
Contact: Assoa Tanoh (a.tanoh@cgiar.org)

40. Name: **FMHS (FARM MANAGEMENT AND HOUSEHOLD SURVEY)**
Location: WARDA Economics Unit
Description: Consists of 3 annual rounds of 22 questionnaires administered between 1993 and 1995 to more than 120 households growing rice in 3 agroecological regions of Côte d'Ivoire.
Contact: Kamara D. Mylène (k.mylene@cgiar.org)

41. Name: **BOUNDIALI DATABASE**
Location: WARDA Economics Unit
Description: Contains socioeconomic information on 20 female rice farmers in Boundiali (Côte d'Ivoire).
Contact: Kamara D. Mylène (k.mylene@cgiar.org)

42. Name: **BACTERIA INHABITING THE RICE SEED**
Location: IRRI Entomology and Plant Pathology Division
Description: Bacterial isolates characterized for pathogenicity, antagonism against *Rhizoctonia solani*, *Pyricularia grisea*, *Fusarium moniliforme*, and *Sarocladium oryzae*, and colony morphology. Cultures are maintained as cryovials at –80 °C.
Contacts: Tom Mew (t.mew@cgiar.org) and Helen Barrios (h.barrios@cgiar.org)

43. Name: **CAMBODIAN RICE FARMER SURVEY**
Location: IRRI Entomology and Plant Pathology Division
Description: A comprehensive baseline survey conducted in 1995-96 of about 1,500 lowland, upland, deepwater, rainfed, and irrigated rice farmers in Cambodia to determine their farming practices with special emphasis on pest management and fertilizer use
Contact: Gary Jahn (g.jahn@cgiar.org)

44. Name: **GIS DIGITAL DATABASES FOR CAMBODIA**
Location: IRRI GIS Laboratory
Description: Information on thematic maps and administrative boundaries in Cambodia, with more detailed information on Takeo Province.
Contact: Nestor Fabellar (n.fabellar@cgiar.org)

45. Name: **GIS DIGITAL DATABASES FOR INDIA**
Location: IRRI GIS Laboratory
Description: Thematic maps showing administrative boundaries and large-scale maps on the four states of eastern India: Uttar Pradesh, Assam, Bihar, and Madhya Pradesh
Contacts: S.P. Kam, V.P. Singh, and Gemma Belarmino (v.p.singh@cgiar.org; g.belarmino@cgiar.org)

46. Name: **GIS DIGITAL DATABASES FOR THAILAND**
Location: IRRI GIS Laboratory
Description: Synoptic and agromet weather stations for north and northeast Thailand, with associated daily weather data (1951-94), rice suitability maps, soils, land and use by selected provinces in north and northeast Thailand
Contacts: S.P. Kam and Nestor Fabellar (n.fabellar@cgiar.org)

47. Name: **GIS DIGITAL DATABASES FOR THE PHILIPPINES**
Location: IRRI GIS Laboratory
Description: Detailed information and thematic maps, including administrative boundaries of provinces in the Philippines
Contacts: S.P. Kam and Nestor Fabellar (n.fabellar@cgiar.org)

48. Name: **INTERNATIONAL DIRECTORY OF RICE WORKERS**
Location: IRRI Library and Documentation Services
Description: A worldwide directory of scientists engaged in rice research; updated daily
Contacts: Rene Manlangit (r.manlangit@cgiar.org) and Mila Ramos (m.ramos@cgiar.org) (http://ricelib.irri.cgiar.org)

49. Name: **RICE AND IRRI IN THE NEWS**
Location: IRRI Library and Documentation Services
Description: Current and past newspaper articles about rice or IRRI; updated daily
Contact: Rene Manlangit (r.manlangit@cgiar.org)

50. Name: **IRRI-DESIGNED MACHINERY AND EQUIPMENT**
Location: IRRI Agricultural Engineering Unit
Description: Agricultural machinery designed by IRRI staff and used in rice cultivation from land preparation to crop processing.
Contact: Joseph Rickman (j.rickman@cgiar.org) (http://www.irri.org/aed/aedrd.html)

51. Name: **STRESS GENES DATABASE**
Location: IRRI Plant Breeding, Genetics, and Biochemistry Division
Description: Stress-related sequences from plants; updated often.
Contact: Kenneth McNally (k.mcnally@cgiar.org)

52. Name: **IR64 MUTANT DATABASE**
Location: IRRI Biometrics and Bioinformatics Unit
Description: A database of phenotypic characteristics of the IR64 mutant collection with images. It presents a description of mutants identified from visual inspection of the M3 lines planted in the field and from disease screening. Updated every season.
Contacts: Richard Bruskiewich (r.bruskiewich@cgiar.org), Hei Leung (h.leung@cgiar.org), and Graham McLaren (g.mclaren@cgiar.org) (http://www.irri.org/genomics/database/IR64.htm)

53. Name: **CPS-IRRI MEDIA ASSETS**
Location: IRRI Communication and Publications Services
Description: A searchable database of more than 5,000 rice images and nearly 400 titles of technical books on rice research.
Contact: Gene Hettel (e.hettel@cgiar.org)

Important conversion factors, by country

Country	Conversion factors
Asia	
Bangladesh	
	1 bushel = 0.73 maund = 29.17 seer = 60 lb
	1 maund = 82.29 lb = 37.32 kg
	1 seer = 2.05 lb = 0.93 kg
	1 kg = 2.20462 lb = 1.07 seer
	1 bushel per acre = 67.253 kg per ha
	1 ha = 2.47109 acres
	1 acre = 0.40468 ha
	1 lakh = 100,000
	1 crore = 10,000,000
Cambodia	
	1 picul = 68 kg
	1 mt = 14.7059 picul
China	
	1 mu = 0.067 ha
	15 mu = 1.0 ha
	1 sheng milled rice = 1 liter milled rice = 0.5 kg
	1 dou milled rice = 10 liters milled rice = 5 kg
	1 dan (picu) milled rice = 100 liters milled rice = 50 kg
	20 dan (picul) = 1 mt
	1 dun = 1 mt = 2204.6 lb
	1 dan (picul) = 100 jin
	1 jin (catty) = 0.5 kg = 1.1023 lb
	1 jin/mu = 7.5 kg/ha
Taiwan, China	
	1 kg rough rice = 0.76366 kg brown rice
	1 kg ponlai brown rice = 0.93 kg milled rice
	1 kg chailai brown rice = 0.94 kg milled rice
	1 ha = 1.03 chia
	1 chia = 0.9699 ha
	1 old catty = 0.5968 kg
	1 shih catty = 0.5 kg

India

1 quintal = 100 kg
1 maund = 37.3 kg = 82.29 lb
1 Madras measure rice = 54 oz = 3.375 lb
Bigha is a land measure in rural areas; its definition varies from state to state.
In Gujarat, 4/7 bigha = 1 acre
In Rajasthan, 21/2 bighas = 1 acre
In West Bengal, 3 bighas = 1 acre

Indonesia

1 liter rice = 0.8 kg
1 gantang rice = 8.58 liters = 0.00686 mt
1 mt rice = 145.69 gantang
Dry stalk rough rice (padi) to milled rice (beras) = 52%
Gabah kering (dry rough rice) to milled rice (beras) = 68%
Dry stalk rough rice (padi) to rough rice = 76.47%

Japan

Brown rice x 1.25 = rough rice
Milled rice x 1.37 = rough rice
Brown rice x 0.91 = milled rice
Rough rice x 0.728 = milled rice
1 koku rough rice = 187.5 kg
1 koku brown rice = 150 kg
1 koku milled rice = 136.5 kg
1 sho milled rice = 1.425 kg
1 kan = 3.75 kg
1 kin = 0.6 kg
1 picul = 50 kg
1 cho = 10 tan = 2.45072 acres = 0.99174 ha
1 ha = 10.0833 tan = 1.00833 cho
1 tan = 0.1 cho = 0.9917 ha

Korea, Rep. of

1 danbo = 0.1 jeongbo = 0.099174 ha
1 ha = 1.0083 jeongbo
1 seok milled rice = 144 kg
1 seok brown rice = 155 kg
1 seok rough rice = 100 kg
100 liters milled rice = 79.8264 kg

Malaysia

1 picul brown rice = 133.33 lb = 60.48 kg
1 gantang rough rice = 5.60 lb = 2.54 kg
1 kati = 0.60478 kg

Myanmar

100 measures rough rice = 1 basket
1 basket rough rice = 46 lb = 20.86 kg
1 basket milled rice = 75 lb = 34.02 kg
1 bag milled rice = 225 lb = 102.06 kg
1 pyi milled rice = 4.69 lb = 2.13 kg
1 maund = 0.037 mt
1 mt = 26.792 maunds

Nepal

1 khet = 1.3 ha
1 bigha = 0.67 ha (Terai)
1 matomuri = 0.13 ha = 0.25 ropani
1 ropani = 0.05 ha (Hills) = 4 muris
1 muri = 0.013 ha
1 seer = 0.80 kg (Hills)
1 seer = 0.93 kg (Terai)
1 mana = 0.3 kg rough rice
1 mana = 0.454 kg rice
1 maund = 37.32 kg rough rice (Terai)

Pakistan

1 kg = 2.2046 lb = 1.0716 seer
1 quintal = 100 kg = 1.96841 cwt = 2.679 maunds
1 metric ton = 1,000 kg = 0.98421 long ton = 26.79 maunds
100 kg per ha = 1.4869 bushels per acre = 1.09 maunds per acre
1 bushel = 0.73 maunds = 29.17 seers = 60 lb
Before 1980, 1 maund = 37.324 kg
After 1980, 1 maund = 40 kg

Philippines

1 cavan rough rice = 50 kg
1 cavan milled rice = 50 kg
1 ganta milled rice = 2.24 kg
Before 1973
1 ganta = 3 liters
1 cavan rough rice = 44 kg
1 cavan milled rice = 56 kg

Sri Lanka

1 bushel rough rice = 46 lb = 20.86 kg
1 bushel rough rice = 30.69 milled rice = 14 kg milled rice
1 bushel milled rice = 64 lb = 32 measures of rice
1 measure milled rice = 2 lb = 0.907 kg

Thailand

1 picul = 60 kg
1 catty = 600 g = 6 kg
1 kwein = 2,000 liters
1 ban = 1,000 liters
1 sat = 20 liters
1 thanan = 1 liter
1 kwein rough rice = 1 mt rough rice
1 rai = 0.16 ha = 0.395 acre

Other countries

Australia

1 bushel rough rice = 42 lb = 19.05 kg

Brazil

1 bushel rough rice = 45 lb = 20.41 kg
1 sack (bag) rough rice = 110.23 lb = 50 kg
1 sack (bag) milled rice = 88.18 lb = 40 kg

Egypt

1 sack (bag) milled rice = 220.11 lb = 99.84 kg

Ghana

1 sack (bag) milled rice = 240 lb = 108.86 kg

Guyana

1 sack (bag) rough rice = 140 lb = 63.50 kg
1 sack (bag) milled rice = 180 lb = 81.65 kg

Malawi

1 sack (bag) rough rice = 150 lb = 68.04 kg
1 sack (bag) milled rice = 200 lb = 90.72 kg

Mexico

1 carga rough rice = 304.24 lb = 138 kg
1 carga milled rice = 352.74 lb = 160 kg

Panama

1 lata milled rice = 24.99 lb = 11.33 kg
1 lata rough rice = 35.99 lb = 16.32 kg

Sierra Leone

1 bushel rough rice = 60 lb = 27.21 kg
1 bushel milled rice = 84 lb = 38.10 kg

Swaziland

1 pocket milled rice = 2 lb = 0.91 kg

United States

1 bushel rough rice = 45 lb = 20.41 kg
1 sack (bag) rough rice = 100 lb = 45.36 kg
1 sack (bag) milled rice = 100 lb = 45.36 kg
1 barrel rough rice = 162 lb = 73.48 kg

Uruguay

1 bolsa brown rice = 110.23 lb = 50 kg

Sources: For Asia: Rose B. 1985. *Appendix to the Rice Economy of Asia.* Resources for the Future, Inc., Washington, D.C. For other countries: FAO Rice Report 1974-75. 1975. Rome.

Rice facts

GENERAL INFORMATION

Selected rice-consuming and -producing countries	Population (million)			Infant mortality (per 1,000 live births)	Life expectancy at birth (years)	GNI[a] per capita	
		Projected				Per capita 2000 ($)	Av annual growth[b] (%)
	1999	2020	2050	1998	1998		1990-99
Asia	**3,634.3**	**4,545.2**	**5,268.5**	–[c]	–	–	–
Bangladesh	126.9	170.2	212.5	79	58.6	1,590	3.1
Bhutan	2.1	3.5	5.7	84	61.2	1,440	2.1
Cambodia	10.9	15.5	20.7	104	53.5	1,440	1.8
China[d]	1,274.1	1,462.7	1,484.9	38	70.1	3,920	8.9
India	998.1	1,272.2	1,528.9	69	62.9	2,340	3.8
Indonesia	209.3	262.3	311.9	40	65.6	2,830	2.2
Japan	126.5	123.9	104.9	4	80.0	27,080	1.0
Korea, DPR	23.7	28.4	30.8	–	63.0	–	–
Korea, Rep. of	46.5	51.9	51.3	5	72.6	17,300	4.6
Lao PDR	5.3	8.8	13.3	96	53.7	1,540	3.6
Malaysia	21.8	29.3	37.0	9	72.2	8,330	3.9
Myanmar	45.1	56.0	64.9	80	60.6	–	–
Nepal	23.4	35.5	49.3	72	57.8	1,370	2.4
Pakistan	152.3	244.2	345.5	95	64.4	1,860	1.6
Philippines	74.5	102.4	130.9	32	68.6	4,220	1.1
Sri Lanka	18.6	22.8	25.9	17	73.3	3,460	3.8
Thailand	60.9	71.0	74.2	30	68.9	6,320	3.4
Vietnam	78.7	102.5	126.8	31	67.8	2,000	5.0
Latin America	**511.3**	**665.1**	**808.9**	**32**	**69.7**	–	–
Brazil	168.0	209.7	244.2	36	67.0	7,300	0.7
Colombia	41.6	56.6	71.6	25	70.7	6,060	1.0
Cuba	11.2	11.7	11.1	7	75.8	–	–
Dominican Rep.	8.4	10.7	12.3	43	70.9	5,710	3.7
Ecuador	12.4	16.9	21.2	30	69.7	2,910	–0.6
Guyana	0.9	1.0	1.2	58	64.8	3,670	8.0
Peru	25.2	33.8	42.3	43	68.6	4,660	2.4
Suriname	0.4	0.5	0.6	28	70.3	3,480	–
Uruguay	3.3	3.8	4.4	16	74.1	8,880	2.9
Africa	**766.6**	**1,187.4**	**1,766.1**	–	–	–	–
Côte d'Ivoire	14.5	21.8	30.5	90	46.9	1,500	0.8
Egypt	67.2	90.5	114.8	51	66.7	3,670	2.7
Guinea	7.4	11.5	16.3	124	46.9	1,930	1.7
Liberia	2.9	5.9	10.0	–	47.0	–	–
Madagascar	15.5	26.2	40.4	95	57.9	820	–1.1
Mali	11.0	18.9	31.4	144	53.7	780	1.1
Mozambique	19.3	27.8	42.9	129	43.8	800	3.5
Nigeria	108.9	168.2	244.3	112	50.1	800	0.4
Senegal	9.2	15.2	23.1	70	52.7	1,480	0.7
Sierra Leone	4.7	7.4	11.0	182	37.9	480	–6.3
Tanzania	32.8	52.5	80.6	91	47.9	520	0.4
Europe	**728.9**	**711.9**	**627.7**	–	–	–	–
France	58.9	61.5	59.9	5	78.2	24,420	1.3
Italy	57.3	52.9	41.2	6	78.3	23,470	1.3
Spain	39.6	37.6	30.2	6	78.1	19,260	1.8
Australia	**18.7**	**22.3**	**25.8**	**5**	**78.3**	**24,970**	**2.0**
USA	**276.2**	**317.1**	**349.3**	**7**	**76.8**	**34,100**	**1.8**
World	**5,978.4**	**7,501.5**	**8,909.1**	**58**	**66.9**	–	–

INFORMATION ABOUT AGRICULTURE

Selected rice-consuming and -producing countries	Arable land[e] per capita	Freshwater resources per capita (m^3)	Annual freshwater withdrawal for agriculture (%)	Agriculture's share (%)		Average annual growth rate		
				Labor force	GDP[g]	GDP[g]	Agriculture value added	Rice production
countries	1998	1996	1980-97[f]	1998	1998	1990-98	1990-98	1990-2000
Asia	**0.14**	**–**	**–**	**57**	**–**	**–**	**–**	**1.6**[h]
Bangladesh	0.06	19,065	96	58	22.2	4.8	1.5	2.5
Bhutan	0.07	–	–	94	38.2	–	–	2.0
Cambodia	0.35	47,530	94	71	50.6	5.5	2.2	6.0
China[d]	0.10	2,282	87	68	18.4	11.1	4.3	0.7
India	0.16	2,167	93	61	29.3	6.1	3.4	2.0
Indonesia	0.09	12,625	76	50	19.5	5.8	2.8	1.2
Japan	0.04	4,338	50	5	1.7	1.3	–2.0	–0.8
Korea, DPR	0.07	–	–	32	–	–	–	–10.1
Korea, Rep. of	0.04	1,438	46	11	4.9	6.2	2.1	–0.4
Lao PDR	0.15	55,679	82	77	52.6	6.7	4.5	4.3
Malaysia	0.09	21,046	47	20	13.2	7.7	2.0	0.5
Myanmar	0.21	24,651	90	71	53.2	6.3	5.0	3.5
Nepal	0.13	7,616	95	93	40.5	4.8	2.3	2.4
Pakistan	0.14	3,256	97	48	26.4	4.1	3.8	4.8
Philippines	0.08	4,373	61	41	16.9	3.3	1.5	1.8
Sri Lanka	0.05	2,329	96	46	21.1	5.3	1.5	0.9
Thailand	0.28	2,954	90	58	11.2	7.4	3.1	2.8
Vietnam	0.07	4,902	78	68	25.7	8.6	5.1	5.4
Latin America	**0.27**	**27,386**	**77**	**21**	**7.8**	**3.7**	**2.6**	**2.9**
Brazil	0.32	42,459	59	18	8.4	3.3	3.1	1.4
Colombia	0.05	26,722	43	22	13.5	4.2	1.6	1.1
Cuba	0.33	–	–	15	–	–	–	–0.3
Dominican Rep.	0.13	2,467	89	18	11.6	5.5	3.6	1.7
Ecuador	0.13	26,305	90	27	12.9	2.9	2.7	3.9
Guyana	0.56	–	–	19	34.7	–	–	12.2
Peru	0.15	1,641	72	31	7.1	5.9	5.5	8.0
Suriname	0.14	–	–	20	–	–	–	-2.3
Uruguay	0.38	37,966	91	13	8.5	3.9	4.2	11.3
Africa	**0.24**	**–**	**–**	**59**	**–**	**–**	**–**	**3.3**
Côte d'Ivoire	0.21	5,468	67	51	26.0	3.5	2.4	6.4
Egypt	0.04	966	86	35	17.5	4.2	2.9	5.7
Guinea	0.12	32,661	87	85	22.4	5.0	4.4	5.9
Liberia	0.07	–	–	69	–	–	–	7.1
Madagascar	0.17	23,819	99	75	30.6	1.3	1.5	0.3
Mali	0.43	9,718	97	82	46.9	3.7	3.3	9.3
Mozambique	0.17	12,989	89	81	34.3	5.7	4.8	13.5
Nigeria	0.27	2,375	54	35	31.7	2.6	2.9	1.3
Senegal	0.25	4,482	92	74	17.4	3.0	1.4	-8.2
Sierra Leone	0.11	33,698	89	63	44.2	-4.7	1.5	-0.9
Tanzania	0.12	2,842	89	81	45.7	2.9	3.7	1.6
Europe	**0.40**	**–**	**–**	**9**	**–**	**–**	**–**	**2.7**[h]
France	0.31	3,029	15	4	2.3	1.5	0.4	–1.5
Italy	0.14	2,903	59	6	2.6	1.2	1.3	0.9
Spain	0.36	2,398	62	8	3.5	1.9	–2.5	6.0
Australia	**2.90**	**18,508**	**33**	**5**	**3.2**	**3.6**	**1.1**	**4.7**
USA	**0.65**	**9,259**	**42**	**2**	**1.7**	**2.9**	**2.0**	**2.1**
World	**0.23**	**8,338**	**69**	**46**	**4.8**	**2.4**	**1.2**	**1.7**

RICE CONSUMPTION

Selected rice-consuming and -producing countries	Total rice consumption (rough equivalent) (×10³ t)		Milled rice consumption per capita (kg/cap/yr)		Daily calorie supply per capita (no.)		Rice in total calorie supply (%)	
	1990	1999	1990	1999	1990	1999	1990	1999
Asia	**414,121**	**465,438**	**89**	**86**	**2,550**	**2,723**	**35**	**32**
Bangladesh	25,946	33,946	157	168	2,082	2,201	75	76
Bhutan	–	–	–	–	–	–	–	–
Cambodia	2,258	3,161	156	165	1,830	2,000	79	76
China[d]	162,785	172,449	93	90	2,713	3,045	35	30
India	97,480	110,421	77	74	2,292	2,417	33	30
Indonesia	40,407	48,330	148	154	2,624	2,931	55	52
Japan	11,946	11,388	64	60	2,822	2,782	24	23
Korea, DPR	2,256	2,499	75	75	2,529	2,100	31	35
Korea, Rep. of	6,674	6,554	104	94	3,037	3,073	37	33
Lao PDR	1,060	1,325	171	171	2,159	2,152	70	70
Malaysia	2,289	2,888	86	88	2,758	2,946	30	29
Myanmar	12,760	14,875	210	211	2,620	2,803	78	73
Nepal	2,987	3,151	110	93	2,481	2,264	41	38
Pakistan	3,433	3,018	21	15	2,412	2,462	9	6
Philippines	9,160	11,088	100	100	2,364	2,357	41	41
Sri Lanka	2,486	2,789	97	99	2,203	2,328	43	38
Thailand	8,937	9,373	109	101	2,142	2,411	51	42
Vietnam	15,382	19,692	155	169	2,219	2,564	71	65
Latin America	**17,000**	**20,098**	**26**	**26**	**2,683**	**2,847**	**10**	**9**
Brazil	9,158	10,154	41	40	2,755	3,012	15	14
Colombia	1,641	1,818	31	29	2,419	2,567	13	12
Cuba	754	773	47	46	3,076	2,490	15	18
Dominican Rep	515	530	49	43	2,225	2,333	21	18
Ecuador	648	940	42	51	2,498	2,679	17	18
Guyana	101	97	93	85	2,342	2,569	34	31
Peru	1,333	1,920	41	51	1,946	2,621	22	19
Suriname	54	48	90	78	2,449	2,604	34	28
Uruguay	48	60	10	12	2,534	2,862	4	4
Africa	**15,144**	**20,752**	**16**	**18**	**2,337**	**2,411**	**7**	**7**
Côte d'Ivoire	970	1,748	51	74	2,395	2,582	20	26
Egypt	2,682	4,061	32	41	3,176	3,323	11	13
Guinea	582	747	63	62	1,988	2,133	32	30
Liberia	290	223	90	55	2,100	2,089	44	27
Madagascar	1,782	2,116	99	91	2,139	1,994	47	46
Mali	325	806	25	49	2,313	2,314	11	21
Mozambique	188	202	9	8	1,840	1,939	5	4
Nigeria	2,962	3,876	23	23	2,376	2,833	10	8
Senegal	735	980	67	71	2,316	2,307	28	30
Sierra Leone	606	635	100	99	1,986	2,016	49	48
Tanzania	714	736	18	14	2,144	1,940	8	7
Europe	**2,820**	**4,772**	**4**	**4**	**3,381**	**3,236**	**1**	**1**
France	306	413	4	5	3,505	3,575	1	1
Italy	415	511	5	6	3,591	3,629	1	2
Spain	400	452	7	8	3,248	3,353	2	2
Australia	**170**	**224**	**7**	**8**	**3,218**	**3,150**	**2**	**2**
USA	**2,595**	**3,778**	**7**	**9**	**3,487**	**3,754**	**2**	**3**
World	**454,796**	**515,851**	**58**	**58**	**2,711**	**2,808**	**21**	**21**

RICE PRODUCTION

Selected rice-consuming and -producing countries	Production ($\times 10^3$ t)		Area ($\times 10^3$ t)		Yield (t/ha)		Area planted to modern varieties (%)
	1990	2000	1990	2000	1990	2000	
Asia	**479,421**	**545,477**[i]	**132,342**	**137,601**[i]	**3.62**	**3.96**[i]	**74**
Bangladesh	26,778	35,821	10,435	10,700	2.57	3.35	65
Bhutan	43	50	26	30	1.65	1.67	100
Cambodia	2,500	3,762	1,740	1,873	1.44	2.01	11[j]
China[d]	191,615	190,168	33,519	30,503	5.72	6.23	100[k]
India	111,517	134,150	42,687	44,600	2.61	3.01	73
Indonesia	45,179	51,000	10,502	11,523	4.30	4.43	77
Japan	13,124	11,863	2,074	1,770	6.33	6.70	100
Korea, DPR	3,570	1,690	650	535	5.49	3.16	100
Korea, Rep. of	7,722	7,067	1,244	1,072	6.21	6.59	100
Lao PDR	1,508	2,155	664	690	2.27	3.12	2
Malaysia	1,885	2,037	681	692	2.77	2.94	90[l]
Myanmar	13,972	21,324	4,760	6,302	2.94	3.38	72
Nepal	3,502	4,030	1,455	1,550	2.41	2.60	36
Pakistan	4,891	7,000	2,113	2,312	2.32	3.03	42
Philippines	9,885	12,415	3,319	4,037	2.98	3.08	89
Sri Lanka	2,538	2,767	828	871	3.06	3.18	91
Thailand	17,193	23,403	8,792	10,048	1.96	2.33	68
Vietnam	19,225	32,554	6,028	7,655	3.19	4.25	80
Latin America	**15,568**	**22,990**	**6,185**	**6,388**	**2.52**	**3.60**	**40**
Brazil	7,421	11,168	3,947	3,672	1.88	3.04	25
Colombia	2,117	2,100	521	440	4.06	4.77	87
Cuba	474	369	155	113	3.06	3.27	100
Dominican Rep.	428	527	89	129	4.78	4.08	81
Ecuador	840	1,520	269	380	3.12	4.00	59
Guyana	156	600	51	145	3.03	4.14	4
Peru	966	1,665	185	300	5.23	5.55	73
Suriname	196	175	52	50	3.77	3.50	92
Uruguay	347	1,175	78	185	4.45	6.35	47
Africa	**12,346**	**17,190**	**6,037**	**7,776**	**2.05**	**2.21**	**–**
Côte d'Ivoire	660	1,162	572	750	1.15	1.55	–
Egypt	3,167	5,997	436	660	7.27	9.09	–
Guinea	424	750	436	500	0.97	1.50	–
Liberia	180	200	175	155	1.03	1.29	–
Madagascar	2,420	2,300	1,165	1,207	2.08	1.91	–
Mali	282	810	197	350	1.44	2.31	–
Mozambique	96	158	110	136	0.87	1.16	–
Nigeria	2,500	3,277	1,208	2,061	2.07	1.59	–
Senegal	181	240	73	96	2.48	2.50	–
Sierra Leone	504	199	393	183	1.28	1.09	–
Tanzania	740	379	385	504	1.92	0.75	–
Europe	**2,404**	**3,103**[i]	**449**	**614**[i]	**5.35**	**5.05**[i]	**–**
France	121	107	20	19	5.95	5.76	–
Italy	1,291	1,300	214	221	6.03	5.89	–
Spain	571	798	90	115	6.32	6.93	–
Australia	**924**	**1,400**	**105**	**145**	**8.84**	**9.66**	**–**
USA	**7,080**	**8,669**	**1,142**	**1,232**	**6.20**	**7.04**	**100**
World	**519,936**	**598,852**	**146,886**	**153,766**	**3.54**	**3.89**	**–**

RICE TRADE

Selected rice-consuming and -producing countries	Milled rice (000 t)					
	Exports			Imports		
	1991-95 Average	1996-2000 Average	2000	1991-95 Average	1996-2000 Average	2000
Asia	11,590	16,848	16,574	7,220	11,655	10,925
Bangladesh	0	0	0	223	1,012	502
Bhutan	0	0	0	24	17	10
Cambodia	0	3	2	63	33	36
China	1,045	2,210	3,071	503	352	245
India	1,566	2,658	1,533	39	11	13
Indonesia	113	1	1	919	2,299	1,355
Japan	3	116	42	542	567	656
Korea, DPR	4	0	0	261	362	400
Korea, Rep. of	1	0	0	1	107	172
Lao PDR	0	0	0	14	20	10
Malaysia	1	1	0	400	618	612
Myanmar	386	88	142	0	0	0
Nepal	0	12	0	34	94	195
Pakistan	1,317	1,829	2,016	3	1	1
Philippines	9	0	0	93	1,096	642
Sri Lanka	15	2	1	119	235	206
Thailand	5,106	6,108	6,140	0	1	1
Vietnam	1,734	3,659	3,477	4	2	5
Latin America	1,052	1,712	1,587	2,439	2,787	2,422
Brazil	8	22	26	820	911	660
Colombia	18	0	0	98	138	59
Cuba	0	0	0	318	356	393
Dominican Rep.	0	0	0	17	60	82
Ecuador	14	60	12	1	22	6
Guyana	135	260	252	0	0	0
Peru	0	2	3	324	247	88
Suriname	75	71	49	0	0	0
Uruguay	395	662	699	4	1	2
Africa	248	362	415	3,939	4,423	4,519
Côte d'Ivoire	5	2	1	341	419	441
Egypt	177	332	393	1	2	1
Guinea	0	0	0	242	169	125
Liberia	0	0	0	109	34	35
Madagascar	1	1	1	46	65	94
Mali	0	0	0	28	49	55
Mozambique	0	0	0	75	43	34
Nigeria	0	0	0	329	603	688
Senegal	0	0	0	390	536	537
Sierra Leone	0	0	0	167	243	243
Tanzania	0	5	5	69	70	47
Europe	1,315	1,449	1,493	2,423	2,873	2,975
France	72	71	66	287	370	417
Italy	620	635	666	48	78	137
Spain	188	270	300	131	105	104
Australia	510	613	622	31	44	52
USA	2,598	2,691	2,736	202	315	305
World	17,319	23,676	23,428	16,802	22,686	21,788

Note: Regional totals include countries not shown. [a]Gross national income in purchasing power parity. [b]In per capita gross national income in constant 1995 dollars. [c]- = data not available. [d]Including Taiwan. [e]Excluding permanent crops. [f]Data are for single years during 1980-97. [g]Gross domestic product. [h]Since 1992, Asia includes former USSR Asia and Europe includes former USSR Europe. Former Asia growth rate = 1.5, former Europe growth rate = 2.5. [i]Former Asia and former Europe rice production are 544,982 and $2,573 \times 10^3$ t, respectively; former Asia and former Europe rice harvested area are 137,368 and 413×10^3 ha, respectively; former Asia and former Europe yields are 3.97 and 6.22 t/ha, respectively. [j]Wet season only. [k]Hybrid rice is 40%. [l]West Malaysia only. Sources: Human Development Report, 2000. World Bank Development Report 1999-2000. FAOSTAT, 2001. IRRI World Rice Statistics.